# Springer Tracts in Mechanical Engineering

*About this Series*

Springer Tracts in Mechanical Engineering (STME) publishes the latest developments in Mechanical Engineering - quickly, informally and with high quality. The intent is to cover all the main branches of mechanical engineering, both theoretical and applied, including:

- Engineering Design
- Machinery and Machine Elements
- Mechanical structures and stress analysis
- Automotive Engineering
- Engine Technology
- Aerospace Technology and Astronautics
- Nanotechnology and Microengineering
- Control, Robotics, Mechatronics
- MEMS
- Theoretical and Applied Mechanics
- Dynamical Systems, Control
- Fluids mechanics
- Engineering Thermodynamics, Heat and Mass Transfer
- Manufacturing
- Precision engineering, Instrumentation, Measurement
- Materials Engineering
- Tribology and surface technology

Within the scopes of the series are monographs, professional books or graduate textbooks, edited volumes as well as outstanding PhD theses and books purposely devoted to support education in mechanical engineering at graduate and post-graduate levels.

More information about this series at http://www.springer.com/series/11693

Giovanni Straffelini

# Friction and Wear

## Methodologies for Design and Control

Springer

Giovanni Straffelini
Industrial Engineering Department
University of Trento
Trento
Italy

ISSN 2195-9862 ISSN 2195-9870 (electronic)
Springer Tracts in Mechanical Engineering
ISBN 978-3-319-35931-1 ISBN 978-3-319-05894-8 (eBook)
DOI 10.1007/978-3-319-05894-8

Springer Cham Heidelberg New York Dordrecht London

Softcover reprint of the hardcover 1st edition 2015


Printed on acid-free paper

Springer International Publishing AG Switzerland is part of Springer Science+Business Media (www.springer.com)

*To my wife Fabiana, and my daughter Beatrice*

# Preface

Mechanical components, manufacturing tools and the majority of engineering systems with mating parts in relative motion may face serious failures due to wear or unacceptable energy dissipations due to friction. The resulting costs are very high, being a significant fraction of the GNPs of an industrialized nation, even if it is a general feeling that large savings could be achieved with the adoption of already existing design solutions, such as adequate choice of materials, appropriate surface treatments, and scheduled maintenance programs.

The present text aims to provide an up-to-date overview of several aspects of tribology and a theoretical framework to explain friction and wear-related problems, together with practical tools for their solution. The basic concepts of contact mechanics, friction, lubrication, and wear mechanisms are introduced, providing simplified analytical relationships, useful for quantitative assessments. Subsequently, the main wear processes are revised and guidelines on the most suitable designing solutions for each specific application are outlined. The final part of the text is devoted to the description of the main materials and surface treatments specifically developed for tribological applications. Tribological systems of particular engineering relevance are also presented.

The text is intended for students of undergraduate and graduate courses in the fields of engineering, applied physics, and materials science, who must develop a sound understanding of friction, wear, lubrication, and surface engineering. Also technicians, researchers and professionals may find useful indications to solve tribological problems related to their own work.

I would like to thank very much my colleagues from the Laboratory of Metallurgy and Tribology at the Department of Industrial Engineering of the University of Trento. Special thanks go to Stefano Gialanella, Vigilio Fontanari, Ulf Olofsson and Luca Fambri for their invaluable help in discussing and preparing the manuscript.

# Contents

# Chapter 1
# Surfaces in Contact

Friction and wear depend on the characteristics of the mating surfaces. The difficulty to explain and to predict with high accuracy such phenomena reflects the complex nature of the surfaces, which is determined by the material properties (such as the microstructure), the geometrical irregularities, the presence of oxides due to the interaction with the surrounding atmosphere, and the presence of organic molecules, water vapour or other impurities adsorbed from the environment. Therefore, when two bodies are brought into close contact, the relevant features of their surfaces determine the nature of the interaction, which has a mechanical character, with the formation of a stress and strain field in the contact region, and a physical-chemical nature, with the establishment of physical or chemical bonds.

To quantitatively evaluate the contact stresses, it is convenient to introduce the concept of *smooth surface*, i.e., of a surface free from geometrical irregularities. This is obviously an ideal vision since it is impossible to produce smooth surfaces at a molecular level. Using the contact mechanics and, in particular, the theoretical analysis developed by Hertz for linear elastic bodies under this assumption, useful relations for the contact stresses and deformations can be obtained. They can be profitably employed when the bodies are in elastic and frictionless contact, with the assumption that the radius of the contacting bodies is large compared to the contact zone size.

This chapter firstly introduces the main concepts of contact mechanics and the types of material response to the contact stresses. Subsequently, the micro-geometrical characteristics of real surfaces are illustrated, showing how they affect the contact at the microscopic level. The chapter is completed with the analysis of the phenomenon of adhesion, which depends on the physical-chemical interactions in the contact region.



G. Straffelini, *Friction and Wear*, Springer Tracts in Mechanical Engineering,
DOI 10.1007/978-3-319-05894-8_1

## 1.1 Contact Between Ideal Surfaces

### *1.1.1 Elastic Contact*

From a geometrical point of view, the contact between two bodies may be conformal or non-conformal (Fig. 1.1). A conformal contact occurs when the mating surfaces fit nearly together. Such a contact occurs, for example, in sliding bearings (between bearing and shaft) or in drawing processes (between wire and tool). If the contacting profiles are rather different, the contact is non-conformal, and it theoretically occurs at a point or along a line. A point contact is present, for example, in rolling bearings (between ball and seat), whereas a line contact occurs in gears (between tooth and tooth).

In the case of conformal contact, the nominal area of contact ($A_n$) has a finite extension and its determination is straightforward. Also in the case of non-conformal contact, the nominal area of contact is finite because of the local deformations. Figure 1.2a schematically shows the contact between two spheres of radius $R_1$ and

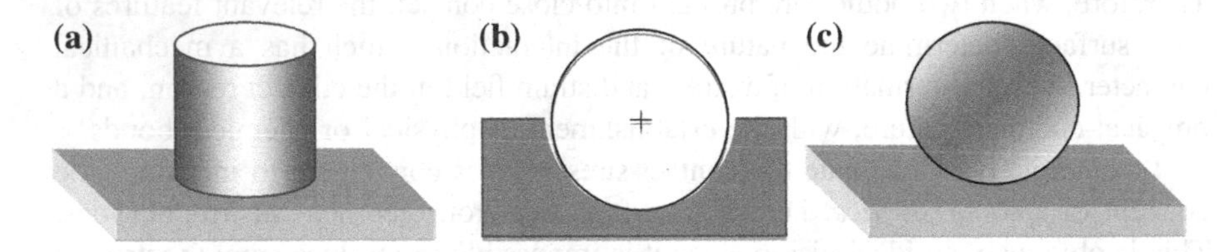


**Fig. 1.1** Contacts between ideal surfaces. Conformal contact between the base of a cylinder and a plane (**a**), and in a sliding bearing (**b**). Non-conformal contact between a sphere and a plane (**c**)

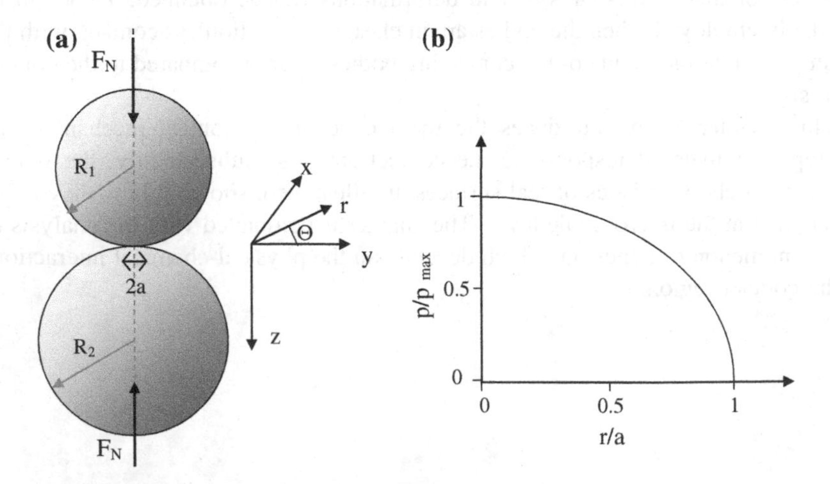


**Fig. 1.2** **a** Point contact between two spheres and definition of the coordinates (the x-y plane is the contact plane, and the z axis lies along the load line directed positively into the lower sphere); **b** variation of contact pressure as a function of distance r from the centre of the contact area (for z = 0)

$R_2$ (the subscripts 1 and 2 refer to the two bodies). In this case, the contact is theoretically in a point. Indicating with $E_1$ and $E_2$ the elastic moduli of the materials of the two spheres, the nominal (or apparent) area of contact can be evaluated from the following relationship:

$$A_n = \pi \cdot a^2 \tag{1.1}$$

where a is the radius of the circular contact region, given by:

$$a = \left(\frac{3 \cdot F_N \cdot R'}{2E'}\right)^{1/3} \tag{1.2}$$

where $F_N$ is the applied normal force, R′ and E′ are the reduced radius of curvature and the effective modulus of elasticity, respectively. They are defined as such:

$$\frac{1}{R'} = \frac{1}{R_1} + \frac{1}{R_2} \tag{1.3}$$

$$\frac{1}{E'} = \frac{1}{2} \cdot \left(\frac{1 - v_1^2}{E_1} + \frac{1 - v_2^2}{E_2}\right) \tag{1.4}$$

where $v_1$ and $v_2$ are the Poisson's ratios of the materials of the two spheres.

Figure 1.2b shows the contact pressure distribution (for z = 0). According to the Hertz theory, it is semi-elliptical:

$$p = -\sigma_z(z = 0) = p_{\max} \cdot \sqrt{1 - \left(\frac{r}{a}\right)^2} \tag{1.5}$$

and the maximum value, which occurs at the centre of the contact, i.e., at r = 0, and is known as the *Hertzian pressure*, is given by:

$$p_{\max} = \frac{3F_N}{2\pi a^2}. \tag{1.6}$$

The evolution of the corresponding surface stresses, in polar coordinates, is illustrated in Fig. 1.3a for $v_1 = v_2 = 0.3$. Note that a tensile radial stress develops at the edge of the nominal area of contact. Its maximum value (for r = a) is given by the following:

$$\sigma_r = p_{\max} \frac{1 - 2v}{3} \tag{1.7}$$

The stresses along the load line (z axis) and for r = 0 are shown in Fig. 1.3b. For symmetry reasons, they are also principal stresses. At the surface: $\sigma_z = -p_{max}$ and

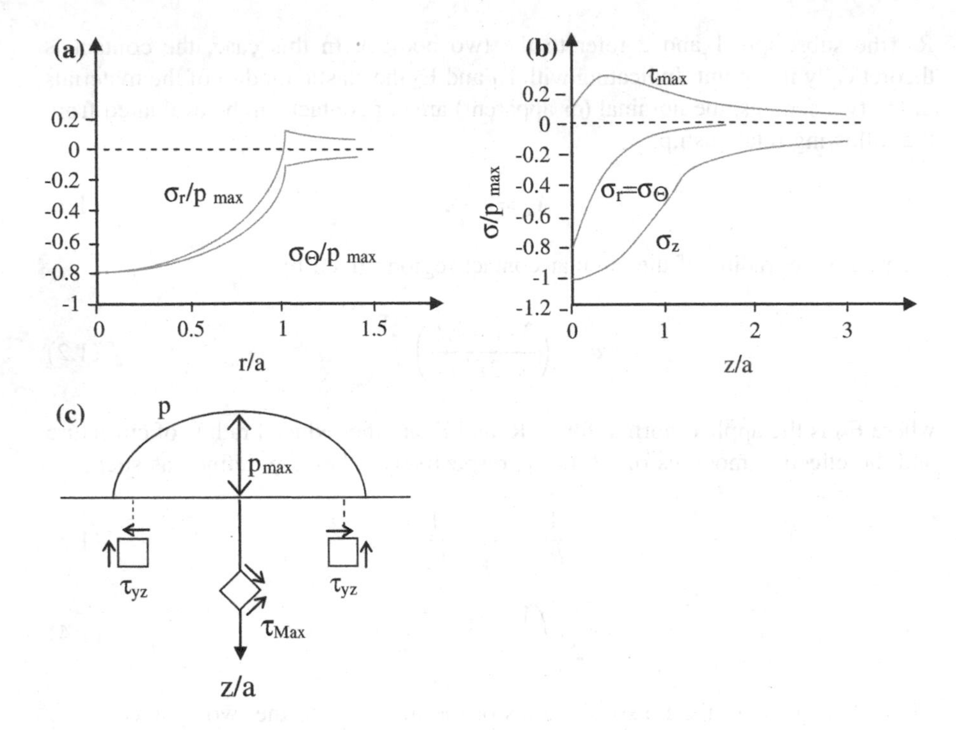


**Fig. 1.3** Elastic contact of spheres ($\nu = 0.3$): **a** distribution of normalized Hertzian stresses $\sigma_r$ and $\sigma_\Theta$ at the surface ($z = 0$); **b** distribution of normalized stresses $\sigma_r$, $\sigma_\Theta$, $\sigma_z$ and $\tau_{max}$ along the z axis, i.e., moving inside one of the two spheres; **c** schematization showing the maximum values of $\tau_{max}$ and $\tau_{yz}$

$\sigma_r = \sigma_\Theta = (1/2 - (1 + \nu))\, p_{max} = -0.8\, p_{max}$ for $\nu = 0.3$. In Fig. 1.3b the *maximum shear stress* ($\tau_{max}$) distribution along the z-axis is also shown. $\tau_{max}$ is defined by:

$$\tau_{max} = \frac{1}{2}|\sigma_z - \sigma_r| \tag{1.8}$$

It is oriented at 45° with respect to the contact surface and reaches its maximum value, equal to $\tau_{Max} = 0.31 p_{max}$ at a distance $z_m = 0.48a$ from the surface (these values are obtained with $\nu = 0.3$).

The stress field far from the z-axis is characterized by the presence of stresses whose modulus is lower than at the load line. Of particular importance is the occurrence of a shear stress, $\tau_{yz}$ (i.e., normal to the z and y axes), which is due to the lateral displacement of material beneath the flattened contact area. It is parallel to the contact surface and is maximum at a depth of 0.5a and at a distance of ± 0.87a from the z-axis. Its maximum value is $0.25 p_{max}$. See the schematization in Fig. 1.3c.

The parameters a, $p_{max}$, $\tau_{Max}$, $z_m$, and the nominal pressure $p_o$ (defined by $F_N/A_n$), for the configurations sphere/sphere (point contact) and cylinder/cylinder (line contact) are listed in Table 1.1. In the table, the relations for the mutual

**Table 1.1** Equations for the calculation of the contact parameters for elastic solids

| | Sphere/sphere (point contact) | Cylinder/cylinder (line contact) |
|---|---|---|
| a | $\left(\frac{3F_N R'}{2E'}\right)^{1/3}$ | $\sqrt{\frac{8F_N R'}{\pi E' L}}$ |
| δ | $a^2/R'$ | $\frac{2F_N}{\pi L}\left\{\begin{array}{c}\frac{1-\nu_1^2}{E_1}\left[\ln\left(\frac{4R_1}{a}\right)-\frac{1}{2}\right]+\\ \frac{1-\nu_2^2}{E_2}\left[\ln\left(\frac{4R_2}{a}\right)-\frac{1}{2}\right]\end{array}\right\}$ |
| $p_{max}$ | $3F_N/2\pi a^2$ | $2F_N/\pi aL$ |
| $p_o$ | $0.67p_{max}$ | $0.78p_{max}$ |
| $\tau_{Max}$ | $0.31p_{max}$ | $0.3p_{max}$ |
| $z_m$ | 0.48a | 0.786a |

The parameters $\tau_{Max}$ and $z_m$ in case of the sphere/sphere contact are determined considering $\nu = 0.3$. The cylinder/plane and sphere/plane contacts are special cases of the cylinder/cylinder and sphere/sphere contacts occurring when $R_1$ (or $R_2$) tends to infinity

displacements, δ, along the load line are also reported. The parameters a and δ are representative of the local elastic strains (for further details on the contact stresses and displacements see, for example, Refs. [1–3]).

For a conformal contact, the nominal contact pressure is simply given by the ratio between the applied normal force and the nominal area of contact ($A_n$). The stress distribution in the contact area, however, is strongly affected by the stress concentration exerted by the edges. If the radius of curvature of the edge tends to zero, the pressure at the edge tends to infinity, as schematically shown in Fig. 1.4a, for a contact between a punch and a plane. Such stress intensification can be alleviated by rounding the edges. Figure 1.4b schematically shows the contact between a punch with rounded corners and a plane. A method to calculate the contact pressures in this case is reported in Ref. [4].

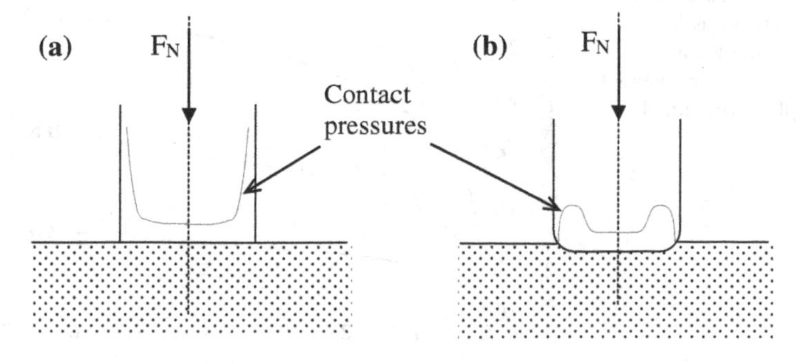


**Fig. 1.4** Pressure distribution for a contact between a punch and a plane. In **a** the radius of curvature of the edges tends to 0, while in **b** the effect of a rounding of the edges is shown

### 1.1.2 Viscoelastic Contact

Some materials, like polymers, may display a particular deformation behaviour that is affected by elastic, viscoelastic and plastic processes. Following the application of a stress $\sigma$, the total deformation $\varepsilon_t$, is thus given by the sum of three terms: the instantaneous elastic deformation, $\varepsilon_1$, the viscoelastic deformation, $\varepsilon_2$, and the plastic deformation, $\varepsilon_3$:

$$\varepsilon_t = \varepsilon_1 + \varepsilon_2 + \varepsilon_3 = \frac{\sigma}{E} + \frac{\sigma}{E_r} \cdot \left(1 - e^{\frac{E_r \cdot t}{\eta_r}}\right) + \frac{\sigma \cdot t}{\eta_0} \tag{1.9}$$

where E is the elastic modulus, $E_r$ is the viscoelastic modulus, $\eta_r$ is a damper parameter, $\eta_o$ is the viscosity parameter and t is time. As a consequence, the nominal area of contact is greater than that predicted by the Hertzian theory, and it increases with time. As an example, Fig. 1.5 shows the experimental dependence of the contact deformation displacement, $\delta$, as a function of time in the case of a polypropylene (PP) sphere pressed against an optically transparent plane [5]. In polymers the viscoelastic contribution is particularly marked if the material contains an amorphous phase and is at a temperature above its glass transition temperature, $T_g$. Polypropylene is a semi-crystalline polymer (with about 60 % of amorphous phase) and has a $T_g$ of −14 °C (see Sect. 6.8 for further details on the properties of polymers for tribological applications).

Similarly to elastic deformations, even viscoelastic deformations are recoverable, although not instantaneously but over a period of time after unloading. In addition, energy losses are associated to the viscoelastic loading and unloading cycles. Such energy dissipation may produce a noticeable material heating (especially in the case of cycling loading) because of the very low thermal conductivity of polymers. On the contrary, plastic deformations are permanent. The viscoelastic and plastic processes strongly depend on temperature and their intensity increases as temperature is increased, especially above the $T_g$-temperature of the polymer.

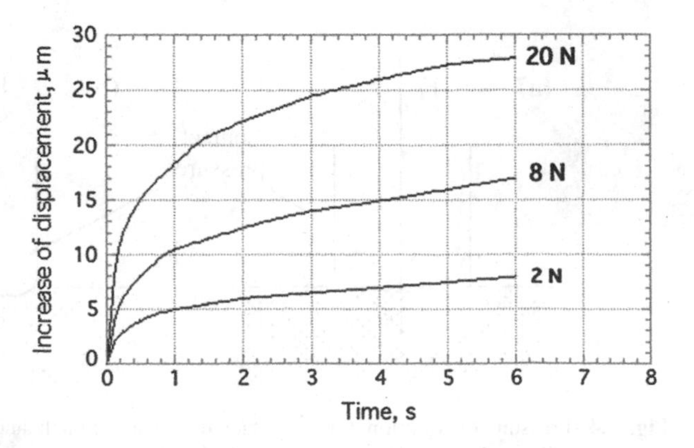


**Fig. 1.5** Variation of load line displacement as a function of time for a polypropylene sphere pressed against a plane (modified from [5])

### 1.1.3 Elastic-Plastic and Fully Plastic Contacts

If a material behaves in a ductile manner, the applied contact force may induce a localized plastic deformation if the equivalent stress at the most critical point reaches the *uniaxial yield stress* of the material, indicated with $\sigma_Y$. In such a case, the contact is no longer elastic but *elastic-plastic* [6]. In the case of conformal contact, yield starts at the surface first, possibly at the edges. In case of non-conformal contact, yield starts first at the depth $z_m$, when $\tau_{Max}$ reaches the *shear yield stress* $\tau_Y$ given by $\sigma_Y/2$ (following the Tresca yield criterion). This means that subsurface localized yield starts when the Hertzian pressure, $p_{max}$, becomes equal to $1.61\sigma_Y$ for a point contact, and to $1.67\sigma_Y$ for a line contact. The schematic of Fig. 1.6a illustrates the condition of elastic-plastic contact in the case of a sphere/plane contact (where the plane has a lower hardness than the sphere). By further increasing the applied load, the size of the plastic zone also increases. If the applied load is removed when the contact pressure is below a specific limit, the additional loads, of the same magnitude, which are possibly applied, give rise to elastic deformations only. This phenomenon is also defined by the term *elastic shakedown* and will be further considered in the next chapter.

If the applied load is high to the point that the plastic deformation reaches the surface, the contact becomes *fully plastic* (Fig. 1.6b). The increase in the plastic zone size from first yield is made difficult by the local stress triaxiality. Fully plastic contact is therefore achieved when the nominal pressure ($p_0$) reaches a critical value, called the *yield pressure* ($p_Y$) that is greater than the uniaxial yield stress. Typically, $p_Y = b\sigma_Y$, where b is a constant greater than unity, which takes into account the difficulty of the spreading of plasticity (it depends on the geometry of the contact, the applied load, and the materials properties). As an example, in the case of a sphere in contact with a plane, it was obtained that b = 2.8 using the slip-line field theory.

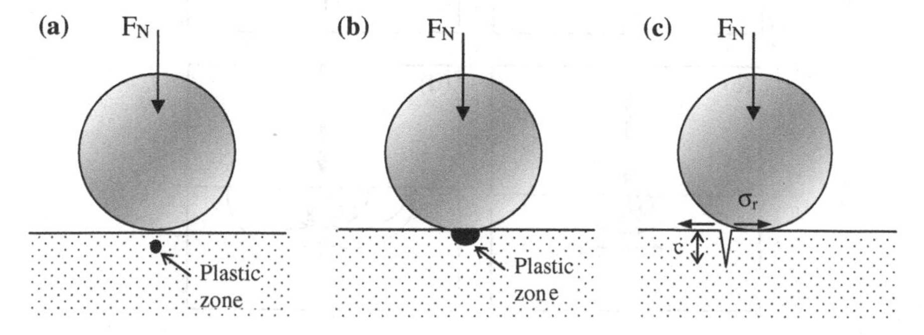


**Fig. 1.6** Elasto-plastic (**a**) fully plastic (**b**) and brittle contact (**c**) between a sphere and a plane

### 1.1.4 Brittle Contact

If the yield strength of a material is high and its fracture toughness is low, the increase in the applied force may lead to brittle fracture at the contact surface. The contact, in this case, is *brittle* [7]. A brittle contact may take place when a micro-crack is present on the surface of one of the two mating bodies and such micro-crack is subject to a critical opening tensile stress. Consider, for example, Fig. 1.6c that displays a sphere in contact with a plane that contains a surface micro-crack, right in correspondence with the outer edge of the nominal area of contact. Brittle contact occurs if the local radial tensile stress ($\sigma_r$), given by the relation 1.7, is greater than the critical value, $\sigma_F$, given by the following relation:

$$\sigma_F = \frac{K_{IC}}{1.12\sqrt{\pi c}} \tag{1.10}$$

where c is the length of the micro-crack and $K_{IC}$ is the fracture toughness of the material of the plane. Surface cracks formed in this way are often called *C-cracks*, since their shape is conical. A brittle contact may take place even in the absence of a surface micro-crack if the load is applied in a very concentrated point, as is the case of the angular contact of a ceramic particle. Figure 1.7 shows the mechanism proposed by Lawn and Swain [8]. The applied load may promote the formation of a

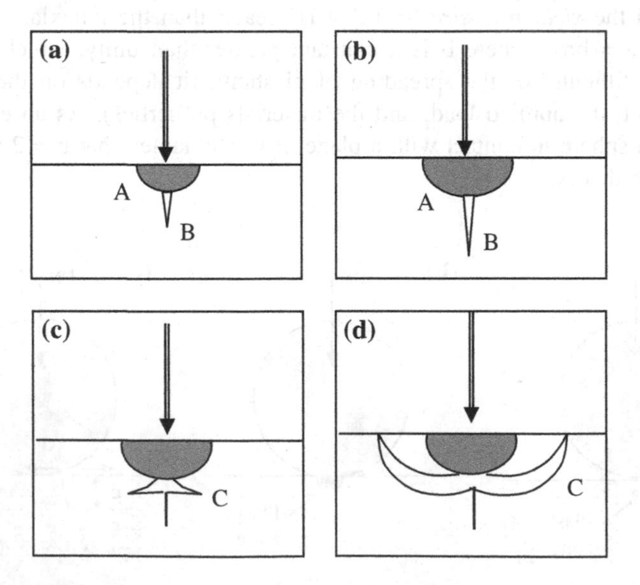


**Fig. 1.7** Formation of radial and lateral cracks by the Lawn and Swain mechanism (modified from [8]). **a** Plastic deformation at the angular contact (*A*) with the formation and opening of a crack perpendicular to the surface (*B*); **b** propagation of the crack with increasing the applied load; **c** load removal followed by closing of the perpendicular crack and formation of new radial cracks (*C*) by the action of local residual stresses; **d** spalling when the radial cracks reach the surface

very concentrated plastic deformation due to the large and compressive local stresses. If it is increased over a critical value, a crack is formed, and it propagates towards the interior of the plane, perpendicularly to the surface. If the load is then removed, the local residual stresses may induce the formation of radial cracks that are parallel to the surface. Such cracks are formed if the initially applied load exceeds a critical value, which is proportional to the so-called *brittleness index*, given by the ratio $H/K_{IC}$, where H is the material hardness.

### 1.1.5 Materials Response to the Contact Stresses

The materials constituting the bodies in contact can differently deform to the applied stresses. The kind of response depends on the applied force, the properties of the materials' involved, and the geometry of the bodies in contact. The most important materials' properties are the elastic properties, such as the longitudinal modulus of elasticity and Poisson's ratio, the yield strength, hardness and fracture toughness. Table 1.2 shows such properties for some engineering metals, ceramics, and polymers.

**Table 1.2** Selected mechanical properties of metals, ceramics and polymers

| Materials | E (GPa) | $\sigma_Y$ (MPa) | H (kg/mm$^2$) | $K_{IC}$, (MPa m$^{1/2}$) | $E/\sigma_Y$ | E/H | $E/K_{Ic}$ (m$^{-1/2}$) | $H/K_I$ (m$^{-1/2}$) |
|---|---|---|---|---|---|---|---|---|
| *Metals* | | | | | | | | |
| Ferritic-pearlitic steel | 207 | 400 | 200 | 140 | 517.5 | 103.5 | 1478 | 14.3 |
| Heat treated steel | 207 | 1200 | 430 | 80 | 172.5 | 49 | 2600 | 54 |
| Phosphorus bronze | 110 | 350 | 120 | 70 | 314.3 | 91.7 | 1571 | 17.1 |
| Aluminum alloys 6061 T6 | 70 | 275 | 100 | 25 | 254.5 | 71.3 | 2800 | 39.2 |
| *Ceramics* | | | | | | | | |
| Glass | 72 | 3600 | 500 | 0.7 | 20 | 14.4 | 102,800 | 7140 |
| Alumina | 380 | 5230 | 1400 | 4 | 72.6 | 27.1 | 95,000 | 3500 |
| Si nitride | 310 | 4250 | 1800 | 4 | 72.9 | 17.2 | 77,500 | 4500 |
| Si carbide | 410 | 10,000 | 3000 | 4 | 41 | 13.7 | 102,500 | 7500 |
| *Polymers* | | | | | | | | |
| Nylon | 3 | 60 | 12 | 3 | 50 | 25 | 1000 | 40 |
| HD PE | 1 | 30 | | 2 | 33 | | 500 | |
| PMMA | 3 | 70 | 30 | 1 | 95.5 | 28.6 | 3000 | 300 |

For ceramic materials $\sigma_Y$ was obtained from hardness measurements using the relationship: $\sigma_Y = H/3$ [9]. Hardness, H, is defined by the resistance to surface penetration under a given force. In the text, it is always expressed in kg/mm$^2$, as obtained from Brinell or Vickers tests

The onset of the elastic-plastic contact is relatively easy in materials with high values of the ratio $E/\sigma_Y$ (or E/H), which is called the *plasticity index*. A brittle contact, if surface micro-cracks are present, is rather easy in materials characterized by high values of the ratio $E/K_{Ic}$, while the brittle contact due to action of a concentrated point load is easy in materials with high values of the ratio $\sigma_Y/K_{Ic}$ (both ratios are *brittleness indices*). In Table 1.2, the calculated indices of the selected materials are also reported. The data show that metallic materials would easily provide elastic-plastic (and fully plastic) contact, while a brittle contact is more common for ceramics. Polymeric materials do not provide easily both plastic and brittle contacts, although they are characterized by a low yield strength and a low toughness. Polymers may give a viscoelastic contact if the temperature is greater than $T_g$.

## 1.2 Surface Roughness

The concepts and relationships reported in the previous paragraph are strictly valid for ideally smooth surfaces. However, a close look at the real surface of a solid body (Fig. 1.8, first zoom) shows that it is not really smooth but consists of asperities and valleys of variable height, typically between 0.1 and a few micrometres. Furthermore, a surface is usually characterized by the presence of defects such as scratches, holes, cracks, and inclusions, having dimensions up to 10 μm or more. Thus each solid surface is characterized by a certain *roughness*.

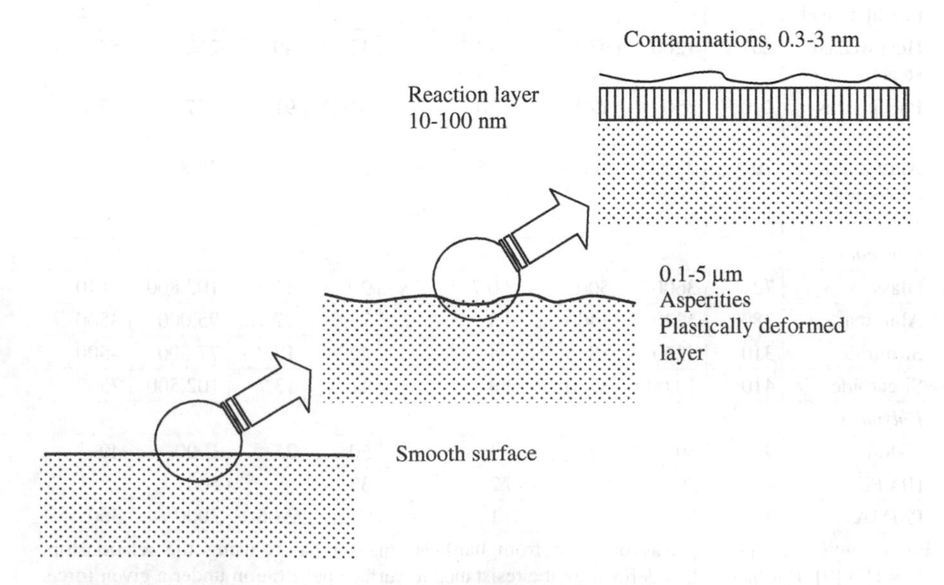


**Fig. 1.8** Schematic of the microstructural characteristics of a material surface

The surface topography is quantified by different geometric parameters. One parameter, which is widely used in engineering applications, is the *average roughness* (also known as centre line average—CLA), which is indicated with $R_a$ and is defined by the following relation:

$$R_a = \frac{1}{n}\sum_{i=1}^{n} |y_i| \qquad (1.11)$$

where $y_i$ are the distances from the mean line of n of points of the roughness profile, ideally obtained by a surface normal section (Fig. 1.9). Another important engineering parameter is the *root-mean-square roughness*. It is indicated with $R_q$ (or with RMS) and is defined by the following relation:

$$R_q = \sqrt{\frac{1}{n}\sum_{i=1}^{n} y_i^2}. \qquad (1.12)$$

For a Gaussian distribution of the distances (or, heights) from the mean line, the ratio $R_q/R_a \approx 1.25$. The parameters $R_a$ and $R_q$ are typically expressed in micrometres and are often evaluated from data obtained using a *stylus profilometer*. With this method, a diamond stylus is moved along the surface and the vertical movement is measured. Several non-contact methods have been also developed, such as optical and capacitive methods, which are better suited for soft materials. As an example, Fig. 1.10 shows the roughness profiles of an AISI D2 tool steel surface polished to obtain two different finishing levels. As shown in Table 1.3, the surfaces with higher roughness are those obtained by foundry processes and hot working processes. After cold working (often in the presence of lubrication) and machining, surfaces with lower surface roughness are obtained (see, for example, Ref. [10]). With super finishing processes $R_a$-values down to 0.025 μm (or even less) may be achieved [11].

The height readings obtained using the stylus profilometer or other methods can be utilized to calculate several roughness parameters according to international standards. This is often done with the measuring instruments built in software. In addition to *amplitude parameters*, such as $R_a$ and $R_q$ (and relevant standard deviations), *spacing parameters*, such as the mean spacing of adjacent asperities, or *hybrid parameters* can be calculated [12]. The hybrid parameters are a combination

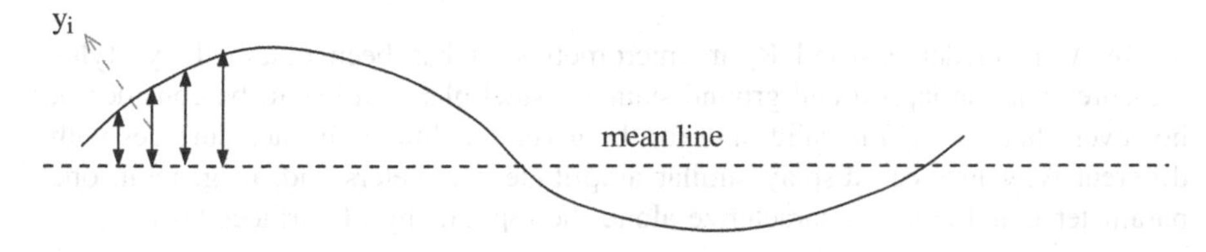


**Fig. 1.9** Schematic of roughness profile for the evaluation of parameters $R_a$ and $R_q$

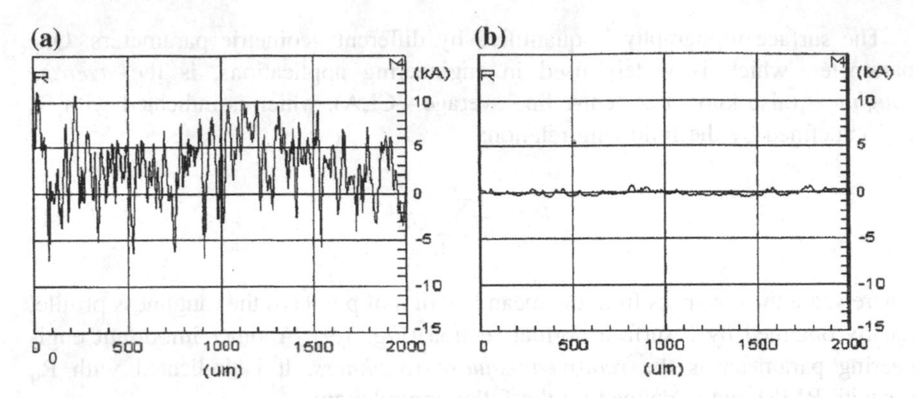


**Fig. 1.10** Roughness profiles of an AISI D2 tool steel surface metallographically polished to obtain two different finishing levels: **a** $R_a = 0.22$ μm; **b** $R_a = 0.02$ μm (mirror finish). Please note that the x- and y-scales are very different, and the graphs thus give a distorted view of roughness. The asperities and valleys are not so sharp as they appear in the roughness profiles!

**Table 1.3** Typical roughness values, $R_a$ in micrometres, of surfaces obtained with different production methods

| Foundry operations | | Cold working processes | |
|---|---|---|---|
| Sand casting | 8–25 | Stamping | 0.6–5 |
| Shell casting | 1.5–4 | Rolling | 0.16–2 |
| Die casting | 0.8–1.6 | Drawing | 0.5–3 |
| *Hot working processes* | | *Machining* | |
| Forging | 4–15 | Turning | 0.4–3 |
| Rolling | 10–25 | Grinding | 0.1–1.2 |
| Extrusion | 0.8–4 | Lapping | 0.05–0.4 |

of amplitude and spacing properties. An example is the *mean slope of the profile* that varies on average from 1° and 5° and is generally less than 10°. It may be evaluated from surface topography measurements data by calculating all slopes between each two successive points of the profile and then making their average. If obtained in this way, it is also indicated with $\Delta_a$. An experimental relationship between $\Delta_a$ and $R_a$ is [13]:

$$\Delta_a = 0.108 \cdot R_a^{1.165} \tag{1.13}$$

where $\Delta_a$ is in degrees and $R_a$ in micrometres. It has been obtained by stylus measurements on lapped and ground stainless steel plates. It has to be considered, however, that Eq. 1.13 is valid only for the given conditions. In fact, surfaces with different $\Delta_a$-values can display similar amplitude parameters and, in general, one parameter is not able to characterize alone the topography of surfaces [14].

If the surface of an engineering material is observed even closer (Fig. 1.8, second zoom), the existence of different surface layers can be recognized:

(1) A deformed layer;
(2) A reacted layer;
(3) A contaminated layer;

The characteristics of the plastically deformed layer depend on the material and the manufacturing processes. Typically, in metals a work-hardened layer is formed with possible local microstructural modifications (such as the formation of white layers that will be considered in Sect. 6.1.1). The reacted layer is formed spontaneously due to exposure to the surrounding environment. In metals, an oxide layer is formed after exposure in air. In ferrous alloys, a mixture of $Fe_3O_4$ and $Fe_2O_3$ is observed at the top surface layer. In addition, FeO may be present in an intermediate layer above the bulk. A very thin and compact layer of chromium oxide, $Cr_2O_3$, covers stainless steel. On the surface of aluminium alloys a thin layer of amorphous $Al_2O_3$ oxide is present, possibly covered by a thicker and porous layer of hydrate oxide. On the surface of copper two oxide layers may be present: an innermost $Cu_2O$ covered by CuO. The surface oxides have a typical thickness of 10 nm. They are well bonded to the metal if the ratio between their specific volume and that of the underlying metal is greater than 1. This happens in most metals such as iron, aluminium, copper, nickel, cobalt, zinc, chromium, molybdenum (and their alloys). In titanium, however, this ratio is less than 1. Surface oxides may be also present on non-oxide ceramics such as carbides or nitrides. In polymers, no oxides are present on their surface.

The outermost contaminated layer is made up of water vapour, hydrocarbons, and gases and has a typical thickness of 2 nm. The contaminants are mainly physically adsorbed on the surface through rather weak van der Waals forces. The adsorption of hydrocarbons is particularly important if the component is operating close to lubricated machinery, since oils tend to vaporize.

## 1.3 Real Area of Contact

When two surfaces are brought into contact only few asperities actually touch each other, as shown schematically in Fig. 1.11. The *real area of contact* ($A_r$) is therefore given by the sum of the individual areas ($A_i$) that form at each contact spot:

$$A_r = \sum_{i=1}^{N} A_i \quad (1.14)$$

where N is the number of contacting asperities. $A_r$ is therefore smaller than the nominal area of contact, $A_n$ (it is typically $10^{-2}$–$10^{-6}$ $A_n$).

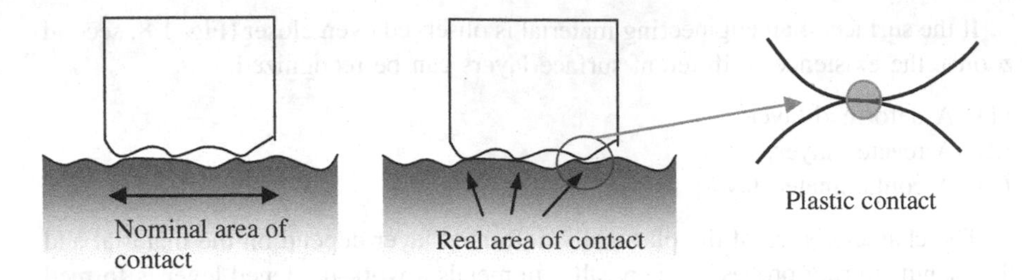


**Fig. 1.11** Definition of nominal and real area of contact and lay out of the junction between two plastic asperity contact

The contact at the asperities may be elastic, plastic, or mixed. Greenwood and Williamson proposed a simplified model assuming the contact between spherically shaped asperities with the same radius of curvature and that follow a Gaussian height distribution [15]. They proposed the following index:

$$\Psi = \frac{E^*}{H}\sqrt{\frac{\sigma_s}{R_s}} \tag{1.15}$$

where $E^* = E'/2$, H is the hardness of the softest material in contact, $\sigma_s$ is the composite standard deviation of the asperity height distribution, and $R_s$ is the composite asperity radius ($1/R_s = 1/R_1 + 1/R_2$). Both $\sigma_s$ and $R_s$ can be obtained from profilometry data. Greenwood and Williamson obtained that when $\Psi < 0.61$ the elastic contact at the asperity tips dominates, whereas when $\Psi > 1$ the plastic contact dominates. Therefore, the kind of contact depends on the plasticity index and the roughness characteristics of the surfaces. With reference to the data listed in Table 1.2, it can be then argued that in metals the contact at the asperities is almost always plastic since the plasticity index is quite high. On the contrary, in ceramics or polymers the asperity contacts may be prevailing elastic if their surface roughness is sufficiently low. In practice, when engineering polymers and ceramics are involved, the contact at the asperities may be assumed to be mixed.

A simple relation for $A_r$ can be obtained in the case of plastic contacts at the asperities (Fig. 1.11). At equilibrium:

$$F_N = \sum_{i=1}^{N} p_Y A_i = p_Y A_r \tag{1.16}$$

where N is the number of asperity junctions. Then

$$A_r = \frac{F_N}{p_Y} \tag{1.17}$$

The yield pressure ($p_Y$) is given by $b\sigma_Y$ as described in Sect. 1.1.3, and b depends on the geometry of the asperities. To a first approximation $p_Y = H$, where

H is the hardness of the material or, better, its microhardness. In the case of elastic contacts a relation for estimating $A_r$ can be obtained by the Greenwood and Williamson model:

$$A_r \cong \frac{3.2F_N}{E^* \sqrt{\frac{\sigma_s}{R_s}}} \tag{1.18}$$

In either plastic or elastic contacts, $A_r$ is thus independent from $A_n$ and it is proportional to the normal load, $F_N$. In plastic contacts, $A_r$ decreases as the hardness of the softest mating material is increased, whereas in elastic contacts, $A_r$ decreases as $E^*$ is increased and the composite roughness is decreased (in particular, as the radius if the asperity tips is increased and the scatter in the height distribution is decreased). It is argued that as normal load is increased, the number N of asperity contacts is increased as well, whereas the average contact size is almost independent from load. In fact, as load is increased the size of pre-existing contact spots increases, but new contacts of smaller size also form. N can be estimated by assuming that each junction is circular in shape, with a mean radius r:

$$N = \frac{A_r}{\pi r^2} = \frac{F_N}{p_Y \pi r^2} \tag{1.19}$$

This relation is valid if the nominal pressure is low (roughly less than $\sigma_Y/2$). In fact, if the applied load becomes so high that $A_r$ tends to $A_n$, the average size of the contact spots increases and N decreases (at the limit, when $A_r = A_n$ only one macro junction is formed).

The experimental evaluation of r is quite difficult. In the case of metals, typical values reported in the literature are of the order of $10^{-5}$–$10^{-6}$ m, and it is common that r is inversely proportional to $p_Y$ [16].

## 1.4 Adhesion Between Surfaces in Contact

If two bodies are brought into contact with an applied normal force, a force may be required to pull the surfaces apart after removal of the normal force (Fig. 1.12a). This force is called *adhesion force*, since the phenomenon of adhesion at the asperity contacts is responsible for its appearance. As an example, Fig. 1.12b shows the measured force between a bioskin probe and different surfaces. The maximum negative force can be taken as the adhesion force.

The adhesion between two surfaces may be due to mechanical, chemical, or physical interactions at the areas where the materials are in intimate contact, possibly favoured by local intense plastic deformations (Fig. 1.11). In the contact of engineering surfaces, the adhesion at these *junctions* is mainly due to the formation

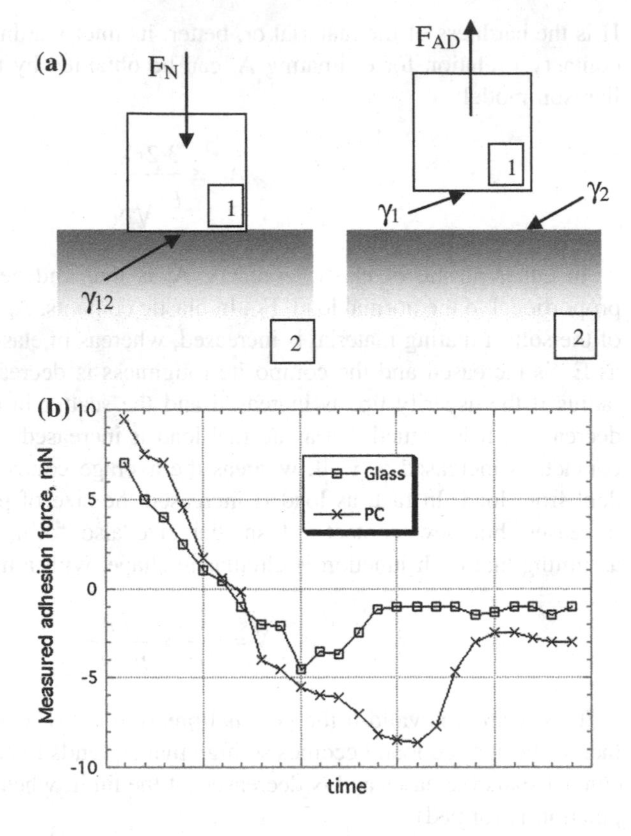


**Fig. 1.12** **a** Adhesion between two bodies in contact and definition of adhesion force, $F_{AD}$. **b** Adhesion test between a bioskin probe and two different surfaces. The probe was firstly pushed against the surfaces with a force of 1 N. Then it was retracted and the force between the probe and the surface was recorded. More negative force shows that the probe was more stuck to the surface due to the adhesion (courtesy of Duvefelt and Olofsson)

of rather weak *van der Waals bonds* and also *hydrogen bonds* in polymers. They involve surface atoms that have free unsaturated bonds, or dipole-dipole interactions between polar molecules. Strong interactions, characterized by chemical bonding or interdiffusion phenomena, can occur during the deposition of thin coatings on various substrates, as discussed in Sect. 7.5.2.

A theoretical evaluation of the adhesion forces is quite difficult. Because of this, the thermodynamic concept of the *work of adhesion* per unit area has been introduced. It is usually indicated with $W_{12}$ (where the subscripts refer to the two materials in contact) and represents the energy that must be theoretically supplied to separate two surfaces in contact. It is defined as the following:

$$W_{12} = \gamma_1 + \gamma_2 - \gamma_{12} \tag{1.20}$$

where $\gamma_1$ and $\gamma_2$ are the *surface energies* of the two bodies (more precisely, the surface energies of the interfaces with the surrounding environment) and $\gamma_{12}$ is the surface energy of the interface that the two bodies form when they are in contact. Surface energies vary between 1 and 3 J/m$^2$ for clean metals, between 0.1 and 0.5 for ceramics and are lower than 0.1 J/m$^2$ for polymers [17].

**Table 1.4** Values of surface energy for different materials, the data relate to clean surfaces and are taken from Refs. [17, 18]

| Metal | γ (J/m²) | Ceramic | γ (J/m²) | Polymer | γ (J/m²) |
|---|---|---|---|---|---|
| Fe | 1.5 | $Al_2O_3$ | 0.8 | HDPE | 0.035 |
| Cu | 1.1 | $ZrO_2$ | 0.53 | PMMA | 0.045 |
| Al | 0.9 | TiC | 0.9 | PA 6 | 0.05 |
| Ni | 1.7 | ZrC | 0.6 | PVC | 0.045 |
| Ag | 0.9 | | | PTFE | 0.018 |
| Pb | 0.45 | | | | |
| Cr | 1 | | | | |

In Table 1.4, typical values of the surface energy for some materials are listed. In the case of metals, the reported values refer to clean surfaces but with the presence of an unavoidable (albeit thin) native oxide layer. The possible presence of organic contaminants on the surfaces strongly affects their surface energy values and, in general, tends to reduce them.

Since the determination of $\gamma_{12}$ is quite difficult, the work of adhesion is better estimated by the following relationship:

$$W_{12} = c(\gamma_1 + \gamma_2) \tag{1.21}$$

where c is a constant that is 1 for the contact between identical materials, and decreases as the *tribological compatibility* between the materials is decreased. The definition of tribological compatibility is not simple. Following Rabinowicz [18], two metals can be considered *compatible* when their phase diagram shows that they have high mutual solubility (>1 %) and are capable of forming intermetallic compounds. Two metals are *partially compatible* if they show a limited mutual solid solubility, between 0.1 and 1 %. They are *partially incompatible* if their mutual solid solubility is less than 0.1 %. Finally, two metals are *incompatible* if their mutual solubility is negligible. Following this approach, Rabinowicz has determined the compatibility chart shown in Fig. 1.13. With reference to experimental data, the *compatibility parameter*, c, is then set to 1 for identical metals, 0.5 for compatible metals, 0.32 for partially compatible metals, 0.2 for partially incompatible metals and 0.12 for incompatible metals. In the contact between ceramics, c can be set to 0.6 for compatible ceramics (such as two oxides or two nitrides), and 0.36 for incompatible ceramics. In the contact between polymers, c typically ranges from 0.8 to 0.95 [19]. The contacts between metals and ceramics, metals and polymers and ceramics and polymers can be assumed to be tribologically incompatible, and c can be set to 0.12 in every case.

In real contacts the adhesion force, $F_{AD}$, is expected to be proportional to the product of the work of adhesion and the real area of contact, $A_r$, since the van der Waals bonds form only at the asperity contacts. We may thus write: $F_{AD} \approx W_{12} A_r$. Considering Eq. 1.17 we can thus obtain an expression for the so-called *adhesion*

| | W | Mo | Cr | Co | Ni | Fe | Nb | Pt | Zr | Ti | Cu | Au | Ag | Al | Zn | Mg | Cd | Sn | Pb | In |
|---|---|---|---|---|---|---|---|---|---|---|---|---|---|---|---|---|---|---|---|---|
| **In** | | | | ⍰ | ⍰ | □ | | △ | △ | ⍰ | ⍰ | △ | △ | □ | ⍰ | △ | △ | △ | △ | ■ |
| **Pb** | ⍰ | ⍰ | □ | □ | □ | □ | □ | △ | △ | △ | □ | ⍰ | ⍰ | □ | □ | ⍰ | ⍰ | △ | ■ | |
| **Sn** | ⍰ | | □ | ⍰ | ⍰ | ⍰ | ⍰ | △ | ⍰ | △ | ⍰ | △ | △ | ⍰ | ⍰ | ⍰ | △ | ■ | | |
| **Cd** | | | ⍰ | ⍰ | ⍰ | ⍰ | | △ | △ | ⍰ | ⍰ | △ | △ | □ | △ | △ | ■ | | | |
| **Mg** | | ⍰ | | ⍰ | ⍰ | ⍰ | □ | | △ | ⍰ | △ | △ | △ | △ | ⍰ | ■ | | | | |
| **Zn** | | ⍰ | △ | △ | △ | △ | ⍰ | △ | ⍰ | ⍰ | △ | △ | △ | △ | ■ | | | | | |
| **Al** | △ | ⍰ | △ | ⍰ | △ | △ | △ | ⍰ | ⍰ | △ | △ | △ | △ | ■ | | | | | | |
| **Ag** | □ | ⍰ | □ | □ | □ | □ | ⍰ | △ | ⍰ | △ | ⍰ | △ | ■ | | | | | | | |
| **Au** | △ | ⍰ | △ | ⍰ | △ | △ | | △ | △ | ⍰ | △ | ■ | | | | | | | | |
| **Cu** | ⍰ | □ | □ | △ | △ | ⍰ | ⍰ | △ | △ | ⍰ | ■ | | | | | | | | | |
| **Ti** | △ | △ | △ | ⍰ | △ | △ | △ | △ | | ■ | | | | | | | | | | |
| **Zr** | ⍰ | △ | ⍰ | △ | ⍰ | ⍰ | △ | ⍰ | ■ | | | | | | | | | | | |
| **Pt** | △ | △ | △ | △ | △ | △ | △ | ■ | | | | | | | | | | | | |
| **Nb** | △ | △ | △ | △ | △ | △ | ■ | | | | | | | | | | | | | |
| **Fe** | △ | △ | △ | △ | △ | ■ | | | | | | | | | | | | | | |
| **Ni** | △ | △ | △ | △ | ■ | | | | | | | | | | | | | | | |
| **Co** | △ | △ | △ | ■ | | | | | | | | | | | | | | | | |
| **Cr** | △ | △ | ■ | | | | | | | | | | | | | | | | | |
| **Mo** | △ | ■ | | | | | | | | | | | | | | | | | | |
| **W** | ■ | | | | | | | | | | | | | | | | | | | |

□ *Incompatible*
○ *Partially incompatible*
⍰ *Partially compatible*
△ *Compatible*
■ *Identical*

**Fig. 1.13** Compatibility chart for metals (modified from [18])

*coefficient*, given by the ratio $F_{AD}/F_N$, where $F_N$ is the load applied to establish the contact:

$$\frac{F_{AD}}{F_N} \propto \frac{W_{12} \cdot A_r}{p_Y \cdot A_r} \cong \frac{W_{12}}{H} \tag{1.22}$$

where H is the hardness of the softest material in contact (if the two materials are different). In Table 1.5, the ratio $W_{12}/H$ is calculated for some material pairs. It can be noted that iron and silver are incompatible and the relative $W_{12}/H$-value is lower than that corresponding to the Fe–Fe pair. However, iron and lead are also incompatible, but their $W_{12}/H$-ratio is quite high. This is because of the low hardness of lead that gives rise to a large real area of contact. The Fe-polymer

**Table 1.5** Calculation of the $W_{12}/H$-values for some metal pairs. H is the hardness of the softer metal in contact

| | c | $W_{12}$ (J/m$^2$) | H (kg/mm$^2$) | $W_{12}/H$ ($10^{-7}$ m) |
|---|---|---|---|---|
| Fe-Fe | 1 | 3 | 80 | 0.038 |
| Fe-Ag | 0.12 | 0.29 | 50 | 0.006 |
| Fe-Pb | 0.12 | 0.23 | 4 | 0.058 |
| Cu-Cu | 1 | 2.2 | 80 | 0.028 |
| Fe-Cu | 0.32 | 0.83 | 80 | 0.01 |
| Fe-Polymer | 0.12 | <0.2 | 10 | <0.002 |

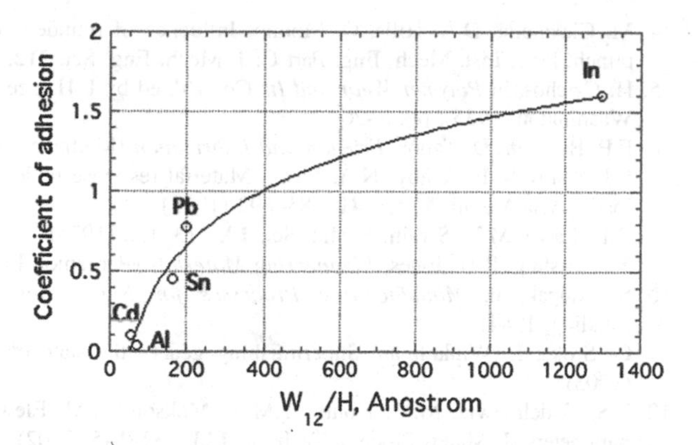


**Fig. 1.14** Coefficient of adhesion versus $W_{12}/H$ (H is the hardness of the soft metal) obtained by McFerlane and Tabor (modified from [18, 22])

contact is characterized by a low $W_{12}/H$-value because of the very low work of adhesion given by the low $\gamma$-values of polymers.

The experimental verification of relation 1.22 by tensile pulling on the interface is quite difficult for different reasons. First of all, it is common experience that adhesion is usually very poor in ordinary conditions. This is because surfaces are usually very contaminated and $W_{12}$ is thus quite low. In order to carry out *pull-off force* measurements with available instruments, the surfaces have to be very clean. But despite this, two more and interrelated effects may render the experimental determination of $F_{AD}$ very difficult. They are connected to the elastic deformations at the contact asperities. After unloading, the elastic part of the local deformation is released and the spring-back favours the detachment of the junctions. Such an effect is particularly pronounced when an elastic contact at the asperities is prevailing and when the dispersion in asperity heights (given by $\sigma_s$) is quite large [20]. The highest adhesion and the highest $F_{AD}$-values are thus achieved when the contacting materials are soft and the local elastic deformations are low, and when all asperities are of the same height and thus the junctions broke simultaneously during unloading [21].

In Fig. 1.14, the classic results of McFerlane and Tabor are shown [18, 22]. The graph shows the experimental coefficient of adhesion of a clean steel ball pressed for 1000 s against different clean planes made of soft metals. The experimental data highlight an increasing trend of the coefficient of adhesion with the ratio $W_{12}/H$, in good agreement with Eq. 1.22.

## References

1. K.L. Johnson, *Contact Mechanics* (Cambridge University Press, Cambridge, 1985)
2. J.E. Shigley, C.R. Mischke, R.G. Budynas, *Mechanical Engineering Design*, 7th edn. (McGraw-Hill, New York, 2004)
3. G.W. Stachowiak, A.W. Batchelor, *Engineering Tribology* (Elsevier, New York, 1993)

4. M. Ciavarella, D.A. Hills, G. Monno, Influence of rounded edges on indentation by a flat punch. Proc. Inst. Mech. Eng. Part C: J. Mech. Eng. Sci. **212**, 319–328 (1998)
5. H. Czichos, in *Polymer Wear and Its Control*, ed by L.H. Lee (American Chemical Society, Washington, 1985), pp. 3–26
6. F.P. Bowden, D. Tabor, *Friction and Lubrication* (Methuen and Co. Ltd., London, 1956)
7. S.J. Sharp, M.F. Ashby, N.A. Fleck, Material response under static and sliding indentation loads. Acta Metall. Mater. **41**, 685–692 (1993)
8. B.R. Lawn, M.V. Swain, J. Mat. Sci. **10**, 113–122 (1975)
9. M.F. Ashby, R.H. Jones, *Engineering Materials* (Pergamon Press, Oxford, 1980)
10. S. Kalpakjian, *Manufacturing Processes for Engineering Materials* (Addison-Wesley, Reading, 1984)
11. G. Sroka, L. Winkelman, Superfinishing gears—the state of the art. Gear Technol. **20**, 6 (2003)
12. E.S. Gadelmawla, M.M. Koura, T.M.A. Maksoud, I.M. Elewa, H.H. Soliman, Roughness parameters. J. Mater. Process. Technol. **123**, 133–145 (2002)
13. D. Dawson et al., in *Polymer Wear and Its Control*, ed. by L.H. Lee (American Chemical Society, Washington, 1985), pp. 171–187
14. P.J. Blau, *Friction Science and Technology* (Marcel Dekker Inc., New York, 1996)
15. J.A. Greenwood, J.B.P. Williamson, Contact of nominally flat surfaces. Proc. R. Soc. (Lond) A, **295**, 295–300 (1966)
16. S.C. Lim, M.F. Ashby, Wear-mechanism maps. Acta Metall. Mater. **35**, 1–24 (1987)
17. D.S. Ramai, L.P. DeMeio, K.L. Mittal, *Fundamentals of Adhesion and Interfaces*, VSP (1995)
18. E. Rabinowicz, *Friction and Wear of Materials*, 2nd edn. (Wiley, New York, 1995)
19. K.H. Czichos, K.H. Habig, *Tribologie Handbuch, Reibung und Verlschleiss* (Vieweg, Braunschweig, 1992)
20. K.L. Johnson, Mechanics of adhesion. Tribol. Int. **31**, 413–418 (1998)
21. M. Benedetti, D. Bortoluzzi, M. Da Lio, V. Fontanari The influence of adhesion and subnewton pull-off forces on the release of objects in outer space. Trans. ASME **128**, 828–840 (2006)
22. J.S. McFarlane, D. Tabor, Adhesion of solids on the influence of surface films. Proc. R. Soc. A **202**, 224–243 (1950)

# Chapter 2
# Friction

A tangential force, $F_T$, is required to start the relative motion between two bodies kept in close contact by a normal force $F_N$ (Fig. 2.1a). Such a force is called *friction force* and is required to overcome the *static* friction force, $F_f$, which opposes the motion. The ratio $F_T/F_N = \mu_s$ is the *coefficient of static friction*. In general, $\mu_s$ is independent from the nominal area of contact while it can depend on $F_N$ and on the properties of the mating materials and their surfaces.

To keep the two bodies moving one with respect to the other with a given sliding speed, it is necessary to apply a tangential force to counteract the *kinetic* friction force. In this case, the ratio $F_T/F_N = \mu$ is the *coefficient of kinetic* (or *dynamic*) *friction*. This coefficient too is generally independent from the nominal area of contact while it may depend on $F_N$, the sliding speed (specially at high speeds), on the materials in contact and on their surfaces. The coefficient of static friction is often higher than the kinetic one, as depicted in Fig. 2.1b, although it is not always so.

In this chapter, the influence of friction on the contact stresses and the surface plastic deformations will be firstly outlined. Then, the adhesive interactions between the asperities and the local plastic deformations will be considered together. This will allow us to obtain a unified view of the origin of friction, highlighting the role of the work of adhesion, the transfer phenomena and the initial surface roughness. The phenomenon of stick-slip, the rolling friction, the influence of abrasion on friction and the friction induced surface heating will be finally treated.

## 2.1 Influence of Friction on the Contact Stresses

The presence of friction between two surfaces in contact and in relative motion changes the stress field in the contact region with respect to the frictionless contact described in the previous chapter. Consider the elastic contact between two smooth ideal surfaces. Figure 2.2 shows the contact stresses in the case of an elastic cylinder sliding (from left to right) over an elastic-plastic flat surface, with a



G. Straffelini, *Friction and Wear*, Springer Tracts in Mechanical Engineering,
DOI 10.1007/978-3-319-05894-8_2

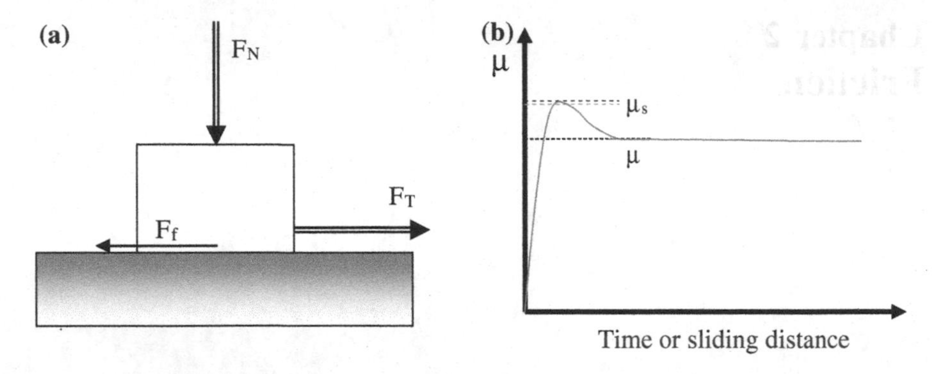


**Fig. 2.1** Schematic showing the forces needed to move two bodies in contact, with relevant definition of coefficients of static and kinetic friction

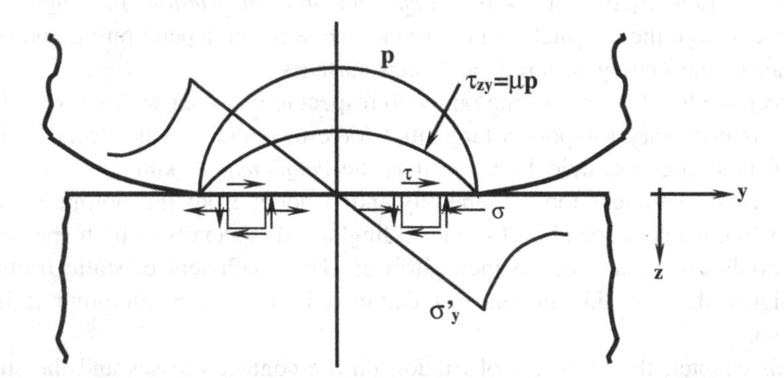


**Fig. 2.2** Cylinder sliding on a flat surface with friction (from *left* to *right*). Stress distribution on the flat surface in the area of contact

coefficient of friction μ.[1] The stresses at the flat surface are indicated. The presence of friction has two consequences [1, 2]:

(1) It induces a shear stress, $\tau_{zy}$, given by $\tau_{zy} = \mu p$;
(2) It establishes a stress in the y direction ($\sigma'_t$), which is a compression stress where the contact begins and a tensile stress at the end of contact.

Therefore, if we consider a material element at the flat surface, the sliding of a cylinder against it induces, in addition to the Hertzian stress p due to the applied load, a stress field characterized by an initial compression ($\sigma'_t < 0$), followed by shearing ($\tau_{zy}$) and then by a tensile stress ($\sigma'_t > 0$). Conversely, an element on the surface of the cylinder and located at the contact front experiences a tensile stress

[1]If not otherwise indicated, the friction coefficient is the kinetic one.

($\sigma'_t > 0$), an element on the load line experiences a shear stress ($\tau_{zy}$) and an element on the rear of the contact experiences a compressive stress ($\sigma'_t < 0$).

For the calculation of the maximum values of $\sigma'_t$ the following relations can be used [3, 4]:

$$\sigma'_{t(\max)} = \frac{4+\nu}{8}\pi \cdot \mu \cdot p_{\max} \qquad \text{(point contact)} \qquad (2.1)$$

$$\sigma'_{t(\max)} = 2 \cdot \mu \cdot p_{\max} \qquad \text{(line contact)} \qquad (2.2)$$

In Fig. 2.3 the stress evolution at the load line ($y = 0$) for a cylinder/cylinder contact with sliding is shown. The cylinders are in relative motion and the role of three values of the friction coefficient is considered. The figure shows the evolution of $\tau_{max}/p_{max}$ as a function of the ratio z/a, i.e., of the normalized depth. It can be noted that with a friction coefficient slightly higher than about 0.2 the peak shear stress at the surface reaches the maximum recorded value in the subsurface region, i.e., at $z = 0.786a$. With friction values greater than about 0.25, the maximum shear stress at the surface exceeds the subsurface value. This has an important consequence on the elastic-plastic behaviour of the materials. In fact, as $\mu$ is increased over 0.25, plastic deformation starts at the surface and at applied loads that are lower than in the frictionless contact.

In Fig. 2.4 the influence of friction coefficient on the onset of yielding (first yielding) for a line contact is shown. As long as $\mu < 0.25$, first yielding takes place in the subsurface region and when $p_{max}/\tau_Y = 3.1$ (see Sect. 1.1.3), whereas for $\mu > 0.25$ first yielding takes place at the surface.

An important situation is that of repeated contacts, like repeated sliding on a plane, or in the case of rolling-sliding. As already introduced in Sect. 1.1.3, if first yielding occurs but $p_{max}$ is below a specific limit, *elastic shakedown* occurs. This means that after first yielding a system of residual stresses is generated that allows the load that is successively applied to be carried out entirely in an elastic manner. The situation is depicted in Fig. 2.5a. In a frictionless contact, the elastic shakedown limit is given by $p_{max}/\tau_Y = 4$. When friction is present, the shakedown limit is decreased as shown in Fig. 2.4. Two boundaries are here shown. The first one pertains to materials that behave in an elastic-perfectly plastic manner and the second to materials that show a kinematic-hardening behaviour. In this latter

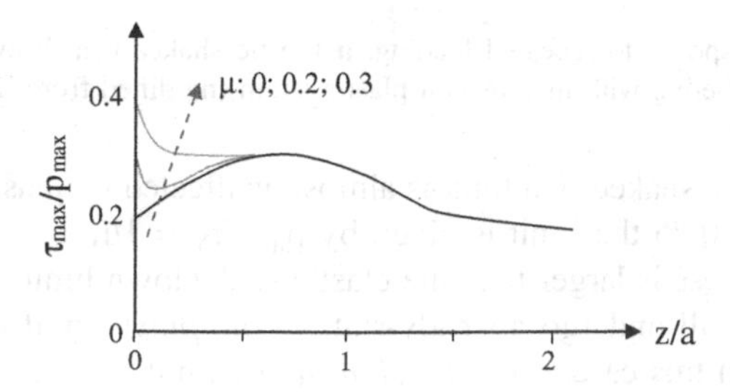


**Fig. 2.3** Evolution of the maximum shear stress ($\tau_{max}$) along the z-axis for a cylinder/cylinder elastic contact, under sliding and for different values of the friction coefficient

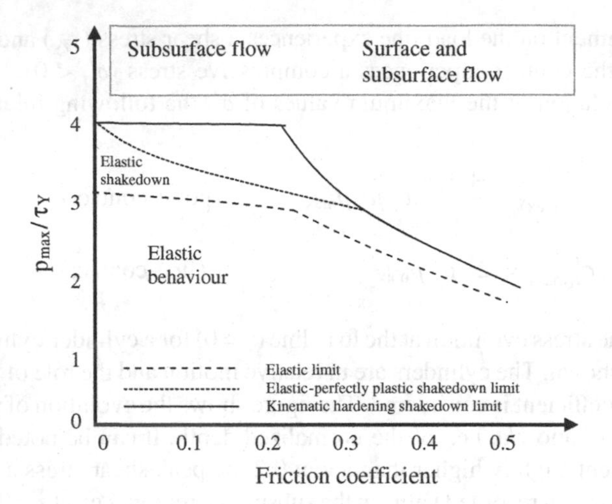


**Fig. 2.4** Influence of friction on the first yielding and elastic shakedown limit for a Hertzian line contact (modified from [1, 2])

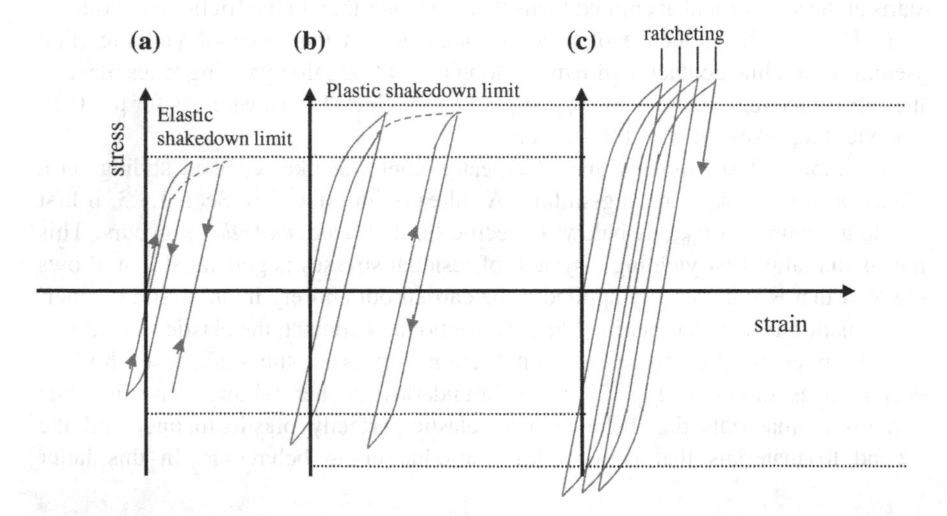


**Fig. 2.5** Materials response to repeated loading. **a** Elastic shakedown; **b** cyclic plasticity, with hysteresis loop; **c** ratcheting with incremental plastic strain (modified from [2])

condition, the elastic shakedown limit is almost unaffected by μ as long as $\mu < 0.25$, whereas when $\mu > 0.25$ the limit is given by $p_{max}/\tau_Y = 1/\mu$.

If the applied stress is larger than the elastic shakedown limit, the materials will strain harden and will undergo a steady-state *cyclic plasticity* if repeatedly loaded (Fig. 2.5b). Also in this case, a *plastic shakedown* limit can be recognized, above which the material undergoes *ratcheting* with an incremental accumulation of plastic strain (Fig. 2.5c).

## 2.2 Friction and Plastic Deformation at the Asperities

As examined in Chap. 1, real surfaces are not perfectly flat and when they are pressed together contact spots occur at their asperities. In most cases, like in metals, such contacts are predominantly plastic. The presence of friction has important consequences that will be analysed using a simple approach, with reference to the situation in which $A_r/A_n \ll 1$, i.e., $p_0 \ll p_Y$.

If two bodies in contact are subjected to sliding with a velocity v, the junctions tend to continuously break off and then reform in other points. A dynamic equilibrium is thus reached characterized by the same rate of formation and breaking of the junctions. The detachment of each junction takes place at a *critical shear stress* ($\tau_m$), acting at the interface between two asperities in contact. Such a shearing process is necessary because of adhesion and it is also associated to the plastic deformation at the asperities. In the simple scheme in Fig. 2.6, each volume element at the asperity junction is submitted to a compressive stress, $\sigma_c$, given by $\sigma_c = F_N/A_r$, and at a shear stress $\tau_m$, given by $\tau_m = F_T/A_r$. Due to the shear stress, yielding is achieved when:

$$\sigma_c^2 + \alpha\tau_m^2 = \sigma_Y^2 \tag{2.3}$$

where $\alpha$ is a constant equal to 4 if the Tresca yield criterion is adopted. Usually, $\sigma_Y$ is replaced with $p_Y$ and the constant $\alpha$ assumes values larger than 4. On the bases of experimental data, some Authors proposed $\alpha = 9$, others $\alpha = 12$ or $\alpha = 25$. Considering $\alpha = 12$ [5, 6], Eq. 2.3 becomes:

$$\sigma_c^2 + 12\tau_m^2 = p_Y^2 \tag{2.4}$$

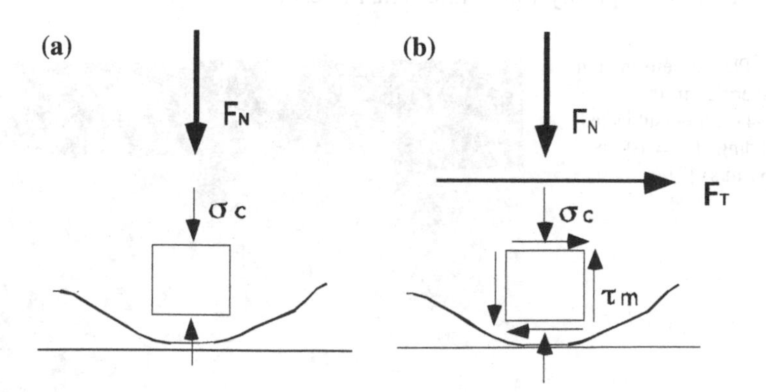


**Fig. 2.6** Schematic showing the stresses acting on each asperity junction. **a** Static contact; **b** sliding contact

A consequence of the greater ease of plastic deformation at the junctions is the increase of the real area of contact. In fact, with appropriate substitutions, Eq. 2.4 becomes:

$$A_r = \frac{F_N}{p_Y}\sqrt{1 + 12\mu^2} \tag{2.5}$$

As an example, Fig. 2.7 shows a cross section of the surface of a Ti-6Al-4V specimen after dry sliding against a counterface of the same alloy. The maximum contact pressure was 325 MPa (the applied load was 100 N), and the sliding speed was 0.063 m/s. In these conditions a friction coefficient of about 0.5 was recorded [7]. The micrograph shows a large shear deformation of the grains located at the surface in the direction of sliding. The extent of plastic deformation decreases moving towards the inside of the material for a thickness depth of around 20–25 μm, which is of the same order of the plastic junctions size. We can therefore say that friction induced a considerable surface microplastic deformation by shear, which is confined at the asperities (the comparison with the undeformed grains allows an estimation of surface plastic deformation of more than 200 %), with an estimated increase in $A_r$ from 100/(350 × 9.81) = 0.0291 mm$^2$ after the initial static contact, to $0.0291\ \sqrt{1 + 12 \cdot 0.5^2} = 0.0582$ mm$^2$ during sliding (with an increase of 100 %).

Considering that the uniaxial yield strength of this titanium alloy was about 980 MPa, it turns out that $\tau_y \approx 490$ MPa (following the Tresca yield criterion) and $p_{max}/\tau_y \approx 0.66$. From Fig. 2.4, it would seem that during sliding surface yielding is not occurring. The discrepancy with the experimental result is due to the fact that the relations given in Sect. 2.1 refer to ideally flat surfaces and thus give information on the *macroscopic behaviour* of the solids in contact. However, at asperity contact points, plastic yielding is certainly achieved. In addition, during sliding the asperities may undergo a cyclic loading due to the repeated contacts, with a continuous accumulation of plastic deformation by ratcheting, which, together with the local compressive state of stress, leads to plastic deformations that are much greater than those typically found after tensile testing.

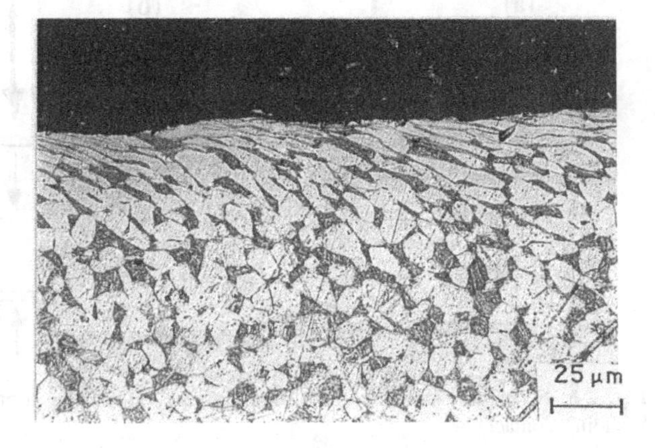


**Fig. 2.7** Plastic deformation at the asperities in the Ti- 6Al -4V alloy, subjected to dry sliding ($F_N$ = 100 N, v = 0.063 m/s) [7]

In general, if the coefficient of friction is low, i.e., typically less than 0.15, the additional plastic contribution by shearing is limited. But if friction coefficient is comparatively high, in excess of 0.3, plastic deformation at the asperities can be quite pronounced since local ratcheting effects become important. Such deformations involve a depth comparable to the junction size and induce a considerable junction growth. They will also possibly promote local damage effects, contributing to wear processes, as will be discussed in the next chapters. In the engineering evaluations, the hypothesis of smooth surfaces that is the basis of the evaluations described in the previous paragraph, has to be used with caution when friction coefficient is greater than 0.3.

In addition to plastic deformation, the asperities in contact undergo local heating and may also interact with the counterface and the surrounding environment. All these phenomena, which will be considered in more detail in the next paragraphs, cause the formation of a surface *tribological layer* (or *friction layer*) whose characteristics contribute to determine the *system response* (in terms of friction and wear) to tribological loading.

## 2.3 The Adhesive Theory of Friction

The theory in the first place considers that the tangential force, $F_T$, required to maintain a sliding velocity, v, between two bodies in contact, on which a normal force, $F_N$ is acting, is due to the critical shear stress ($\tau_m$) required to separate the asperities in contact [8]:

$$F_T = \tau_m A_r \tag{2.6}$$

Using Eqs. 2.5 and 2.6 along with the definition of friction coefficient, the following relationship is obtained [9]:

$$\mu = \frac{\tau_m}{p_Y} \frac{1}{\sqrt{1 - 12\left(\frac{\tau_m}{p_Y}\right)^2}} \tag{2.7}$$

which is graphically shown in Fig. 2.8. It can be seen that μ tends to very high values as the $\tau_m/p_Y$ ratio is increased (μ tends to infinity if $\tau_m/p_Y$ tends about 0.29, which is the maximum allowed value; such a value would be 0.5 if α is set to 4 according to the Tresca yield criterion). On the contrary, when the ratio $\tau_m/p_Y$ tends to low values, the coefficient of friction tends to 0. When it becomes less than, say, 0.15, the contribution of the junctions' growth may be neglected and the relation 2.7 can be simplified in the following way:

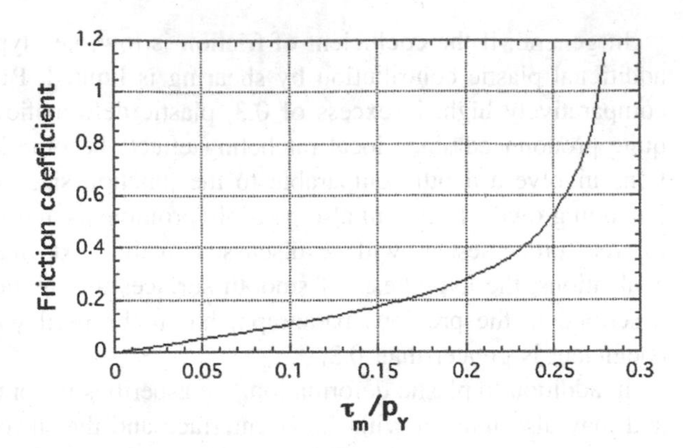


**Fig. 2.8** Graphical representation of Eq. 2.7, which illustrates the dependence of friction coefficient on the ratio $\tau_m/p_Y$

$$\mu = \frac{\tau_m}{p_Y} \tag{2.8}$$

In the case of prevailing elastic contacts at the asperities, the use of Eqs. 1.17 and 2.5 is not valid. The real area of contact can be expressed by Eq. 1.18. Hence, combining Eq. 1.18 with Eq. 2.6, a relation for μ can be obtained:

$$\mu = c \cdot \frac{\tau_m}{E^*} \tag{2.9}$$

where $c = 3.2\sqrt{\frac{R_s}{\sigma_s}}$.

To explain the magnitude of friction that is generated between two surfaces in contact and in mutual sliding, it is therefore necessary to understand the meaning of $\tau_m$. The adhesive theory of friction assumes that friction is due to the adhesive interactions at the asperities that form junctions during the contact time (roughly given by 2r/v, where r is the average junction radius and v is the sliding velocity). Following the ideas exposed in Sect. 1.4, it can be then proposed that $\tau_m$ is a function of $W_{12}$, which has the dimension of a force per unit length. In a very simplified approach, $\tau_m$ is taken to be directly proportional to $W_{12}$. This means that in the case of plastic junctions:

$$\mu \propto \frac{W_{12}}{H} \tag{2.10a}$$

whereas in the case of elastic junctions:

$$\mu \propto \frac{W_{12}}{E^*} \tag{2.10b}$$

As a matter of fact, the measurement of friction coefficient is often considered as a quick and simple way for evaluating the adhesion (and the work of adhesion) between two surfaces [10]. In this respect, a few significant examples are reported

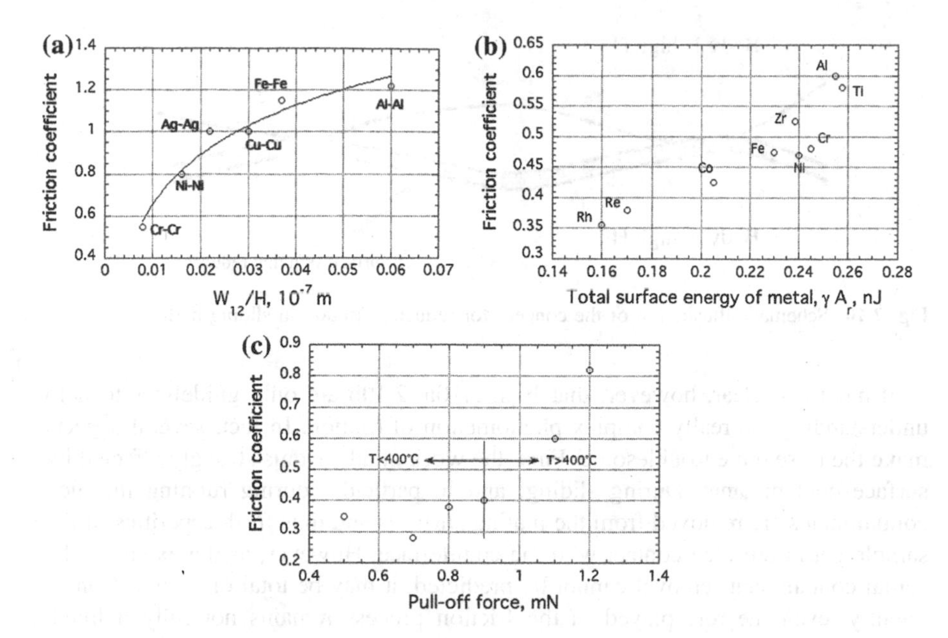


**Fig. 2.9** **a** Friction coefficient versus $W_{12}/H$ for some metals sliding against themselves (modified from [11]); **b** friction coefficient versus $\gamma$ $A_r$ for some metals sliding against SiC; **c** friction coefficient versus pull of force for low frequency-plasma-deposited $Si_3N_4$ films in contact with monolithic $Si_3N_4$ pins (modified from [10])

in Fig. 2.9. Figure 2.9a, shows the experimental relationship between $\mu$ and $W_{12}/H$ for different metal pairs [11]. Figure 2.9b displays a similar concept. It shows $\mu$ as a function of $\gamma A_r$ for different metals, in the case of dry sliding against SiC [10]. Here $\gamma$ is the surface energy of the metal, and $A_r$ is the real area of contact evaluated by means of Eq. 1.17. The product $\gamma A_r$ is thus proportional to $W_{12}/H$. By comparing Fig. 2.9a, b, it is further observed that aluminium, iron and nickel display a higher friction coefficient when sliding against themselves than against a ceramic (SiC) counterface. This can be explained by considering that $W_{12}$ against SiC is lower. Figure 2.9c shows the friction coefficient versus the pull-off force for low frequency-plasma-deposited $Si_3N_4$ films in contact with monolithic $Si_3N_4$ pins [10]. The adhesion and friction tests were carried out at temperatures ranging from room temperature up to 700 °C. The parameters show a clear proportionality because both depend on adhesion. In particular, the Author stated that the increase in temperature induced an increase in adhesion due to the removal of a surface contamination layer. Such an effect was particularly effective for temperatures in excess of 400 °C.

Another interesting case from a design point of view concerns the sliding contact between metals with high $W_{12}$ values and under vacuum conditions, where most of the contaminants are removed from the surfaces. In such cases, very high experimental values for $\mu$ are often recorded, up to 10 or more for some pure metals, in agreement with Eqs. 2.7 and 2.10a.

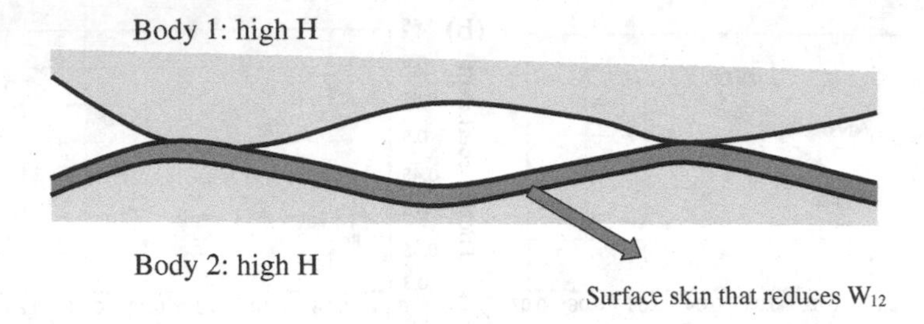


**Fig. 2.10** Schematic illustration of the concept for reducing friction in sliding bodies

It has to be clear, however, that Eqs. 2.10a, 2.10b are only guidelines to help understanding the really complex phenomenon of friction. In fact, several aspects make their use quite troublesome. First, the work of adhesionis strongly affected by surface contaminants. During sliding, and in particular during running in, such contaminants are removed from the mating surfaces and thus fresh asperities of the sample get in repeated contact with the counterface. However, as the extent of the actual contaminant removal cannot be predicted, it may be total or partial. Consequently, even the role played in the friction process remains not fully defined. Second, adhesion is affected by local plasticity that develops during each contact, as shown in the previous section. Such an effect induces an adhesion hysteresis in which the work of separating the contacts is greater than that for approaching it [12]. The thermodynamic work of adhesion given by Eq. 1.20 simply provides a broad picture of the adhesion phenomena during sliding.

Keeping in mind such limitations, the use of Eqs. 2.10a, 2.10b (together with Eq. 2.7 in the case of plastic contacts at the asperities) can be nevertheless very useful, because it provides a powerful tool for explaining the experimental results, as shown by the examples in Fig. 2.9 and by others that will be shown in the next sections. Equations 2.8 (or 2.7) and 2.9 also suggest an important way for reducing friction. First, it is necessary to reduce the real area of contact and thus increase hardness H (of the softest material in contact) and/or increase E*. Second, it is necessary to decrease $\tau_m$, and thus $W_{12}$. Both tasks should be accomplished together, by increasing the hardness and/or the stiffness of the materials in contact, and realising a surface skin able to reduce $W_{12}$. A sketch of this concept is shown in Fig. 2.10. It forms the basis for the development of the so-called *solid lubricants* (see Sect. 3.1).

## 2.4 Friction Between Metals, Ceramics and Polymers

In this paragraph, the concepts of the adhesive theory of friction outlined in the previous section will be used to provide a general explanation of the friction coefficients that are recorded in the case of dry sliding (i.e., without any lubricant) between metals, polymers and ceramics. The influence of normal load will be also considered.

## 2.4.1 Metals

The asperity contact in metals is typically plastic. As predicted by Eq. 2.10a, friction is expected to be proportional to the $W_{12}/H$ ratio. As seen, the experimental data in Figs. 2.9a confirms this statement in the case of different pure metals. The comparison with some data reported in Fig. 2.9b also shows the role of the counterface in modifying $W_{12}$ and thus friction. A particular behaviour is displayed by metals such as cobalt, titanium and magnesium with a hexagonal closed packed (hcp) crystal lattice. When sliding against themselves they provide friction coefficients around 0.5, i.e., much lower than those predicted by the experimental plot of Fig. 2.9a. For example, the pair Ti/Ti is characterised by a ratio $W_{12}/H$ that is similar to that of the Fe/Fe pair, which however displays a friction coefficient greater than 1. This behaviour can be explained by considering that hcp metals possess a reduced ability to deform plastically (at least at relatively low temperatures). Therefore, during sliding the asperity junctions deform with lower intensity and the adhesion forces at the junctions are not able to develop fully [13].

In the case of sliding between metal alloys, friction coefficient turns out to be lower than in the corresponding pure metals. For example, in dry sliding between two bronzes (Cu-8 % Sn) a value of $\mu$ equal to 0.6 has been recorded, against a typical value of 1 for the Cu/Cu pair [6]. This is mainly due to an increase in hardness of the alloy. In the case of bronzes, for example, if we set $W_{12} = 1$ J/m$^2$ and consider a typical hardness of 115 kg/mm$^2$, from the data of Fig. 2.9a a value of 0.6 for $\mu$ is obtained in agreement with the experimental result [6]. Another example concerns steels, which display $\mu$-values in the 0.6–0.8 range, i.e., much lower than the typical values of the Fe/Fe pairs (Fig. 2.9a).

In the case of coupling between different metals, the friction coefficient is determined by the ratio $W_{12}/H$ as far as *transfer phenomena* are not triggered during sliding, as it will be illustrated in the next section. Friction coefficient is thus expected to decrease as the tribological compatibility is reduced, and the hardness of the softer metal (which determines the extent of the real contact) is increased. For example, in the case of the Cu/Fe pair (that is partially compatible, see Fig. 1.13), $W_{12} = 0.8$ J/m$^2$ (from the data listed in Table 1.4) and then a value of $\mu$ around 0.5 may be predicted from the data in Fig. 2.9a, in good agreement with the experimental data. In the case of bronze sliding against steel, following the same calculation, a value of $\mu$ definitely below 0.4 can be inferred. As it will be shown in the next paragraph, for this coupling an experimental value of 0.18 has been recorded.

Metal surfaces may easily undergo oxidation during sliding, with the formation of an oxide layer (or scale) that may reduce friction coefficient, thus acting as a lubricating skin (as schematised in Fig. 2.10). This occurs for example in iron, copper and nickel alloys. If an oxide layer forms during sliding and covers the metal (the mechanism is illustrated in Sect. 4.2 when dealing with the tribo-oxidative wear), friction coefficient is reduced since the contact is ceramic/ceramic and no more metal/metal, with a consequent reduction in $W_{12}$. As an example, Fig. 2.11 shows the friction evolution with sliding distance for a steel/steel contact [14]. For

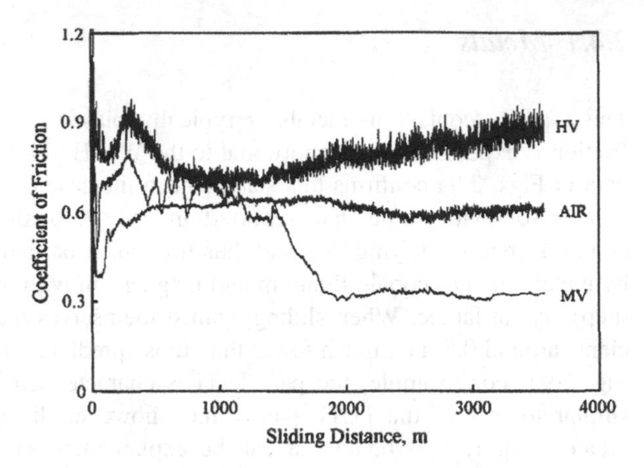


**Fig. 2.11** Evolution of the friction coefficient during tests involving steel on steel sliding in air, medium vacuum (1.3 Pa–MV) and high vacuum (5 $10^{-3}$ Pa–HV) conditions, as specified in the graph legend [14]

the test in air, a friction coefficient of around 0.6 is reached at steady state. During sliding the steel surface becomes covered with hematite ($Fe_2O_3$), as typically reported for this type of testing. For the test in medium vacuum (1.3 Pa), the prevailing oxide is magnetite ($Fe_3O_4$) containing a lower fraction of oxygen. After 2000 m of sliding (probably necessary to generate and compact the oxides), the friction coefficient drops to 0.3 thus highlighting a lubricant capacity of magnetite. If the tests are carried out in high vacuum ($5 \times 10^{-3}$ Pa), i.e., in the absence of oxygen, no oxides are forming during sliding and metal-to-metal contacts still prevails all through the test. As expected, friction coefficient raises to around 0.8.

Equation 2.10a shows that friction in metals should be independent from normal load and this is generally true. An example is shown in Fig. 2.12a where the results of the well-known experiments carried out by Whitehead in 1950 are displayed [8]. They refer to steel sliding against aluminium in air, and show that friction coefficient is around 1.25, in agreement with the data of Fig. 2.9a, and that it is nearly independent from load over a very wide range. However, there are some sliding conditions that change the picture and the normal load may play a role.

A load increase may influence the local hardness of the mating metals in two main ways. It may induce a hardness increase by strain hardening (with a decrease in $\mu$), or it may induce a hardness decrease by thermal softening (with an increase in $\mu$). These two effects may cancel out, or one may be prevailing. Figure 2.12b and c show two cases that may be explained in this way. Figure 2.12b refers to an austempered nodular cast iron sliding against a pearlitic cast iron [15]. The tests were carried out at a sliding speed of 1 m/s, and at two loads. The relevant microhardness of the worn surfaces are also shown in the graph. It can be noted that the surface microhardness is increased as load is increased and correspondingly friction is decreased. Figure 2.12c refers to 36NiCrMo4 steel hardened to yield strength of 900 MPa and dry sliding against a high-speed steel [16]. Friction coefficient is seen to increase with load and this may be attributed to the softening

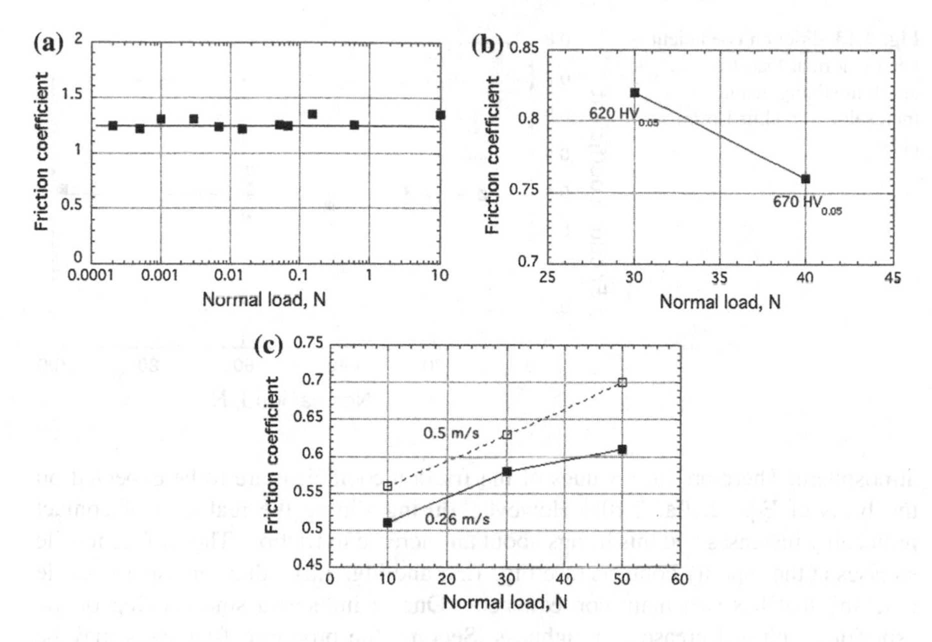


**Fig. 2.12** Effect of normal load on friction coefficient. **a** Steel sliding on aluminium; **b** austempered nodular cast ironsliding against pearlitic cast iron (the microhardness of the worn surfaces are also indicated) (modified from [15]); **c** 36NiCrMo4 steel sliding against a tool steel (modified from [16])

effect due to the heat generated at the contacting asperities. This is further confirmed by the observation that friction increases also with sliding speed that also contributes to local heating.

## 2.4.2 Ceramics

In ceramics the contact at the asperities is typically mixed. It may be completely elastic if the surface roughness is sufficiently low. Otherwise, it may be plastic if surface roughness is high. Equation 2.10b shows that even for elastic contacts, friction coefficient should be independent from the normal load. Figure 2.13 shows an example in the case of alumina balls sliding against an alumina flat sample [17]. It shows that friction is around 0.4. As a matter of fact, friction coefficient in ceramics and in dry conditions is low (around 0.3–0.7), when the normal load is relatively low and temperature is lower than about 200 °C. Such values are actually quite similar to those displayed by metal alloys and this may appear quite surprisingly. In fact, ceramics are characterised by high values of hardness and elastic modulus (see Table 1.2), and low values of the surface energy (Table 1.4). In addition, the surface energy of ceramics is often reduced further by the reaction of the surfaces with water vapour or other substances present in the working

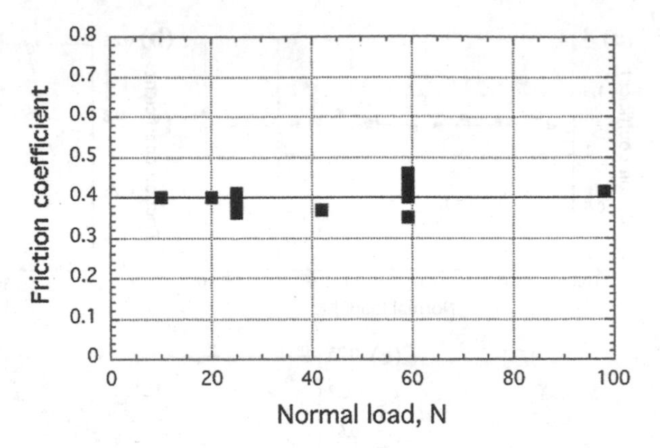


**Fig. 2.13** Friction coefficient versus normal load for alumina sliding against themselves (modified from [17])

atmosphere. Therefore, low values of the friction coefficient are to be expected on the basis of Eqs. 2.10a, 2.10b. However, during sliding the real area of contact noticeably increases and this brings about an increase in friction. The surface tensile stresses at the asperity contacts (see Fig. 1.3a and Fig. 2.2) induce an asperity-scale cracking that has two main consequences. One, it induces a smoothening of the asperities with a decrease in roughness. Second, the produced fragments may be compacted during sliding and form surface scales that support the normal load. An example of this last effect is shown in Fig. 2.14.

In ceramic materials the normal load exerts an important role. If it is increased over a critical value, a macroscopic brittle contact is established and friction coefficient may even reach values in excess of 0.8. As seen in Sect. 1.1.4, brittle

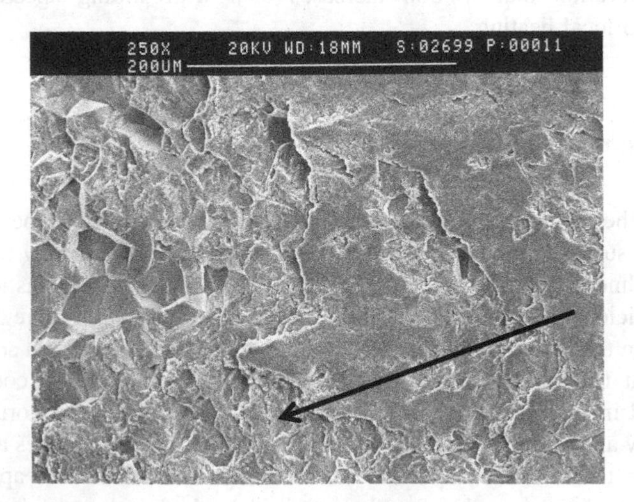


**Fig. 2.14** Wear surface of alumina after dry sliding against a SiC counterface. To the right of the picture, the presence of protective scales that support the applied load may be recognised. On the left the presence of areas with macroscopic brittle contact, displaying intergranular fracture may be also observed (*arrow* indicates the direction of sliding)

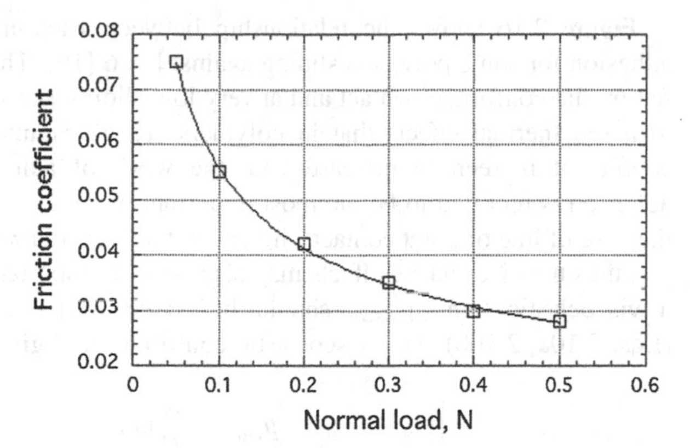


**Fig. 2.15** Friction coefficient versus load for a hemispherical diamond pin sliding on SiC (modified from [10])

contact occurs when the tangential stresses due to friction are high due to the presence of critical surface microcracks. Such cracks are often present on the surfaces of ceramic materials, originating from processing defects, such as flaws, inclusions, and grain boundary porosity. The cracks may also originate from the growth of asperity-scale micro cracks due to repeated contact loading during sliding. Figure 2.14 shows, on the left, some areas of intergranular brittle fracture occurred during dry sliding. The formation of cracks and fragments, and also the comminution of such fragments entrapped in the contact region, all involves energy dissipation and, consequently, an increase in the tangential force, which must be applied to maintain the relative motion at constant sliding speed.

A particular dependency of friction on normal load is observed when $A_r$ is close to $A_n$. In this case, $\mu = \tau_m \cdot A_n/F_N = \tau_m/p_0$ (where $p_0$ is the nominal pressure). A typical situation in which this happens is when a sphere, with a comparatively small radius, is pressed against a plane surface. With reference to the Hertz relations (Table 1.1), it is thus obtained that $\mu \approx \tau_m F_N^{-1/3}$. An example is shown in Fig. 2.15, where friction coefficient for hemispherical diamond pin (radius 0.3 mm) sliding against single-crystal SiC is displayed [10].

### 2.4.3 Polymers

Some polymers, like PTFE (polytetrafluoroethilene), give rise to very low friction coefficients (lower than 0.1) when sliding against themselves or other materials (specially metals), and therefore behave as *solid lubricants* [18]. In general, however, friction coefficient in polymers under dry sliding typically ranges between 0.2 and 1, and therefore is not that different from the values displayed by metals and ceramics. This result can be explained, at least in the first instance, referring to Eqs. 2.10a, 2.10b. In fact, the work of adhesion in polymers is lower than in metals and ceramics (see Table 1.4) but, at the same time, their hardness and stiffness too are lower and the two effects are almost proportional.

Figure 2.16 shows the relationship between friction coefficient and work of adhesion for some polymers sliding against PA 6 [19]. The tests were carried out in flat on flat conformal contact and at very low sliding speed (v = 0.24 μm/s) in order to avoid thermal effects that in polymers are of paramount importance. Friction coefficient is seen to increase with the work of adhesion. In such conditions adhesion is reported to be the most important parameter in determining friction. In the case of line or point contacts, however, the local deformations can be quite large and thus the viscoelastic effects may also play an important role. As a consequence, a viscoelastic term, $\mu_{visco}$, should be added to the adhesive term of friction (Eqs. 2.10a, 2.10b). This viscoelastic contribution is given by [20]:

$$\mu_{visco} = \frac{K}{H} \tan \delta \tag{2.11}$$

where K is a constant, H is the hardness and tanδ is the loss factor, connected to the hysteresis loss during the loading-unloading cycles at the contacts. This term is particularly important when temperature is close to the glass transition temperature, $T_g$, of the polymer, in correspondence of which the loss factor is maximum. An example is shown in Fig. 2.17, where the friction coefficient is plotted as a function of testing temperature in a scratch test on PMMA (polymethylmethacrylate; here friction is depurated from the ploughing contribution, introduced in Sect. 2.8) [21]. The sliding velocity was 16 μm/s to avoid thermal effects. It can be noted that friction reaches a maximum at about at about 120 °C, which is the glass transition temperature of the polymer.

At relatively low loads, friction coefficient is almost independent from load as predicted by Eqs. 2.10a, 2.10b. As load is increased, however, friction coefficient decreases. In fact, if the asperity contacts are mainly elastic (as it occurs quite often at relatively low normal loads), $A_r$ tends to $A_n$ because of the low stiffness of polymers. As shown previously, in this case μ depends on $F_N$ through the relation: $\mu \approx \tau_m F_N^{-1/3}$. In general, however, the following relationship is obeyed: $\mu \approx \tau_m F_N^{-n}$, where n is typically between 0.1 and 0.3. This is because the contacts at the

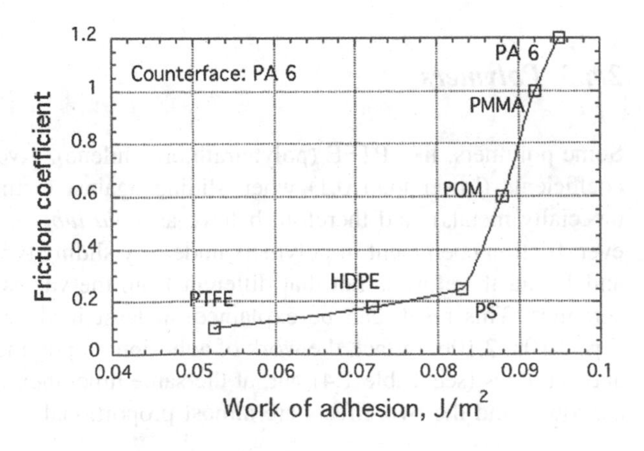


**Fig. 2.16** Friction coefficient versus work of adhesion for some polymers sliding against PA 6 (modified from [19])

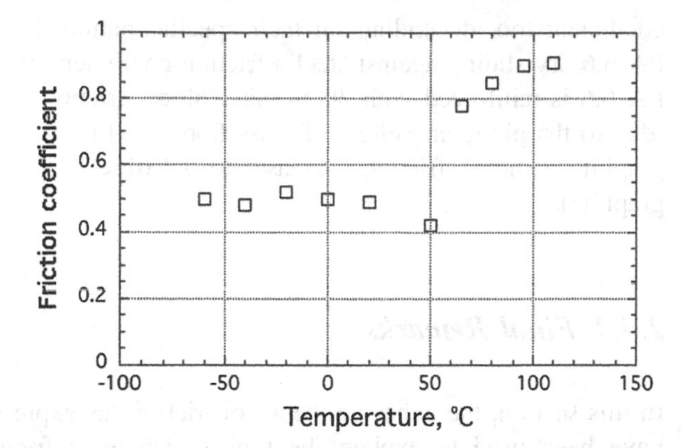


**Fig. 2.17** Friction coefficient (depurated from the ploughing contribution) versus temperature in a scratch test on PMMA (modified from [21])

asperities are not completely elastic and some plastic junctions may still form during sliding. Figure 2.18 shows an example for a steel sphere sliding against two polymers at 250 μm/s [22]. If load increases further, plastic deformation at the asperities gains in importance and friction coefficient may start to increase [23, 24].

Polymers are widely used in coupling with themselves and metals. Metals are much harder than polymers and surface asperities may thus exert a ploughing action that increases friction. This topic will be analysed in Sect. 2.8, but we can say that this contribution is of minor importance if surface roughness is sufficiently low. On the other hand, very important are the *transfer phenomena* from the polymer to counterface. Such transfer of material is favoured by the adhesion bonds between the polar molecules of the polymers and the counterface and also by the mechanical ploughing action of the asperities. The formation of a transfer layer modifies friction coefficient since sliding occurs between similar materials. In the case of PTFE (polytetrafluoroethylene) transfer is quite easy since fluorine interact with the metallic counterface. The formation of a transferred layer made of macromolecules stretched in the direction of sliding induces in this case a decrease in friction.

Quite often, in engineering applications polymers are reinforced with fibres or particles to increase their mechanical strength. Such additions modify the friction

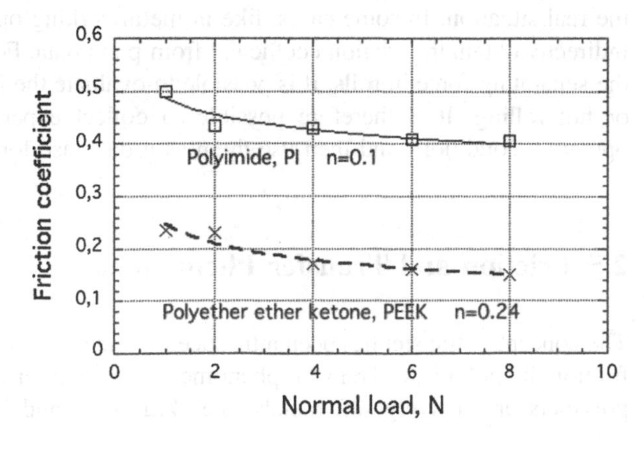


**Fig. 2.18** Dependency of friction coefficient on normal load for a steel sphere sliding against two polymers at 250 μm/s (modified from [22])

coefficient too, depending on their specific nature. For example, in the case of PA 6.6 dry sliding against steel a friction coefficient of 0.28 was obtained [19]. If PA 6.6 is reinforced with 30 % glass fibres, friction coefficient increases to 0.31 (due to the ploughing effect of glass fibres), while if it was reinforced with 30 % graphite particles, friction decreases to 0.2 (due to the solid lubricating action of graphite).

### 2.4.4 Final Remarks

In this section, the adhesive theory of friction, as expressed by Eqs. 2.10a, 2.10b, have been used to explain the typical values of friction coefficient in metals, ceramics and polymers. No lists of friction coefficients have been reported herewith, since friction is a *system property* that strongly depend on the *sliding conditions* [25]. We have introduced the role of oxidation (with the formation of a friction layer), roughness, material transfer phenomena, applied load, thermal heating and so forth. In the next sections, other parameters influencing friction (including lubrication) will be introduced and discussed. Such parameters may also act differently during sliding, giving rise to *transitions* in the friction coefficient. Typically a transition always occurs after the very early stage of sliding, from the so-called *running in stage* (involving the polishing of the surfaces and the removal of dirty and contaminants) to the steady-state stage. Therefore, it cannot be a surprise if in the literature different values of friction coefficients are found for the same sliding materials and data collections may be contradictory or even misleading [26].

The selection of a proper friction coefficient for a specific engineering application is thus a very difficult task. The starting point is the adhesive theory of friction and the relations and arguments outlined in the present section. Literature data are also very helpful but the complete knowledge of the testing conditions is important in order to make the right choice (and to avoid mistakes). Another possibility is to perform laboratory friction tests. In this case, the selection of the testing apparatus is very important since it must simulate the contact conditions as close as possible to the real situation. In some cases, like in metalworking operations, it is possible to indirectly obtain the friction coefficient from plant data. For example, by measuring the separating force in rolls, it is possible to evaluate the friction coefficient in cold or hot rolling. It is therefore possible to collect experimental data in different operating conditions and then use them as a database for successive evaluations.

## 2.5 Friction and Transfer Phenomena

The concept of transfer has been introduced in the previous section when treating the friction in polymers. Transfer phenomena typically involve soft materials, like polymers or relatively soft metals like aluminium and its alloys, forming plastic junctions at the contacting asperities during sliding. In general, a transferred layer of the soft material on the harder counter surface is formed. The mechanism is schematised in Fig. 2.19. The repeated formation of plastic junctions during sliding induces an accumulation of plastic damage at the asperities of the softest material in contact, i.e., of the material with the lowest mechanical strength. This may lead to the formation of a fragment when a critical damage is reached. It may become a wear fragment (and in this case reference is made to the *adhesive wear*), or it may remain attached to the asperities of the counter material, contributing to the transfer layer.

Figure 2.20a shows, as an example, the wear surface of an AISI D2 tool steel (with hardness of 800 kg/mm$^2$) after dry sliding for 5 min against a bronze (Cu-8 % Sn; hardness: 230 kg/mm$^2$) [6]. The presence of many bronze particles firmly transferred to the steel counterface can be clearly appreciated. The degree of transfer was observed to increase with sliding time. At the end of the test (10 min) the degree of counterface coverage with transferred particles was about 30 %. Figure 2.20b shows the friction evolution with time (normal load: 50 N; sliding speed: 0.04 m/s). The results of two tests are reported since for the steel counterface two surface finishing conditions were investigated: $R_a = 0.02$ μm and 0.22 μm (see also Fig. 1.10). It can be noted that there is a first stage characterised by a low friction coefficient, which is about 0.18 for both couplings. After about 1 min of sliding, however, friction starts to increase reaching a value around 0.3 (in both cases) at the end of the test. In the initial stage of sliding the contact between the mating surfaces is actually bronze/steel. The friction coefficient of 0.18 is in good agreement with the arguments developed in the previous section. As the transfer phenomena starts and gradually becomes more important, friction coefficient also increases since part of the contacts change from bronze/steel to bronze/bronze. Friction coefficient may thus be evaluated using a rule of mixtures:

$$\mu = \mu_{11} \cdot \delta + \mu_{12} \cdot (1 - \delta) \tag{2.12}$$

where $\mu_{11}$ is the friction coefficient for the copper alloy sliding against itself (which is about 0.6), $\mu_{12}$ is the friction coefficient for the copper alloy/steel pair, and $\delta$ is the degree of coverage. Setting $\mu_{11} = 0.6$, $\mu_{12} = 0.18$, it is obtained that at the end of the test $\mu = 0.3$ and therefore $\delta = 0.29$, in good agreement with the experimental evaluations.

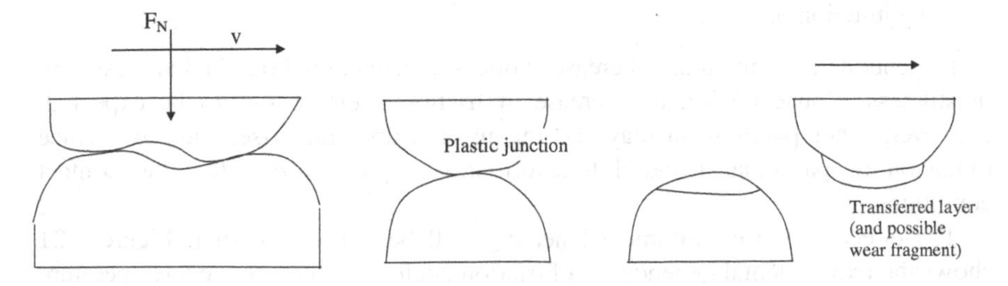


**Fig. 2.19** Schematic showing the formation of the transferred layer Transferred layer

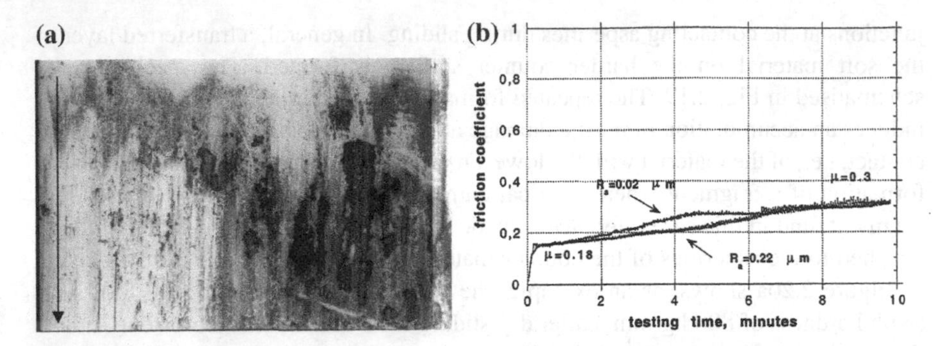


**Fig. 2.20** Cu-8 % Sn alloy dry sliding against AISI D2 tool steel. **a** Appearance of the steel surface. The *arrow* indicates the sliding direction. Note the presence of many black particles that are bronze particles transferred to the steel. **b** Evolution of friction coefficient [6]

The phenomenon of transfer is difficult to predict. In general it starts after an incubation period, whose duration depends on the loading parameters (normal load and sliding velocity) and the type of contact. During sliding, transfer and also back-transfer phenomena may occur in a dynamic equilibrium that involves the formation of a transfer layer (or *third-body*) between the mating surfaces. The characteristics of such transfer layer determine the friction and wear properties of the tribological system. For example, the action and relevant properties of almost all solid lubricants rely on the formation of a transfer film able to reduce the work of adhesion, following the concept outlined in Fig. 2.10.

## 2.6 Effect of Temperature and Sliding Speed

Temperature significantly influences the mechanical properties of materials and therefore the friction coefficient. Two main factors control the temperature in the contact region:

(1) the surrounding environment (consider, for example, the heating of gears located close to a combustion engine, or the cutting tool-piece interface, which is heated by the intense plastic shearing of the chip);
(2) the heating due to frictional energy dissipation (due to the adhesive shearing at the junctions).

In general, a temperature increase induces a reduction both in hardness and in stiffness. Consequently, an increase in friction coefficient is to be expected. However, other phenomena may accompany a temperature rise, such as surface oxidation or particular material transformations, which also affect the contact conditions.

The effect of the environmental heating will be considered first. Figure 2.21 shows the experimental dependence of friction coefficient from the test temperature in the case of different materials dry sliding against themselves:

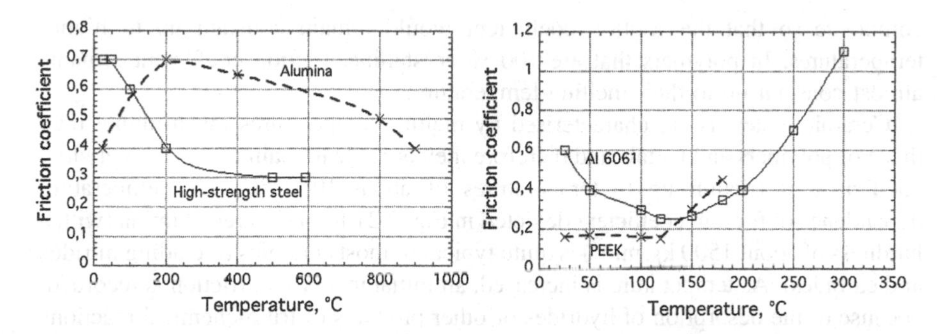


**Fig. 2.21** Dependence of friction coefficient on temperature for different materials (data from [17, 27, 28])

- a high-strength steel (hardened to about 500 kg/mm$^2$; the data are obtained from different literature sources);
- an Al alloy (Al 6061, data from [27]);
- a polymer (PEEK, data from [28]);
- a ceramic (alumina, data from [17])

In the case of the high-strength steel, the friction coefficient decreases with increasing temperature (at least up to the temperature of 600 °C, which corresponds to a homologous temperature of approximately 0.5, given by the ratio between the temperature and the melting temperature of the metal, both in Kelvin). Indeed, the steel is covered with an oxide layer, which is well supported by the underlying high strength material. This implies that the coefficient of friction is relatively low at room temperature. As temperature is increased, the degree of coverage also increases and becomes almost complete at 500 °C. At this temperature the friction coefficient has stabilised at a value of about 0.3.

Even in the case of the aluminium alloy, friction coefficient slightly decreases as temperature is increased, at least up to a temperature of about 150 °C, which, again, corresponds to a homologous temperature of about 0.5. Also for this alloy the decrease in friction coefficient may be attributed to the formation of an oxide layer. For higher temperatures, however, the friction coefficient begins to increase. This behaviour is due to the onset of a noticeable thermal softening, aided by creep phenomena, which renders the alloy unable to support the surface oxide layer.

In the case of the polymer (PEEK: polyetheretherketone), the friction coefficient is substantially unchanged up to about 125 °C and then begins to increase. This behaviour is typical of polymeric materials and is due to their high sensitivity to temperature if they contain an amorphous phase. This polymer is *partly crystalline* and has a glass transition temperature ($T_g$) of about 148 °C. Hence, at about 130 °C it begins to soften and also its loss factor increases, contributing to friction by increasing the viscoelastic dissipation, as outlined in the previous section. Any addition of reinforcing fibres (such as carbon fibres) may increase the $T_g$ of the composite so that the friction coefficient would remain constant up to higher temperatures. In polymers that are 100 % crystalline, friction coefficient remains almost constant up to their melting temperature.

Ceramic materials are characterised by melting temperatures much higher than those of polymers and metals and therefore they are able to maintain a relatively low coefficient of friction up to temperatures of about 1000 °C. The temperature dependence of friction coefficient depicted in Fig. 2.21 for a sintered alumina (with a hardness of about 1500 kg/mm$^2$) is quite typical of most ceramics (including nitrides and carbides). As temperature is increased, an initial increase in friction is recorded, because of the desorption of hydrides or other products of tribo-chemical reactions that causes an increase in the surface energy. Subsequently, the friction coefficient decreases due to the incipient melting, in the contact regions, of the sintering additives that are segregated at the grain boundaries. This process results in the formation in the contact areas of a thin layer of glassy material, which behaves as a lubricant and reduces the shear force necessary to separate the contacting asperities. At very high temperatures (above 1000 °C), melting of the sintering additives at the grain boundaries leads to a massive weakening of the material and the coefficient of friction inevitably increases, reaching values even greater than 1 [29].

As mentioned above, heating in the contact zones can be also induced by an increase in sliding speed and normal load. In Fig. 2.22, the dependence of the friction coefficient from the sliding speed, v, in the case of high strength steel, a polymer (PP: polypropylene) and a ceramic (silicon nitride) is shown (data from [9, 30]). In the case of steel, the friction coefficient is little influenced by v, as long as it is sufficiently low. However, an increase in v results in a decrease in friction coefficient due to two main effects. The first one is a decrease in the contact time at the asperities that reduces the possibility of forming strong junctions by adhesion. Secondly, there is an intensification of the surface oxidation with the formation, at very high speeds of the order of 100 m/s, of a very plastic oxide. If at these sliding speeds the applied load is also very high (with nominal pressures of the order of one

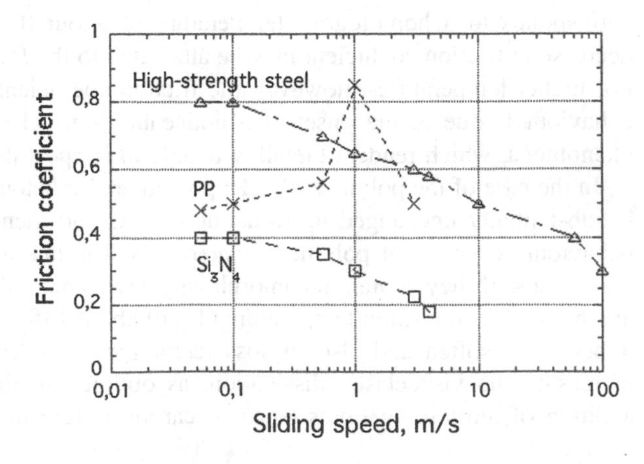


**Fig. 2.22** Dependence of the coefficient of friction on sliding speed in the case of a polymer (PP sliding against steel), a ceramic (silicon nitride sliding against itself) and high strength steel against itself (data from [9, 30])

tenth of the steel hardness) it is possible to reach local melting at the asperities and friction coefficient falls to very low values, even below 0.1, since the liquid metal film acts as a lubricant (the corresponding wear, however, becomes very high) [25]. In the case of polymers, the achievement of the conditions for the asperity melting can take place at speeds (and pressures) much lower than for metals. The reduction of the friction coefficient shown by the PP for sliding speed greater than about 1 m/s is precisely due to the occurrence of asperity melting. At low speeds, polymers in general show a continuous increase in friction coefficient with sliding speed, up to the attainment of the melting conditions at the asperities, since a lubricant layer during sliding is not formed. The reduction of the friction coefficient with sliding speed shown by silicon nitride in Fig. 2.22 is quite a typical behaviour of ceramic materials and is due to increased contact temperature, which causes the formation of interfacial layers with lubricating characteristics.

## 2.7 The Stick-Slip Phenomenon

The tangential force required to trigger the relative motion between two bodies in contact is given by $F_N \mu_s$, where $\mu_s$ is the coefficient of static friction. Typically $\mu_s$ is greater than $\mu$ of about 20–30 %. The reason lies in the fact that during the rest time before the onset of motion, a large stress relaxation at the junctions may occur, which causes an increase in the real area of contact and allows the adhesive forces to fully develop. This effect is particularly important when the contacts are mostly plastic and the mating surfaces are free from contaminants.

When the static friction coefficient is markedly greater than the kinematic one, the phenomenon of *stick-slip* may occur. In this case, the tangential force experiences a periodic instability during sliding, because of an intermittent motion. Consequently, the evolution of the resulting friction coefficient is characterized by large oscillations around the mean value. For example, Fig. 2.23 shows the friction coefficient as a function of time in the case of a Cu-8 % Sn alloy sliding against itself (in the same tribological conditions of the couplings of Fig. 2.20b) [6]. It can be clearly seen that friction fluctuates intensely around the mean value, equal to about 0.6.

The stick-slip phenomenon can be explained in this way. A high work of adhesion and a low sliding speed, which increases the contact time at the junctions, may favour the formation of strong junctions that cause a significant increase in the tangential force. Such conditions may give rise to a period with no relative motion between the surfaces. Hence, $\mu$ approaches $\mu_s$ during this stage (the *stick stage*). Under the action of the tangential force, however, the junctions break off and the sliding speed suddenly increases. The *slip stage* is thus entered. During this stage, the junctions are weak since the adhesion forces have no time to sufficiently develop. As a consequence, friction coefficient drops to a low value (certainly lower than $\mu_s$). At this stage, the system decelerates to keep the sliding speed constant. This slowdown may again result in the formation of strong junctions and the cycle is repeated.

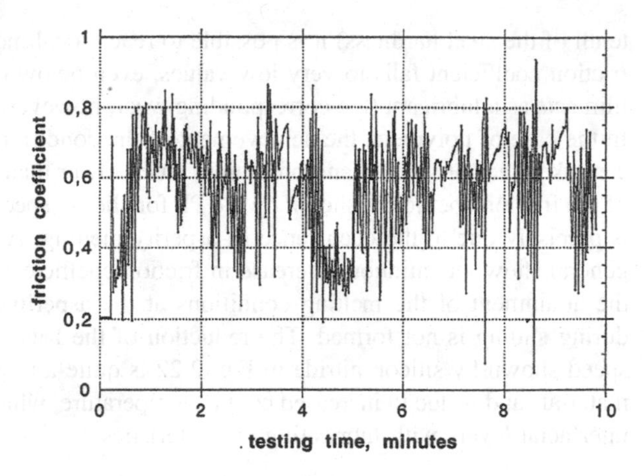


**Fig. 2.23** Friction coefficient as a function of time for the coupling Cu-8 % Sn/Cu -8 % Sn in dry sliding [6]

As seen, the stick-slip is originated from phenomena taking place in the contact region, although it is clearly influenced also by the rigidity of the tribological system, which can accentuate or damp the oscillations. The stick-slip phenomenon produces an intermittent motion that in most cases is not acceptable for the correct operations of mechanical systems. In addition, it produces an audible chatter and squeal that may be quite undesirable too. It can be prevented or minimized by reducing the work of adhesion, so that the coefficient of static friction is lower than the kinematic one. This may be achieved through a correct choice of the materials of the tribological system, including, if necessary, suitable lubricants. Other beneficial effects can be obtained by changing the sliding speed and/or the stiffness of the mating bodies.

The stick-slip phenomenon can be modelled by considering a mass pressed against a block that is restrained by a spring and a dashpot attached to a support. In this way it is possible to understand, for example, if harmonic oscillations may occur at high sliding speeds, and therefore design the tribological system in order to avoid them [30].

## 2.8 Contribution of Abrasion to Friction

Along with the adhesive interaction, two bodies in contact and in mutual sliding may experience an *abrasive interaction*, which is characterized by large surface plastic deformations. Such interaction takes place when a hard body exerts a ploughing or grooving action on the surface of a softer body (in general, a hardness difference of about 20–30 % is necessary to have an abrasive interaction).

As schematized in Fig. 2.24, it is possible to distinguish between *two-body abrasion* and *three-body abrasion*. Two-body abrasion (Fig. 2.24a) takes place when hard particles are firmly fixed in one body and are allowed to plastically penetrate the counterface and plough groove on it. Such hard particles may be part of the material microstructure (such as in ceramic-reinforced composites, or in

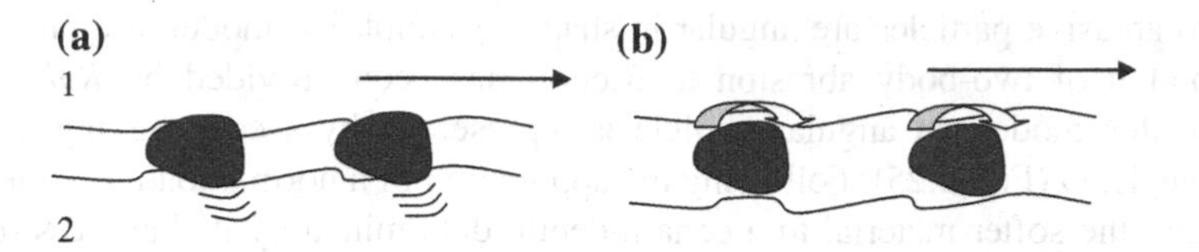


**Fig. 2.24** Scheme of two-body **a** and three-body **b** abrasion

steels or cast irons containing hard carbides), or may come from the surrounding environment (typical examples are the sand particles that contaminate tribological systems). Two-body abrasion may be exerted also by the hard asperities of a surface sliding on a softer counterface. Such particular case will be treated in the next section. Three-body abrasion (Fig. 2.24b) takes place when hard particles trapped between two contacting surfaces are quite free to roll. This typically occurs when both bodies in contact are comparatively hard and the particles are not able to remain fixed in neither of the two. The hard particles may even come from a damaging (wear) action of one of the two bodies.

Hard abrasive particles are typically ceramic particles. The first column of Table 2.1 lists the hardness of some ceramic materials that may have an abrasive action in different tribological systems. For example, sand (silica) particles originate from soil re-suspension and may contaminate the mechanical systems. Alumina ($Al_2O_3$) or silicon carbide (SiC) particles are commonly used as grinding media where the two-body abrasive interaction (that increases friction and, especially, wear) plays in this case a positive role. The second column of Table 2.1 lists some hard carbides or nitrides that may be present in the microstructure of different steels, cast iron or hard metals. In the third column, finally, the hardness values of some metals that may be abraded by the hard particles are listed. It can be noted that even heat-treated steels with a hardness of 900 kg/mm$^2$ can be abraded by sand particles.

**Table 2.1** Typical hardness ranges in different materials (data taken from [9] and other sources of literature)

| Material | Hardness (kg/mm$^2$) | Material | Hardness (kg/mm$^2$) | Material | Hardness (kg/mm$^2$) |
|---|---|---|---|---|---|
| Diamond | 6000–10,000 | Cr carbide ($Cr_7C_3$) | 1600 | Ferritc/pearlitic steel | <350 |
| Boron carbide | 2700–3700 | Cr carbide ($Cr_3C_2$) | 1300 | Heat-treated steel | 500–900 |
| Silicon carbide | 2000–3000 | Mo carbide | 1500 | Al alloy | 100–200 |
| Alumina | 1100–1800 | Ti carbide | 2000–3200 | Cu-Be alloy | 150–400 |
| Si nitride | 1400–2000 | W carbide | 2000–2400 | Hard Cr | 500–1250 |
| Sand (quartz) | 750–1200 | V carbide | 2460–3150 | Au | 30–70 |
| Glass | 500 | Cr nitride | 2200 | Ag | 25–80 |
| | | Ti nitride | 1200–2000 | Pb | 5 |

Hard abrasive particles are angular in shape. A simplified model to quantify the contribution of two-body abrasion to friction has been provided by *Rabinowicz* [13]. In this model, an angular particle is represented by a cone having a given attack angle, $\Theta$ (Fig. 2.25). Following the application of a normal load, $F_N$, the cone penetrates the softer material to a certain depth, determined by its hardness (or the yield pressure). The tangential force, $F_T$, necessary to horizontally move the cone can be expressed by the product of the section of the groove ($A_p$) to the average stress for plastic deformation ($p_Y$). $A_p$ is given by: $A_p = 1/2\ 2r\ r\ tg\Theta$, where r is here the radius of the base of the cone. Therefore:

$$F_T = r^2 tg\Theta p_Y \tag{2.13}$$

since $F_N$ is given by: $(\pi r^2)/2p_Y$ (note that, due to the movement of translation, only the frontal projected area sustains the normal load), the following relation for the abrasive contribution to friction, $\mu_{abr}$, is obtained:

$$\mu_{abr} = \frac{F_T}{F_N} = \frac{2tg\Theta}{\pi} \tag{2.14}$$

In general, the friction coefficient is then given by the sum of two contributions [8]:

$$\mu = \mu_{ad} + \mu_{abr} \tag{2.15}$$

where the subscripts *ad* and *abr* indicate the contributions of adhesion and two-body abrasion. Such contributions have clearly different weights, depending on the specific characteristics of the tribological system. Figure 2.26a shows the results of scratch tests carried out on a number of polymers using hard and undeformable cones (typically made of diamond) with varying attack angles (the experimental data were collected in [31]). In such cases the adhesive contribution may be neglected and the results confirm the substantial validity of Eq. 2.14. In Fig. 2.26b, the friction coefficient expressed by Eq. 2.15 is plotted as a function of $\tau_m/p_Y$ taking

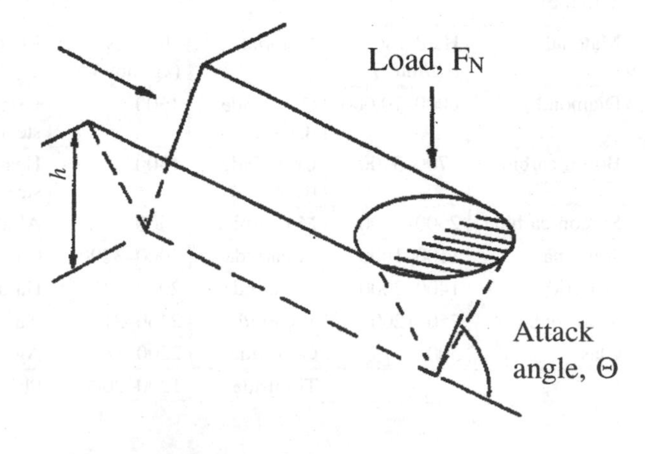


**Fig. 2.25** Scheme of the two-body abrasive action of a particle, which is represented by a cone with attack angle, $\Theta$

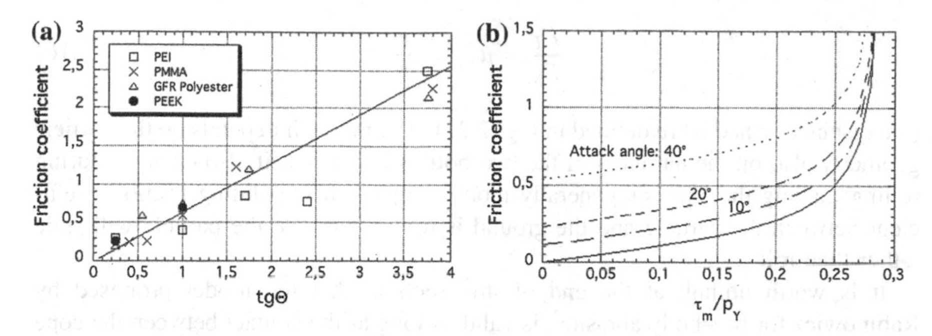


**Fig. 2.26** **a** Friction coefficient versus attack angle (Θ) for a number of polymers (data from scratch tests collected in [31]); the straight line is given by Eq. 2.14; **b** dependence of friction coefficient on the ratio $\tau_m/p_Y$ and the attack angle, Θ (the curves are obtained using relations 2.7, 2.14 and 2.15)

into account both contributions. Note that hard abrasive particles are characterized by Θ-values, which, due to their angularity, typically vary between 5° and 30°, giving rise to $\mu_{abr}$-values ranging from 0.05 to 0.37.

Consider, as an example, the friction data reported in Sect. 2.4.3 regarding the PA 66 composite reinforced with 30 % of glass fibres, dry sliding against a steel. A friction coefficient of 0.31 was recorded. Without reinforcement, friction coefficient was 0.28. How was the abrasive action of the glass fibres? The interaction of glass fibres with steel has an adhesive and abrasive nature. $\mu_{ad}$ may be set to 0.1 since glass and steel are incompatible materials. Hence, considering that 30 % of the contact is between glass and steel, and the remaining 70 % is between polymer and steel, from Eq. 2.15 it is obtained that: $0.31 = 0.28\ 0.7 + 0.1\ 0.3 + \mu_{abr}$. Therefore $\mu_{abr} = 0.084$ and the average attack angle of the glass fibers is 8°.

The evaluation of the abrasive contribution to friction in the case of three-body abrasion is much more difficult. With reference to Fig. 2.27, it is obtained that particles trapped between the two sliding surfaces are allowed to rotate if $F_T\,h > F_N\,e$. Therefore:

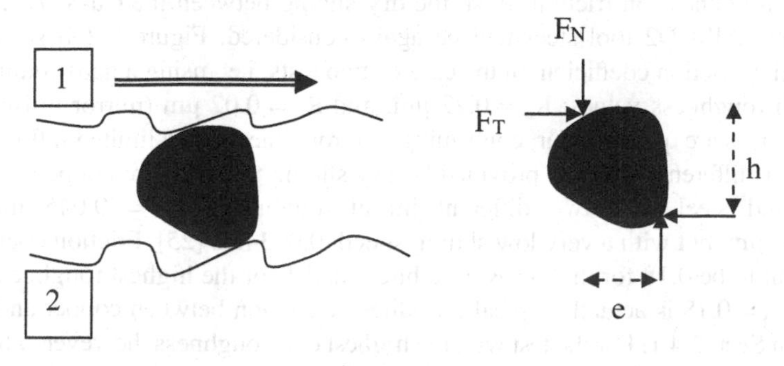


**Fig. 2.27** Scheme of the three-body interaction, and free body picture showing the forces acting on the particle

$$\frac{F_T}{F_N} = \mu_{abr} > \frac{e}{h} \tag{2.16}$$

(the distances e and h are defined in Fig. 2.27). The ratio e/h depends on the particle geometry also on the hardness of the two bodies in contact. It also changes during rolling making the analysis generally more complex. If the sliding friction coefficient between the particle and the ground is lower than e/h, the particle will slide rather than roll.

It is worth noting, at the end of this section, that the model proposed by Rabinowicz for two-body abrasion, is valid as long as the contact between the cone (i.e., the hard particle) and the plane is plastic. If the plane is made by a ceramic material with high hardness and low fracture toughness, the contact can be brittle instead of plastic. In particular, the type of contact may be that described by the Lawn and Swain model (Fig. 1.7), because of the angular nature of the hard particles. In this case the formation of (wear) fragments may take place by lateral cracking. Such fragments may remain trapped between the two contacting surfaces and crushed by repeated brittle contacts. It is clear that all the energy spent in cracking and crushing will increase the friction coefficient even if no ploughing occurs.

## 2.9 Effect of Initial Roughness

As said in the previous section, a two-body abrasive interaction may also occur between the asperities of a hard body and the softer counterface. Equation 2.14 can be then used to evaluate the ploughing contribution to friction. Since the mean slope of the roughness profile, Θ, on average varies between 1° (or less) and 5°, the corresponding $\mu_{abr}$-values are quite low, ranging between 0.010 (or less) and 0.056. It turns out that the initial roughness has a very low (if any) influence on friction coefficient. More advanced models have shown that the abrasive contribution to friction due to surface roughness may be actually greater than that predicted by the Rabinowicz model, although the question is still open.

Some representative examples will be presented herewith on the important role of initial roughness on friction. First, the dry sliding between the Cu-8 % Sn alloy against the AISI D2 tool steel will be again considered. Figure 2.20b shows the evolution of friction coefficient in the case of two tests, i.e., using a hard counterface disc with roughness values: $R_a = 0.22$ μm, and $R_a = 0.02$ μm (mirror finish). The friction traces are quite similar, confirming that roughness has a limited influence on friction. A different picture is provided by dry sliding tests of pure copper against a heat-treated steel with two different initial roughness: $R_a = 0.045$ μm and $R_a = 0.42$ μm, but with a very low sliding speed: 0.003 m/s [25]. Friction coefficient turned out to be 0.18 for the lowest roughness and 1 for the highest roughness. The value of $\mu = 0.18$ is actually typical for adhesive friction between copper and steel, as seen in Sect. 2.4.1. For the test with the highest disc roughness, however, a transfer of copper took place and contact soon became Cu/Cu with a corresponding increase in friction coefficient. The value $\mu = 1$ is actually typical of the Cu/Cu contact. Such a result highlights the role of roughness and sliding conditions (sliding velocity and possibly normal load) on transfer phenomena and thus on friction. It is therefore confirmed that roughness does not directly influence friction by its ploughing action but it may influence it by activating or not the transfer phenomena.

Further information on this topic is provided by an investigation on pure aluminium, dry sliding on harder 080 M40 steel with different initial roughness values [32]. In this work, the Authors produced steel surfaces with different $R_a$-values, adopting different grinding processes. In particular, they obtained a unidirectional surface texture by grinding the steel with emery papers in the same direction; they obtained an 8-ground surface texture by moving the steel plate with the shape of an "8" during grinding; finally, they obtained a random surface texture by random grinding the steel surface. The results of the sliding tests are summarized in Fig. 2.28. It can be noted that friction coefficient is independent on initial roughness for each grinding process. However, it increases, on average, in passing from random grinding to 8-ground, and to unidirectional texturing with sliding perpendicular (indicated with U-PL) and parallel (indicated with U-PD) to the grinding marks. Such transitions can be explained by considering the transfer phenomena. Figure 2.29 shows the sliding surface of the steel plate in two conditions: after the U-PL test and initial roughness of $R_a = 0.22$ μm; after the test on the random grinded surface with $R_a = 0.2$ μm. In the first case, a large transfer of aluminium (the black regions in the backscattered scanning electron micrograph image) took place during sliding and this explains the high resulting values of friction coefficient, around 0.7, since most of the contacts during sliding are between aluminium and aluminium. In the second case the transfer was very limited. Consequently the recorded friction coefficient was very low, around 0.28,

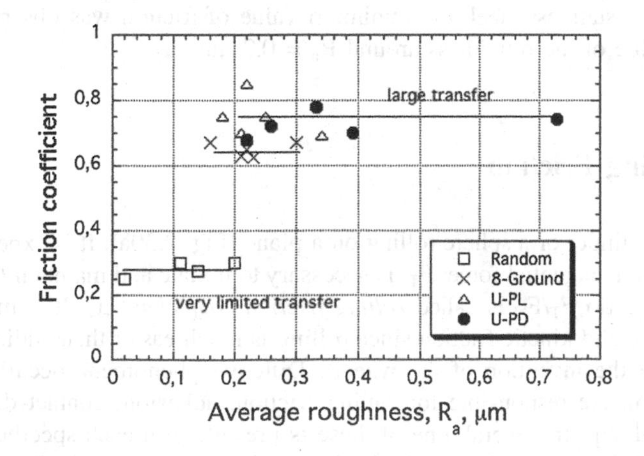


**Fig. 2.28** Variation of friction coefficient with average roughness and grinding process (modified from [32])

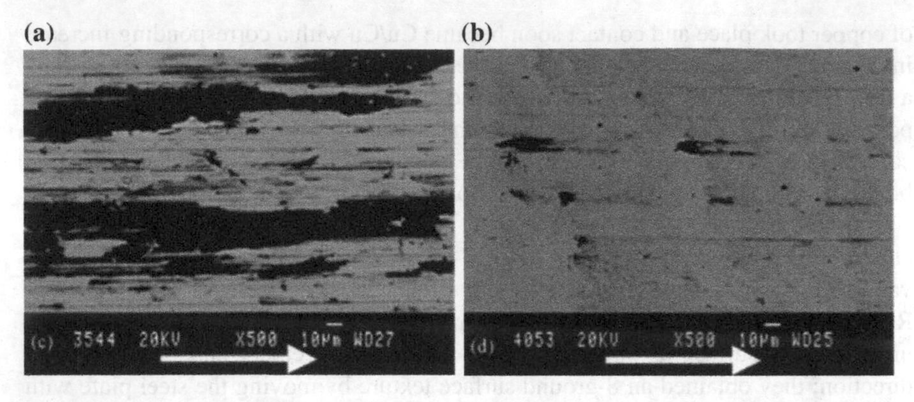


**Fig. 2.29** Backscattered electron SEM images of steel plates after dry sliding tests against pure Al. **a** Initial roughness $R_a = 0.22$ μm, U-PL condition; **b** initial roughness $R_a = 0.2$ μm, random condition [32]

since most of the contacts during sliding are between aluminium and steel. These results confirm that initial roughness does not have a direct influence on friction (by ploughing) but it may influence friction by modifying the transfer phenomena.

The initial roughness is of great importance in the case of polymers sliding against hard counterfaces, such as steels. For polymers the initial roughness of the hard (steel) counterface has to be optimized in order to minimize friction coefficient. Also in this case, friction is controlled by transfer phenomena, since the work of adhesion (and thus friction) is reduced when the polymer is sliding against polymer particles transferred to the counterface with the macromolecules that stretch in the direction of sliding [23]. If the initial surface roughness of steel is too low, the asperities are not able to promote an adequate transfer of the polymer. But if it is too large, friction may be actually increased by the ploughing action. For example, in the case of dry sliding between UHMW-PE (polyethylene with high molecular weight) and a stainless steel, the minimum value of friction was observed for an optimum value of the roughness around $R_a = 0.2$ μm.

## 2.10 Rolling Friction

Consider a cylinder or a sphere rolling on a plane (Fig. 2.30a). It is experimentally observed that a tangential force, $F_T$, is necessary to initiate and maintain *free rolling* motion. The ratio $F_T/F_N$ is called *rolling friction coefficient*, $\mu_v$. It is much lower than the static and kinetic friction since rolling is much easier than sliding (as well known since the invention of the wheel). Different phenomena occurring in the contact region are responsible for rolling friction: adhesion, contact deformation and localized slip. In general, one of these is prevailing in each specific situation.

Figure 2.30b shows a hard cylinder in contact with an elastically deformable plane. Deformation in the contact region is exaggerated for the sake of clarity.

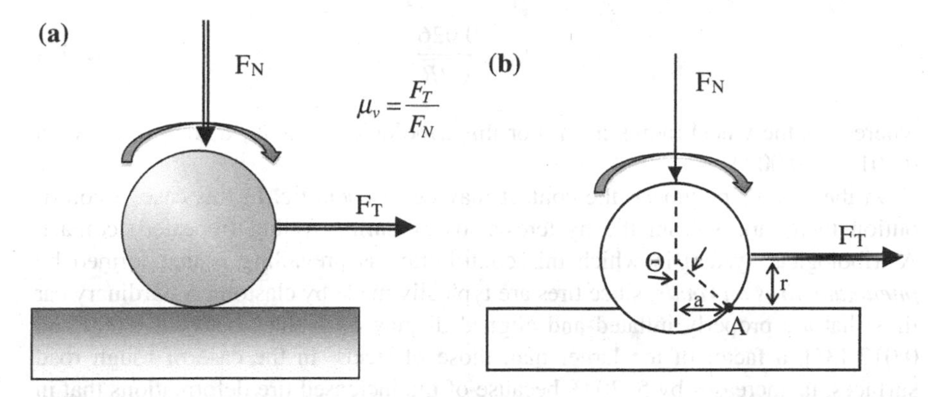


**Fig. 2.30** **a** Definition of rolling friction coefficient; **b** Schematic showing a cylinder rolling on a flat (the elastic deformation is deliberately enhanced for the sake of clarity)

During rolling, the asperities of the two surfaces get in contact and form adhesive junctions. Such junctions are then separated by a tension stress when contact is released, i.e., at the trailing end of the rolling contact. Such a contribution is negligible in lubricated conditions but it is prevailing when conditions involving large $W_{12}$-values are encountered, as in dry contact between metals. In such conditions, $\mu_v$ is proportional to $W_{12}/H$, and therefore to the kinetic friction coefficient. Elastic deformations in the contact region also introduce a resistance to rolling. A tangential force, $F_T$, is in fact required to allow the torque $F_T$ r to overcome the resisting torque $F_N$ a (both calculated with respect to point A in Fig. 2.30b), where a is the half-width of the (rectangular) Hertzian contact (Table 1.1). At equilibrium: $F_N$ a = $F_T$ R cos Θ. Since Θ is very small, cos Θ may be set equal to 1, and therefore $\mu_v$ = a/R. In a line contact $\mu_v$ is thus proportional to $F_N^{1/2}/(R^{1/2}\ E^{1/2})$ whereas in a point contact $\mu_v$ is proportional to $F_N^{1/3}/(R^{2/3}\ E^{1/3})$.

During rolling some localized micro-slip may also take place at the contact region. As seen, a nominal area of contact is established also in the case of non-conformal contacts. In the case of a sphere rolling on a flat, the contact points within this area lie in different planes. As a consequence, pure rolling is attained in some contact points only, whereas in the most part of the area of contact a combination of rolling and micro-slip is established [33].

Several empirical relationships have been proposed for evaluating $\mu_v$. For mechanical rolling parts in point contact, a useful relation is the following [34]:

$$\mu_v = 0.005 \cdot \mu \cdot F_N^{1/3} \tag{2.17}$$

where $F_N$ is in N. For example, in lubricated hardened steel ball bearings on steel, $\mu_v$ ranges between 0.0011 and 0.0015. Empirical relationships were also proposed for the *wheel-rail system* where a theoretical line contact occurs. A general relation for steel wheels free rolling on steel rails is:

$$\mu_v = \frac{0.026}{\sqrt{2R}} \tag{2.18}$$

where R is the wheel radius in m. For this tribological system, $\mu_v$ ranges between 0.001 and 0.0024.

In the case of polymers, the contact may be viscoelastic. In this case, a contribution to $\mu_v$ arises from the hysteresis losses during rolling (repeated) contact. A tribological system in which this contribution is prevailing is that formed by *pneumatic tires on roads*, since tires are typically made by elastomers. Ordinary car tires that are properly inflated and aligned display $\mu_v$ -values between 0.009 and 0.012 [35], a factor of ten larger than those of steels. In the case of rough road surfaces, $\mu_v$ increases by 5–20 % because of the increased tire deformations that in turn increase the hysteresis losses.

A particular situation occurs in case of *tractive rolling*. Figure 2.31a shows two cylinders under rolling contact, where a *torque* is applied to the lower cylinder to drive the upper one at a given angular velocity. If the tangential force, $F_T$, is lower than $\mu_s F_N$ (where $\mu_s$ is the coefficient of static friction), the torque is transmitted. In case of tractive rolling, $\mu_v$ is greater than in free rolling because some slip occurs in the contact region. Therefore, $\mu_v$ is related to the length, c, of the slip area that is established at the trailing edge (Fig. 2.31b) [33]:

$$c = 2 \cdot a \cdot \left(1 - \sqrt{1 - \frac{F_T}{\mu_s \cdot F_N}}\right) \tag{2.19}$$

where a is the half width of the (rectangular) Hertzian contact (see Table 1.1).

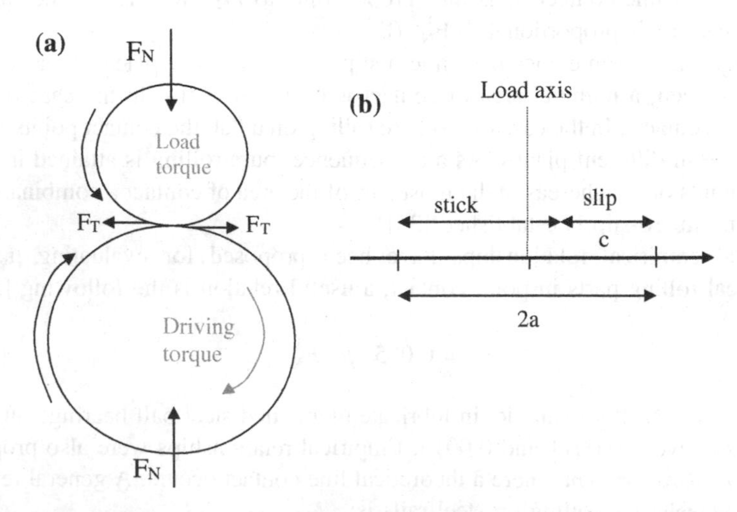


**Fig. 2.31** **a** Two cylinders under tractive rolling with a load torque applied to the driven cylinder; **b** stick and slip areas in the contact region

It has to be noted that in some applications different and more intense sliding conditions may occur in the contact region. An example involves the contacts between gear teeth. Here sliding may be also quite large depending on the distance from the pitch line.

## 2.11 Friction and Surface Heating

Friction is due to dissipative processes occurring at the contacting asperities, namely adhesion and plastic deformations. Most of the dissipated energy (more than 90 %) is lost as heat that increases the temperature of the surfaces in contact. The remaining part is stored in the materials as structural defects. The heat generated by friction per unit sliding time (units: J/s) is given by:

$$q \approx F_T v = \mu F_N v \tag{2.20}$$

Figure 2.32 shows that the heat flux (units: J/s m$^2$) is greater in correspondence of the junctions, where the real section is smaller (here the heat flux is: $\mu F_N v/A_r$, where $A_r$ is the real area of contact defined in Sect. 1.3). On the other hand, it is lower in the sub-surface region where it is: $\mu F_N v/A_n$ ($A_n$ is the nominal area of contact). Therefore, during sliding a higher temperature is reached at the junctions. Such a temperature is called *flash temperature* ($T_f$), since it is attained for a very short time, given by the contact duration of the junctions. The average temperature that is reached just beneath the asperities is generally known as the *average surface temperature* ($T_s$). It is lower than $T_f$ and decreases moving inside of the two bodies in contact down to the reference bulk temperature of the bodies, $T_o$. $T_o$, typically room temperature, is thus the temperature far from the surface where heat flows, i.e., the temperature of the heat sink.

Surface heating can be responsible for microstructural transformations that are very important with regard to the tribological behaviour of the bodies in contact. The flash temperature, for example, can induce the direct oxidation of the asperities or their melting. In the case of steels, it can induce local formation of austenite that,

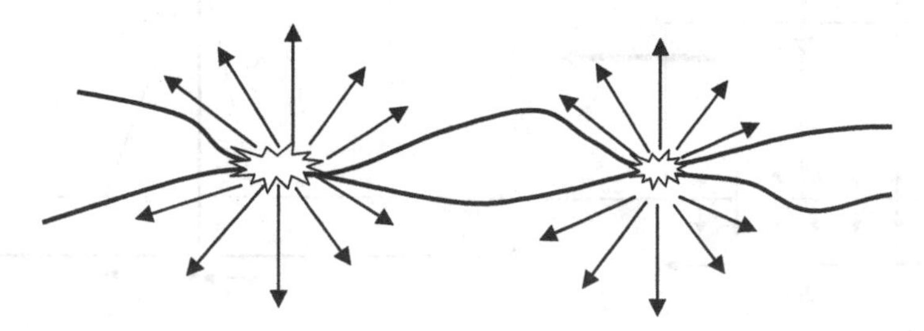

**Fig. 2.32** Schematic diagram of the flow of heat generated by friction in correspondence of the asperities in contact

if followed by rapid cooling, may transform into martensite. The average surface temperature promotes other phenomena, requiring larger thermal activation. For instance, thermal recovery, tempering or even recrystallization that leads to thermal softening, precipitation of secondary phases or their ageing, extended surface melting (when $T_s$ exceeds the melting point). In the case of polymers, surface heating can cause shape distortions, softening or even localized melting. All such phenomena limit the tribological performance of these materials. Eventually, as seen in the next chapter, the surface heating may also play a role in lubricated contacts and must be properly considered in the design of the lubricating systems.

### *2.11.1 Evaluation of the Average Surface Temperature*

The approach that is followed here for the assessment of the average surface temperature has been proposed by M.F. Ashby and co-workers and is essentially based on the classic work of Block and Jaeger [36–38]. Consider the simple geometric configuration of a block (body 1) sliding on a plane (body 2), shown schematically in Fig. 2.33a. The block is the moving body. It is always in contact with the counterface and has a length $2r_0$ in the direction of sliding. The plane is stationary and a surface element is in contact with the counterface only for a limited time interval, given by $2r_0/v$. In Fig. 2.33b, the *equivalent heat diffusion distances* are shown (in the y direction). They are indicated with $l_1$ and $l_2$ and represent the distances between the contact region and the point to which heat flows and where temperature has the reference value, $T_o$. The heat generated at the interface diffuses inside the two bodies in contact. Assuming that heat flows by conduction in the y-direction only, we can write:

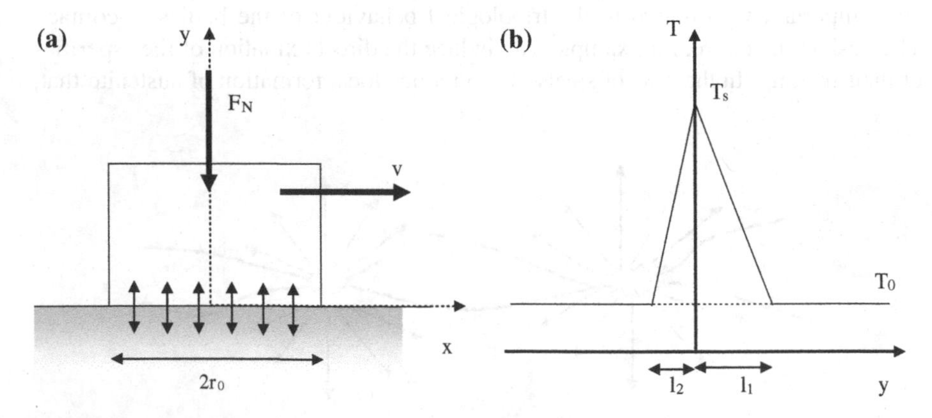


**Fig. 2.33** **a** Schematic of a block sliding on a plane and **b** definition of the equivalent heat diffusion distances Equivalent heat diffusion distances

$$\frac{\mu F_N v}{A_n} = k_1 \frac{T_s - T_0}{l_1} + k_2 \frac{T_s - T_0}{l_2} \tag{2.21}$$

where $k_1$ and $k_2$ are the thermal conductivities of the materials of the two bodies in contact. They depend on temperature (in most cases they decrease as temperature is increased) but they may be considered as constant to a first approximation. From this relation it is obtained:

$$T_s - T_o = \frac{\mu F_N v}{A_n} \frac{1}{\frac{k_1}{l_1} + \frac{k_2}{l_2}} \tag{2.22}$$

which gives the temperature rise and allows the calculation of $T_s$.

This approach can be extended to other geometries and the critical part of the calculation relies in the estimation of the distances $l_1$ and $l_2$, which depend on the system geometry and the thermal properties of the materials. They are linked with the fraction, $\gamma$, of heat that enters body 1 by the following relation:

$$\gamma = \frac{1}{1 + \frac{k_2 \cdot l_1}{k_1 \cdot l_2}} \tag{2.23}$$

(the fraction of heat flowing into the stationary body is: $1 - \gamma$).

In body 1, always in contact with body 2, the attainment of a steady state condition, characterized by a temperature profile that is constant with time, is easy. As a result, $l_1$ is independent from time. In body 2, however, the situation is quite different, since a surface element is in contact with the counterface for a limited time. In this case, it can be assumed that a stationary state is attained if the sliding speed is sufficiently low with respect to the thermal diffusivity, a, of body 2 (given by: $a = k/\rho c$, where $\rho$ is density and c is the specific heat.) A useful adimensional parameter is thus the *Peclet number*, Pe, given by:

$$Pe = \frac{v \cdot r_0}{2 \cdot a} \tag{2.24}$$

If it is sufficiently low, typically less than 0.1, a steady state is reached in both bodies, and in both bodies enters the same fraction of heat. Consequently: $\gamma = 1/2$, $l_1/l_2 = k_1/k_2$ and obviously $l_1 = l_2$ if $k_1 = k_2$. On the other hand, if $Pe > 1$, $\gamma < 1/2$ and most heat enters body 2. As a consequence, the thermal length $l_2$ is time-dependent and lower than that it would be for the steady state condition.

For the simple configuration in Fig. 2.33a, the following relation has been proposed for $l_1$ [36]:

$$l_1 = h + \frac{A_n \cdot k_1}{A_i \cdot h_i} \tag{2.25}$$

where h is the block height, $A_n$ is the nominal area of contact, $A_i$ is the area between the block and the heat sink (which may be, for example, the holder) and $h_i$ is the heat transfer coefficient between the block and the holder. Now, if Pe < 0.1, $\gamma = 1/2$ and $l_2$ is immediately obtained from Eq. 2.23. But if Pe > 1 (i.e., almost always) a relation is needed for obtaining $l_2$. From Eq. 2.21, the specific heat entering a surface element of body 2 is: $k_2 \frac{T_s - T_0}{l_2} \cdot \frac{2r_0}{v}$. To a very first approximation, it may be stated that the heat stored by the same surface element is: $\rho c(T_s - T_0)l_2/2$. Hence:

$$l_2 = 2\sqrt{\frac{a \cdot r_0}{v}} \tag{2.26}$$

Ashby and co-workers, proposed the following relationship [36]:

$$l_2 = \frac{r_0}{\sqrt{\pi}} \tan^{-1}\left[\frac{2\pi a}{r_0 v}\right]^{1/2} \tag{2.27}$$

A useful application of the above relationships is for the *pin-on-disc configuration*, since pin-on-disc testing is very common in the tribological investigation of the sliding behaviour of materials. As shown in Fig. 4.17, in this type of testing a pin (with a diameter which is commonly between 5 and 10 mm) is pressed against a rotating disk. With reference to Fig. 2.33a, the pin may be regarded as body 1, whereas the disc is the stationary plane (body 2). $l_1$ can be calculated using Eq. 2.25. However, a different approach is often used, and $l_1$ is set equal to: $\beta\, r_o$, and $\beta$ is a constant that is obtained from experimental measurements. In general, two thermocouples are placed in small holes drilled in the pin, at different distances from the contact surface. In fact, Fig. 2.33b shows that knowing two values of temperature, $l_1$ and then $\beta$ can be estimated. In general $\beta$ does not depend so much on the testing conditions (load and sliding speed) and therefore the obtained values can be used even for further calculations from Eq. 2.22. Typical examples are the following [39]: Steel: $\beta = 4.6$; 7072 aluminium alloy: $\beta = 5.4$; Ti-6Al-4 V alloy: $\beta = 3.8$.

For the evaluation of $l_2$, the Eq. 2.27 can be used, assuming Pe > 1. However, the use of such equation is possible if a surface element in contact with the pin is able to cool down to $T_o$ before entering the contact again after one revolution of the disc. If this is not happening, the disc heats up to an average temperature T* and the heat flow during contact is now to a sink at T*. In order to account for this in the calculations, it is necessary to increase the actual heat diffusion distance, $l_2$, over the value calculated by Eq. 2.27. Alternatively, it is necessary to estimate T* and use it in place of $T_0$. Again, a fine-tuning with experimental measurements can help in this task.

Let us briefly consider the situation of non-conformal contacts. For a ball sliding on a plane (point contact), an effective $l_1$ value can be obtained by considering that heat flow is not one-dimensional but has a spherical symmetry. The following relation has been proposed [36]:

$$l_1 = \frac{\sqrt{\pi}}{2} \cdot r_0 = 0.89 \cdot r_0 \quad (2.28)$$

where $r_0$ is the radius of the nominal area of contact (the extension of the worn surface has to be considered in the case of dry sliding conditions that typically imply the occurrence of some wear).

For two rolling cylinders (denominated with 1 and 2) that are in contact along a generatrix (line contact), $l_1$ and $l_2$ may be simply evaluated using Eq. 2.27 where $r_0$ is set to the half contact width. For this, the thermal diffusivity has to be calculated using the properties of the cylinder and the tangential surface velocity of the cylinder. The speed v in Eq. 2.22 is the relative (or, sliding) velocity given by $|v_1 - v_2|$. Also in this case, the surfaces may not be able to cool down to $T_o$ after each revolution. As a consequence, $l_1$ and $l_2$ has to be properly increased, or $T^*$ has to be evaluated. More specific relations for evaluating $T_s$ in non-conformal contacts having circular, square and rectangular shapes can be found in [40].

Different experimental techniques have been developed for the measurement of $T_s$. As already seen, thermocouples can be placed close to the contact area in the stationary body (even just beneath the surface), and then it is possible to evaluate $T_s$ by considering the local thermal gradient. Reliable measurements can be also obtained by detecting the emitted thermal radiation, using pyrometers or thermal imaging cameras. Other techniques are summarized in Ref. [30]. It is clear that the availability of experimental data can aid in the calibration of β or the other parameters, and this allows using with greater precision the formulas given above and subsequently performing further assessments without the use of measuring systems.

Table 2.2 shows selected thermal properties of some engineering materials. In the selection of the most suitable materials for contact temperature reductions, it is useful to consider that materials with high thermal conductivities are to be used for the body that is not in continuous contact with the counterface (at least when Pe is greater than 1).

For specific engineering applications, simplified relations are often used to check for the surface heating and avoid that surface temperature exceeds a critical value. As an example, for sliding bearings, the following relation is often used:

$$T_s - T_0 = C\mu F_N v \quad (2.29)$$

where C is an experimental constant, which depends on the materials in contact and the geometry of the tribological system (compare with Eq. 2.22). In the case of dry sliding bearings, C ranges usually between 0.1 and 1 °Cs/Nm. If the bearings are made of a polymeric material and have diameter of 25 mm and height of 25 mm, with a rotating shaft made by steel, C is around 0.5 °Cs/Nm [19]. In several applications, the control of surface heating is simply made by calculating the dissipated thermal power (Eq. 2.20) and by verifying that it is less than an acceptable limit value.

**Table 2.2** Density and thermal properties of some materials (data from [41])

| Material | Properties at 20 °C | | | | Thermal conductivity | |
|---|---|---|---|---|---|---|
| | Density (g/cm$^3$) | Specific heat, (kJ/kg °C) | Thermal conductivity (W/m °C) | Thermal diffusivity (m$^2$/s)10$^5$ | At 100 °C | At 300 °C |
| Al-4 % Cu | 2.79 | 0.883 | 164 | 6.676 | 182 | – |
| Al-20 % Si | 2.63 | 0.854 | 161 | 7.172 | 168 | 178 |
| Fe-0.5 % C | 7.83 | 0.465 | 54 | 1.474 | 52 | 45 |
| Fe-1 % C | 7.8 | 0.473 | 43 | 1.172 | 43 | 40 |
| Fe-18 % Cr-8 % Ni | 7.82 | 0.46 | 16.3 | 0.444 | 17 | 19 |
| Cu-5 % Al | 8.67 | 0.41 | 83 | 2.33 | – | – |
| Cu-25 % Sn | 8.67 | 0.343 | 26 | 0.859 | – | – |
| Polycrystalline diamond | 3.51 | 0.51 | 2000 | 110 | 1300 | – |
| Alumina | 3.9 | 0.752 | 30 | 1.02 | 13 a 400 °C | |
| Si carbide | 3.2 | 0.67 | 50 | 2.33 | 34 a 400 °C | |
| Nylon | 1.14 | 1.67 | 0.25 | 0.013 | – | – |
| Teflon (PTFE) | 2.2 | 1.05 | 0.24 | 0.01 | – | – |
| Glass | 2.2 | 0.8 | 1.25 | 0.08 | – | – |

### 2.11.2 Evaluation of the Flash Temperature

An estimate of the flash temperature, $T_f$, can be made referring to the model of the asperities in contact (Fig. 2.32, which is valid as long as $A_r < A_n$. Applying arguments similar to those disclosed in the previous section, the following relationship can be obtained [36]:

$$T_f - T_s = \frac{\mu F_N v}{A_r} \frac{1}{\frac{k_1}{l_{1f}} + \frac{k_2}{l_{2f}}} \tag{2.30}$$

The distances $l_{1f}$ and $l_{2f}$ and can be simply considered equal to the mean junction radius, r. In case of plastic contacts at the asperities, an approximation for r is: r = 0.01/H, where r is in m, and H is the hardness in kg/mm$^2$ [36]. Hence, Eq. 2.30 can then be recast in the following way:

$$T_f - T_s = 8.8 \cdot 10^4 \frac{\mu}{\sqrt{1 + 12\mu^2}} \frac{v}{k_1 + k_2} \tag{2.31}$$

Again, the critical part of the calculation of the flash temperature relies in the estimation of the radius of the junctions. Equation 2.31 has proved to be relatively correct for many tribological systems. An experimental method that can be used to estimate r a posteriori, is to evaluate, by microscopic observation, the size of the wear fragments transferred to the opposing counterface [42].

When $A_r$ is close to $A_n$, as it occurs in metalworking operations, $T_f$ coincides with $T_s$.

# References

1. K.L. Johnson, *Contact Mechanics* (Cambridge University Press, Cambridge, 1985)
2. K.L. Johnson, Contact mechanics and the wear of metals. Wear **190**, 162–170 (1995)
3. G.M. Hamilton, Proc. Instn. Mech. Engrs. **197**, 53–59 (1983) (point contact)
4. J.O. Smith, C.K. Liu, J. Appl. Mech. **20**, 157–168 (1953) (line contact)
5. S.C. Lim, M.F. Ashby, Wear-Mech. Maps. Acta Metall. **35**, 1–24 (1987)
6. G. Straffelini, A simplified approach to the adhesive theory of friction. Wear **249**, 79–85 (1995)
7. A. Molinari, G. Straffelini, B. Tesi, T. Bacci, Dry sliding wear mechanisms of the Ti-6Al-4 V. Wear **208**, 105–112 (1997)
8. F.P. Bowden, D. Tabor, *Friction and Lubrication* (Methuen and Co., Ltd., London, 1956)
9. I.M. Hutchings, *Tribology* (Edwald Arnold, London, 1992)
10. K. Miyoshi (2001) *Solid Lubrication.* Fundamentals and Applications, Marcel Dekker
11. G. Straffelini, *Attrito e Usura*, Tecniche Nuove (2007)
12. A.E. Giannakopoulos, T.A. Venkatesh, T.C. Lindley, S. Suresh, The role of adhesion in contact fatigue. Acta Mater. **47**, 4653–4664 (1999)
13. E. Rabinowicz, *Friction and Wear of Materials*, 2nd edn. (Wiley, New York, 1995)
14. H. Kong, E.S. Yoon, O.K. Kwon, Self-formation of protective oxide films at dry sliding mild steel surfaces under a medium vacuum. Wear **181–183**, 325–533 (1995)
15. G. Straffelini, M. Pellizzari, L. Maines, Effect of sliding speed and contact pressure on the oxidative wear of austempered ductile iron. Wear **270**, 714–719 (2011)
16. S. Mitrovic, D. Adamovic, F. Zivic, D. Dzunic, M. Pantic, Friction and wear behaviour of shot peened surfaces of 36CrNiMo4 and 36 NiCrMo16 alloyed steels under dry and lubricated contact. Appl. Surf. Sci. **290**, 223–232 (2014)
17. S. Jahanmir (ed.), *Friction and Wear of Ceramics*, (Marcel Dekker, New York, 1994)
18. T. Blanchet, F. Kennedy, Sliding wear mechanism of polytetrafluoroethylene (PTFE) and PTFE composites. Wear **153**, 229–243 (1992)
19. K.H. Czichos e K.H. Habig, *Tribologie Handbuch*, Reibung und Verlschleiss, Vieweg (1992)
20. D.F. Moore, *The Friction and Lubrication of Elastomer*. (Pergamon Press, Oxford, 1972)
21. S. Lafaye, C. Gauthier, R. Schirrer, Analysis of the apparent friction of polymeric surfaces. J. Mater. Sci. **41**, 6441–6452 (2006)
22. L.H. Lee (ed.), *Polymer Wear and its Control*, (American Chemical Society, New York 1985)
23. N.K. Myshkin, M.I. Petrokovets, A.V. Kovalev, Tribology of Polymers: Adhesion, friction, wear and mass-transfer. Tribol. Int. **38**, 910–921 (2005)
24. A. Akinci, S. Sen, U. Sen, Friction and wear of zirconium oxide reinforced PMMA composites. Compos. B **56**, 42–47 (2014)
25. S.C. Lim, M.F. Ashby, J.H. Brunton, The effects of dry sliding conditions on the dry friction of metals. Acta Metal. **37**, 767–772 (1989)
26. P.J. Blau, The significance and use of the friction coefficient. Tribol. Int. **34**, 585–591 (2001)
27. I. Singh, A.T. Alpas, High-temperature wear and deformation processes in metal matrix composites. Metall. Trans. A **27**, 3135–3148 (1996)
28. J. Hanchi, N.S. Eiss, Dry sliding friction and wear of short carbon-fiber-reinforced polyethertherketone (PEEK) at elevated temperatures. Wear **203**(204), 380–386 (1997)
29. C. Melandri, M.G. Gee, G. de Portu, S. Guicciardi, High temperature friction and wear testing of silicon nitride ceramics. Tribol. Int. **28**, 403–423 (1995)
30. B. Bushan, *Introduction to Tribology* (Wiley, New York, 2002)
31. B.J. Briscoe, S.K. Sinha, Scratch Resistance and Localised Damage Characteristics of Polymer Surfaces—a review Mat.-wiss. U. Werkstofftech **34**, 989–1002 (2003)
32. P.L. Menezes, Kishore, S.V. Kailas, Influence of surface texture and roughness parameters on friction and transfer layer formation during sliding of aluminium pin on steel plate. Wear **267**, 1534–1549 (2009)

33. J.A. Williams, R.S. Dweyer-Joyce, Contact between solid surfaces, in *Modern Tribology Handbook Volume One*, ed. by B. Bhushan, (CRC Press, Boca Raton, 2001)
34. A. Over, K. Stevens, Friction: the uniquitous coefficient can it be controlled? Surf. Eng. **9**, 305–311 (1993)
35. Tires and Passenger Vehicle Fuel Economy: Informing Consumers, Improving Performance—Special Report 286 (2006) National Academy of Sciences, transportation Research Board
36. M.F. Ashby, J. Abulawi, H.-S. Kong, *T-MAPS Software and User Manual* (Engineering Department, Cambridge, 1990)
37. H.-S. Kong, F. Ashby *Friction-Heating Maps and Their Applications*. MRS Bull. **16**, 41–48 (1991)
38. J.C. Jaeger, Moving sources of heat and the temperatures ay sliding contacts. Proc. R. Soc. S. W. **56**, 203–224 (1942)
39. G. Straffelini, A. Molinari, Mild sliding wear of Fe-0.2 % C, Ti-6 % Al-4 % V and Al-7072: A comparative study. Tribol Lett **41**, 227–238 (2011)
40. G.W. Stachowiak, A.W. Batchelor, *Engineering Tribology* (Elsevier, New York, 1993)
41. ASM Handbook (1992) Friction, Lubr. Wear Technol. **18**, ASM 621–648 (1992)
42. S. Wilson, A.T. Alpas Thermal effects on mild wear transitions in dry sliding of an aluminium alloy. Wear **225–229**, 440–449 (1999)

# Chapter 3
# Lubrication and Lubricants

A reduction of friction between two sliding surfaces can be obtained by interposing a substance capable of reducing the shear stress, $\tau_m$, necessary to allow the relative motion. This substance, which can be solid, liquid or gaseous, is called *lubricant*.

Solid lubricants are materials able to strongly adhere to one or both surfaces to be lubricated, and to realize in this way sliding planes characterized by a low value of $\tau_m$. Liquid lubricants are interposed to the mating surfaces and their action depends much on the *lubrication regime* that they establish. Lubrication regime can be classified according to the so-called *lambda factor*, $\Lambda$, defined by the following relation:

$$\Lambda = \frac{h_{min}}{\sqrt{R_{q1}^2 + R_{q2}^2}} \tag{3.1}$$

where $h_{min}$ is the minimum lubricant thickness between the surfaces, and $R_{q1}$ and $R_{q2}$ are their root-mean-square roughness values. The meaning of $\Lambda$ is simply illustrated in Fig. 3.1a. Figure 3.1b shows, as an example, the friction coefficient versus the lambda factor in the case of two steel discs under rolling-sliding motion with a mineral oil as a lubricant (the tests were carried out using the apparatus described in Sect. 4.5.3). For high values of $\Lambda$ (greater than 3), friction coefficient is very low, whereas for low values of $\Lambda$ (below about 0.5), friction coefficient is rather high (around 0.1–0.2). Such results are quite general and allow us to identify three *lubrication regimes* as a function of the lambda factor [1, 2]:

(1) *Fluid film* lubrication if $\Lambda$ is greater than 3;
(2) *Boundary* lubrication if $\Lambda$ is less than 0.5;
(3) *Mixed* lubrication in the intermediate region.

In addition to reducing friction, lubricant may have other functions. It may remove heat from the tribological system (a cooling task, which is important, for example, in machining operations), or protect the surfaces against environmental aggression (preventing, for example, oxygen and moisture from reaching the metal surfaces thus causing corrosion phenomena).



G. Straffelini, *Friction and Wear*, Springer Tracts in Mechanical Engineering,
DOI 10.1007/978-3-319-05894-8_3

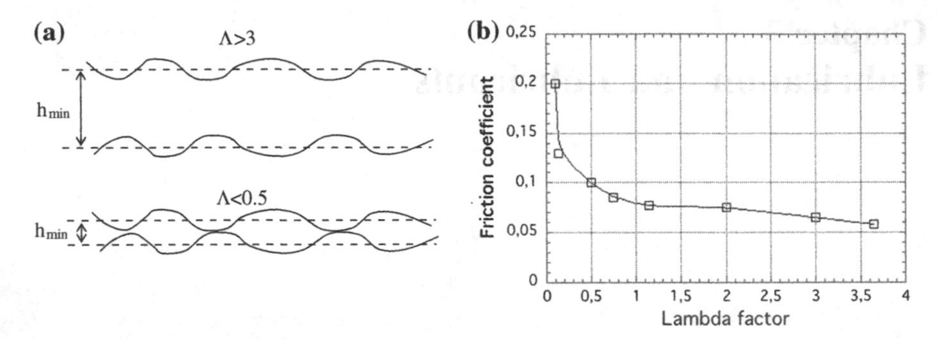


**Fig. 3.1** **a** Scheme showing the meaning of the Λ factor; **b** friction coefficient as a function of the lambda factor in the case of two steel discs under rolling-sliding motion using a mineral oil as a lubricant

In this chapter, the main features of solid lubricants and solid lubrication will be introduced first. Then, the salient features of the main liquid lubricants and of the fluid film lubrication theory will be described. Since gas lubricants (air, $N_2$, $H_2$, He) are used in some special cases only, the gas lubrication will not treated herewith. Subsequently, boundary and mixed lubrication will be considered and the important phenomenon of scuffing described. The chapter ends with an introduction to lubrication when large plastic deformations are involved.

## 3.1 Solid Lubricants

Solid lubricants are usually employed in the following situations [3]:

(1) When the points to be lubricated are inaccessible (such as in equipment working under vacuum or in presence of toxic gases);
(2) When the absence of contaminating substances is mandatory (as in the food-processing machines, optical equipment's, space telescopes, etc.);
(3) When operating temperatures are particularly high (as in hot working of metals) or low (as in cryogenic systems) for liquid lubricants.

Quite often, therefore, solid lubricants are used when liquid lubricants cannot be employed (consider that the typical temperature range for lubricating oils varies from −20 to 120 °C, whereas solid lubricants may operate at temperatures in excess of 250 or 300 °C). Disadvantages related to the use of solid lubricants concern their limited ability of heat removal from the tribological system, and their low wear resistance.

Solid lubricants may be applied to a surface as loose powders (by simply rubbing onto the surface; by dipping or spraying the powder suspended in a suitable medium that will evaporate; by using a binding resin), as powders dispersed in oils and greases, or in the form of coatings, typically obtained using different vapour deposition techniques (see Chap. 7 for details on such technologies). The mechanism of friction reduction by solid lubricants is depicted in Fig. 2.10: the use of hard and stiff materials, required to reduce the real area of contact, is coupled with the formation of a skin (i.e., a powder layer or a coating) able to reduce the work of adhesion (and thus $\tau_m$). In most cases, the solid lubricant is able to transfer to the counterface to establish a sliding plane between the solid lubricant and itself, which results in a large reduction in the work of adhesion.

Friction coefficient is dependent on the *coverage degree*, i.e., on the fraction of the real contact area (δ) covered with lubricant. Hence, the rule of mixture can be used [4]:

$$\mu = \mu_l \cdot \delta + \mu_S \cdot (1 - \delta) \quad (3.2)$$

where $\mu_l$ and $\mu_s$ are the friction coefficients with and without the lubricant, respectively. It is not easy to know in advance the coverage degree that is reached during sliding. In general it depends on the type and amount of lubricant used, the lubricating method, the environmental conditions (mainly temperature) and on the tribological parameters (normal load and sliding velocity).

There are four main classes of solid lubricants [5]:

(1) Carbon-based materials (such as graphite and diamond like coatings, DLCs);
(2) Transition metal dichalcogenide compounds (TMDs, such as $MoS_2$);
(3) Polymers (such as PTFE);
(4) Soft metals (such as Ag, Pb, In, Au).

In the following, the main characteristics of some important solid lubricants will be presented. They are also summarized in Table 3.1.

### *3.1.1 Graphite*

A graphite crystal is made by a large number of parallel planes where carbon atoms are arranged with an hexagonal symmetry (Fig. 3.2a). In the planes, carbon atoms form strong covalent bonds whereas weak van der Waals forces held the planes together. Figure 3.2b shows the morphology of typical graphite particles. By the application of a shear stress, the planes slide quite easily one over the other and then transfer to the counterface. If water vapour and air are present in the surrounding environment, gaseous molecules are adsorbed on the planes (in particular on high-energy edge sites [5, 6]). As a consequence, the work of adhesion between the sliding planes (a plane of graphite against an other plane of graphite) is strongly reduced, thus reducing the friction coefficient to values typically below 0.2. To work as a solid lubricant, graphite thus needs the presence of enough moisture of gas (>100 ppm) to reduce the surface energy of the planes as well as of the edge sites.

**Table 3.1** Main characteristics of some common solid lubricants (modified from [5])

| Methods of use | Common coating thickness range (μm) | Temperature operating range, °C | Typical applications | Comments |
|---|---|---|---|---|
| *Graphite* | | | | |
| • Powder (burnished, sprayed, with binder; as additive in oil, grease and composites)<br>• Coating: vacuum evaporation<br>• Monolithic parts | 0.2–5 | −200 up to 400* | Bearings, seals, pumps and valve parts, electrical contacts | * Increasing friction with increasing temperature<br>Effective only in the presence of water vapour and oxygen |
| *DLC* | | | | |
| Coating (CVD, high energetic PVD techniques) | 0.005–1 | −200 up to 400 | Magnetic media (protect head slider and disk), bearings, bushings, seals, gears, razor blades, car engines (e.g., fuel injectors, camshafts, valve tappets and piston rings), femoral heads for hip implants | Friction coefficients are very dependent on operating conditions |
| *$MoS_2$* | | | | |
| • Powder (like graphite)<br>• Coating (PVD techniques) | 0.2–2 | −150 to 350 ($MoS_2$) and 400 ($WS_2$) in air. In $N_2$ or vacuum up to 800 | Widely used in industry (specially in resin-bonded coatings); as PVD coating is mainly used in aerospace applications | Basically effective in the absence of water vapour and oxygen |
| *PTFE* | | | | |
| • Powder (in composites, co-deposited in Ni layers);<br>• Monolithic (reinforced) parts;<br>• Less used as coatings (by spray, sputtering, CVD) | 2–10 | −70 up to 200 | Bearings; seals; automotive (seat belt clips, fasteners) | $T_g$ around −90 °C |
| *Soft metals (Ag, Pb, In, Au)* | | | | |
| Coating (electrochemical, PVD techniques) | 0.2–1 (or more in some cases) | | Bearings; rolling bearings (in lubricated conditions) | |

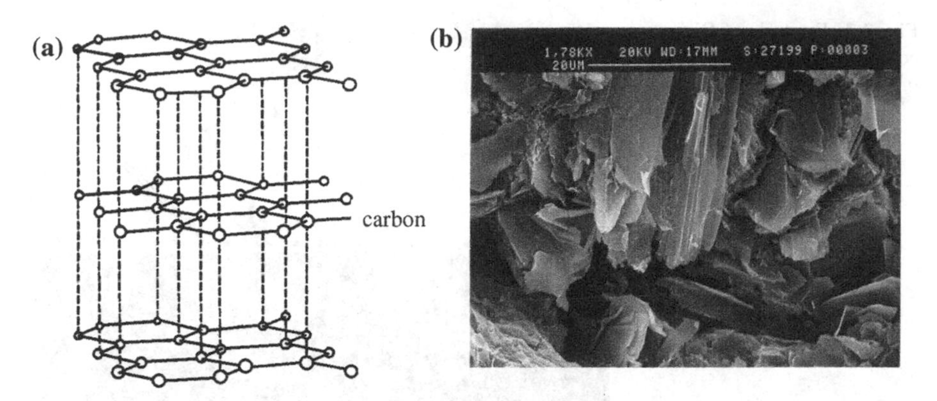


**Fig. 3.2** **a** Crystal structure of graphite; **b** typical morphology of graphite particles

As shown in Table 3.1, graphite can be used as loose particles, thin layers and also as monolithic components, obtained by machining carbon-graphite blocks that are produced following particular processing routes (further details will be given in Sect. 6.7). For example, graphite seals and special braking pads are made in this way.

### *3.1.2 Diamond-Like Coatings*

Diamond-Like Coatings (DLCs) are made of carbon too, but have a different crystal structure than graphite [6]. In fact, they are made of amorphous metastable carbon with variable short-range diamond ($sp^3$-type) and graphitic ($sp^2$-type) hybridization. DLCs display high hardness, high elastic modulus and also excellent tribological properties, i.e., low friction and high wear resistance. In order to optimize their performances in different conditions, they are usually doped with hydrogen. They may be also doped with N, Si, SiO that reduce the residual stresses, which may cause the spallation of the coating. On the basis of hydrogen content, a distinction is made between amorphous DLC (a-C) with the majority of $sp^2$-bonds, and hydrogenated amorphous carbon (a-C:H) with typically 10–50 at. % H that stabilizes the $sp^3$ bonds and prevents them to transform into graphitic bonds during deposition.

The solid-lubricating performance of DLCs during sliding is due to different factors that most probably work together. The first is the $sp^3 \rightarrow sp^2$ transformation at the contacting asperities because of the local temperature rise during the run-in stage. A layer of graphite is thus transferred to the counterface and low friction is attained because of the sliding between graphitic planes. The second is the contamination of the surface and the third is the hydrogenation (if present) of carbon atoms that reduces the surface energy [7, 8]. In this latter case, a low friction is attained without the formation of a transfer layer. An excessive temperature rise, however, may induce a degradation of DLCs and also a release of hydrogen, with a

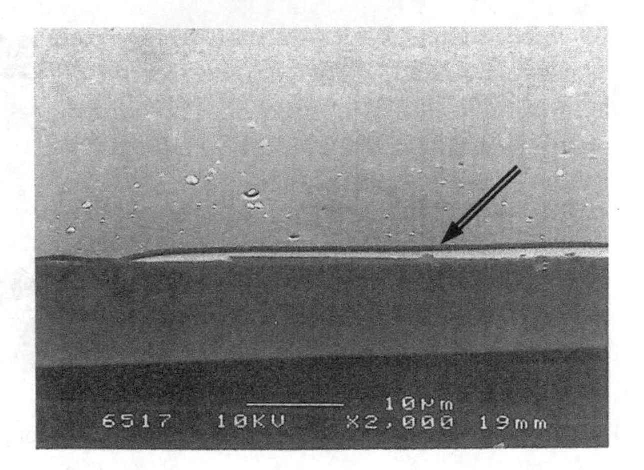


**Fig. 3.3** Example of DLC coating (indicated by the *arrow*) [6]

deterioration of the performances. For temperatures greater than 100 °C friction starts to increase, and above 300 °C the DLCs almost loose their properties [5].

In general, the friction coefficient may range from 0.01 (very low value) to 0.5 (no solid lubricant effect), in dependence of the operating conditions, i.e., contact load, sliding speed, temperature, counterface material and environment. A correct selection and optimization is therefore required for each engineering applications. As an example, H-free DLCs perform better in humid air, while hydrogenated coatings perform better in dry or inert gas environments [5]. Figure 3.3 shows an example of a DLC coating. The very low surface roughness of the coating after deposition can be clearly appreciated.

### 3.1.3 Molybdenum Disulphide

$MoS_2$ has a layered structure very similar to graphite. A sandwich of S-Mo-S hexagonally packed layers is held together by rather weak van der Waals forces. Shear stresses lead the planes to slide and form a transfer layer. No water vapour molecules are required for a solid lubricating effect. $MoS_2$ shows a low friction coefficient in dry environments and even vacuum. Indeed, the presence of oxygen or water vapour may oxidize molybdenum forming $MoO_3$ that increases friction, and $H_2SO_4$ that may also induce corrosive phenomena in the tribological system.

$MoS_2$ can be used in the form of powder, simply burnished onto the component surface or sprayed with a volatile solvent. It can be also deposited using a binder material (Fig. 3.4 shows an example) or used as a sputtered coating (PVD, Sect. 7.5.2) in precision components. The best performances in terms of friction and wear are obtained with $MoS_2$ coatings. Another member of the TMD family is $WS_2$, which displays a similar behavior as $MoS_2$.

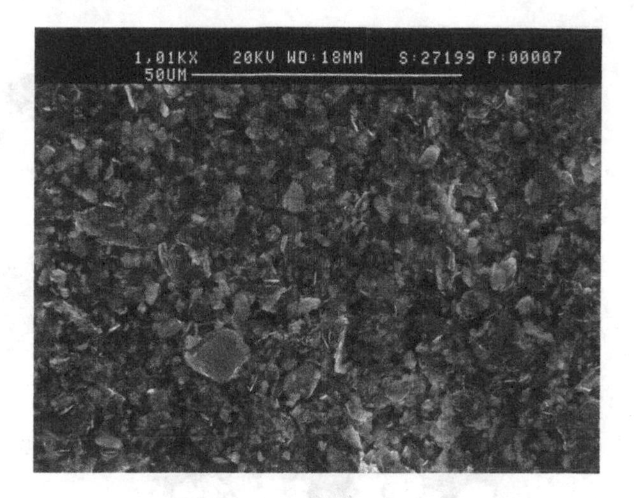


**Fig. 3.4** $MoS_2$ powder bonded with epoxy resin [9]

## 3.1.4 PTFE

Some polymers with unbranched molecular chains, such as PTFE (polytetrafluoroethylene) and some polyamides and acetal resins, constitute another important class of solid lubricants. As already described in Chap. 2, during sliding PTFE forms a thin transfer layer on the counterface (the presence of fluorine atoms greatly facilitates the adhesion to the metal surfaces). Very low friction is attained because of the low surface energy, due to the absence of unsaturated bonds. Unfortunately, the mechanical strength of PTFE is very low as well as its wear resistance. Different approaches are followed to exploit the solid lubricating capacity of PTFE. For example, it is reinforced by the use of suitable fillers or particles. As an example, Fig. 3.5a shows the fracture surface of a PTFE seal reinforced with bronze particles. PTFE particles are also used in polymeric composites or co-deposited in Ni-coatings (Fig. 3.5b). It may be also impregnated in the open porosity of layers made by sintering bronze powders, as shown in Fig. 3.5c. In this case, wear resistance is provided by the bronze skeleton whereas the solid lubricating effect is continuously provided by the PTFE.

## 3.1.5 Soft Metals

Low coefficients of friction (even lower than 0.1) can be obtained with thin coatings (thickness in the 1 μm range are commonly used) of the so-called soft metals, such as lead, silver, gold and indium. The lubricant capacity of such metals is due to their low surface energy. For example, the surface energy of the lead is around 0.45 J/m$^2$. If a thin layer of lead is deposited on a steel with a hardness of 400 kg/mm$^2$, during

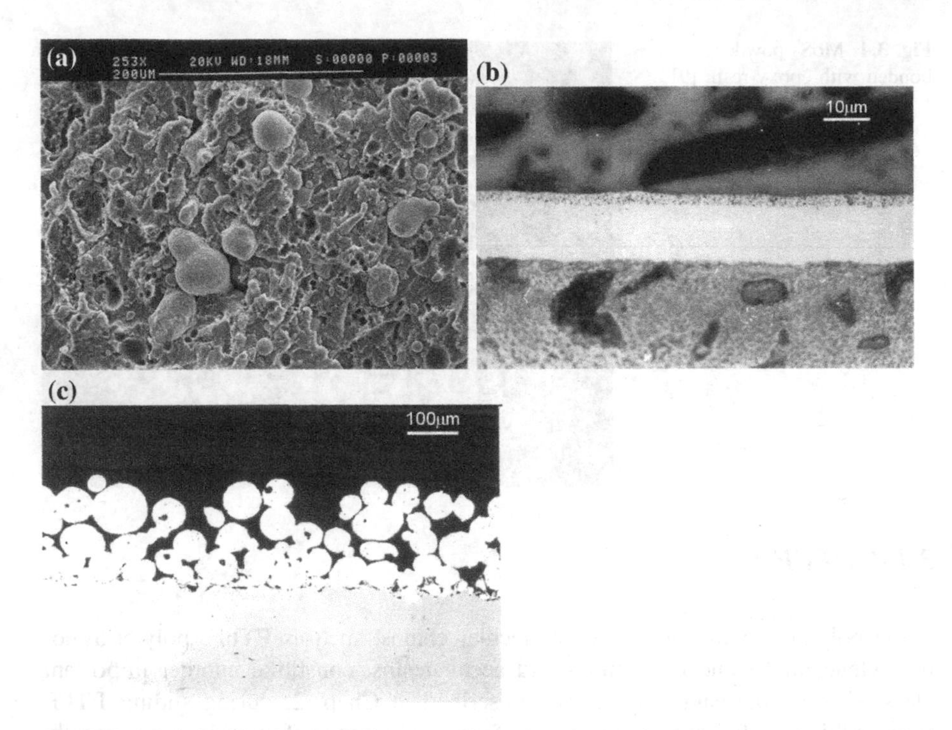


**Fig. 3.5** **a** Fracture surface of a PTFE gasket reinforced with bronze particles; **b** cross section of a nickel coating on steel that contains, in the outer part, a dispersion of PTFE particles; **c** cross section a surface bronze skeleton made by sintered particles with an impregnation of PTFE [9]

sliding Pb–Pb contacts are realized because of transfer, and the ratio $W_{12}/H$ is about $0.0023\ 10^{-7}$ m. Therefore, from Fig. 2.9a friction coefficient lower than 0.2 can be predicted. It is clear that film thickness should be low enough to avoid increasing the real area of contact.

Soft coatings are usually produced with electrolytic techniques and PVD methods, like vacuum deposition and ion plating (see Sect. 7.5.2). Note that soft metals can be easily removed during the tribological contact. Thus, their use is generally restricted to particular applications (such as under boundary lubricated conditions, see Sect. 3.4), and their wear life should be properly assessed. As an example, Fig. 3.6 shows the sliding specific wear rates (see next chapter) vs. friction coefficient for different solid lubricant coatings sliding against AISI 440C stainless steel balls [3]. The tests were carried out in humid air as well as in high vacuum. The lead and silver coatings display the worst behaviour in humid air and an intermediate behaviour under vacuum. As previously stated, the $MoS_2$ coatings display an outstanding behaviour in vacuum, whereas the DLCs show their best performances in humid air.

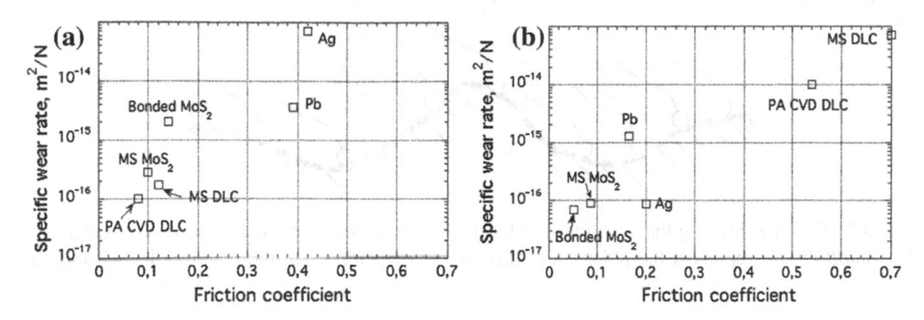


**Fig. 3.6** Specific wear rates versus steady-state friction coefficient for a number of solid lubricant coatings sliding against a stainless steel ball. **a** In humid air; **b** in ultra high vacuum (modified from [3])

## 3.2 Liquid Lubricants

When a liquid lubricant is employed, the knowledge of the lubrication regime attained in service is of paramount importance for a correct design, including a correct selection of the lubricant type. If $\Lambda > 3$ (fluid film lubrication) the lubricant thickness in the mating region is such that no contacts among the asperities are allowed (Fig. 3.1a). The lubricant is under pressure and is thus able to exert a suitable force to maintain the complete separation between the two surfaces. In this case, the physical properties of the lubricant, such as viscosity, play the most important role. If $\Lambda < 0.5$ (boundary lubrication), a thin lubricant film covers the asperities but it does not exert any force to separate the surfaces. In this case, the performance of the lubricant is mainly dependent on its chemical properties, i.e., on its ability to chemically interact with the surfaces and form a tribofilm able to reduce friction and wear.

### *3.2.1 Mineral and Synthetic Oils*

The main lubricating oils are mineral and synthetic. Mineral oils are largely used due to their high performance/price ratio. They account for more than 95 % of all lubricants production [10]. The base oils are derived from refining of petroleum crude oil and contain organic macromolecules and impurities (such as sulphur), depending on the provenance of petroleum. The majority of mineral oils are *paraffinic*. Paraffin's are alkanes ($C_nH_{2n+2}$) with linear or branched chains. Mineral oils may also be *naphthenic* (and thus contain cyclic molecules) or *aromatic* (and thus contain molecules with very stable aromatic rings). Synthetic oils are more expensive than minerals oils. They are obtained by the polymerization of organic molecules and are produced for specific applications (such as high temperature applications). Synthetic oils may be based on hydrocarbons (polyolefins, ester oils

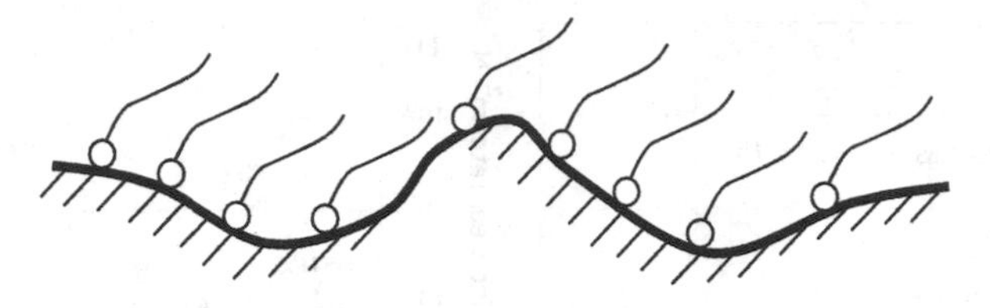

**Fig. 3.7** Scheme showing the adsorption of lubricant macromolecules on the surface of materials. The *small circles* represent the polar ends whereas the *lines* represent the macromolecular chains

or polyglycols) or they may be based on *silicones* (with a –Si–O–Si– chain that is particularly suitable for use at high temperature). Compared with mineral oils, synthetic oils have better thermal properties such as higher stability, higher decomposition temperature and better oxidation resistance.

Special *additives* are added to base oils (in amounts typically ranging from 1 to 5 % by weight) to improve their specific performances. There are additives that improve the wettability of the lubricant on the surfaces, such as *fatty acids*. Most oils contain such molecules but they are also specifically added in lubricants for boundary lubrication where the adhesion of the lubricant to the surfaces is paramount. Such molecules contain polar ends (−OH) that are able to adhere to the metal surface (see the scheme in Fig. 3.7). In the case of steels such adhesion is very strong since a chemisorption process is involved in bond formation among hydroxyl groups and the oxides present on the surface of the steel. The amount of additive should be enough to ensure complete coverage of the surfaces to be wetted by the lubricant.

Other classes of additives for boundary lubrication are the *anti-wear* and the *extreme pressure (EP)* additives [11]. They contain sulphur, lead and zinc compounds that are able to react at the asperities during sliding because of the high local flash temperatures, and thus form protective solid lubricant layers. An important anti-wear additive is *ZDDP* (zinc dialkil dithiophosphate), which actually has multifunctional properties including anti-wear and *anti-oxidant* action [12]. This additive interacts with the steel surface (the oxides) during sliding and forms an oxy-sulphide layer on the top of which a glassy phosphate forms. The whole tribofilm has a thickness ranging between 100 and 1000 nm and acts as a barrier against wear. The anti-oxidant action of ZDDP is also important, since oxidation strongly reduce lubricant performances as temperature is increased. In fact, oxidation induces an increase in viscosity and in acidity that makes the lubricant chemically aggressive against the surface to be lubricated. Different oxidation inhibitors are available to increase to oil life when oil temperature rise may be of great concern. EP-additives mainly contain organic sulphur, phosphorus compounds with a strong affinity to metals (e.g. steels) that during sliding promote formation of compounds, like sulphides, with a strong solid lubricant action.

A very important property of liquid lubricants is *viscosity*, which is a measure of their resistance to flow. If we consider two planes, separated by a liquid lubricant

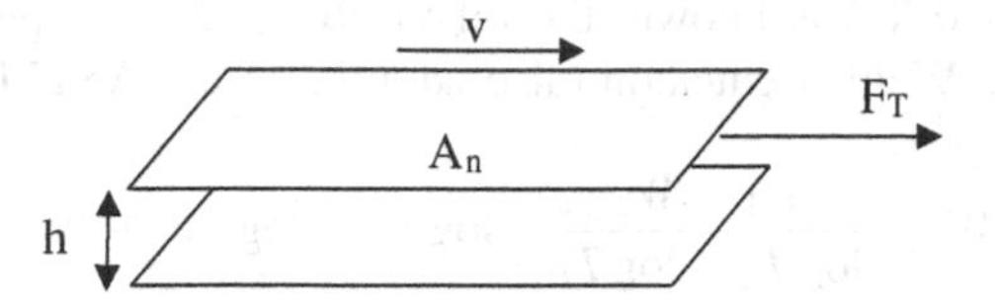


**Fig. 3.8** Schematic for the definition of oil viscosity

with thickness h, and sliding with a relative velocity v (Fig. 3.8), the shear stress, $\tau_m$ (given by $F_T/A_n$) required to maintain the relative motion is given by:

$$\tau_m = \eta \cdot \frac{v}{h} \tag{3.3}$$

where η is the *dynamic viscosity* of the lubricant (units: Pa s; 1 cP = $10^{-3}$ Pa s). The ratio between the dynamic viscosity and the liquid density gives the *kinetic viscosity* that can be measured in laboratory tests (note that oil density is around 0.9 g/cm$^3$). It is commonly indicated with ν, and it is typically measured in mm$^2$/s (1 mm$^2$/s = 1 centi Stokes, cSt). As an example, at room temperature water has a kinematic viscosity of around 1 cSt, whereas automobile engine oils have a kinematic viscosity of around 200 cSt.

Viscosity greatly depends on temperature and, in particular, it decreases as temperature is increased, as shown in Fig. 3.9 for some industrial oils. In certain lubricating oils, viscosity decreases by 80 % for a temperature increase of only 25 °C. The variation in temperature of viscosity may be expressed by an exponential equation:

$$\eta = \eta_r \cdot \exp(-\beta \cdot (T - T_r)) \tag{3.4}$$

where β is the temperature viscosity coefficient and $\eta_r$ is the viscosity at the reference temperature, $T_r$ (typically 20 °C). When the viscosity of a lubricant at two

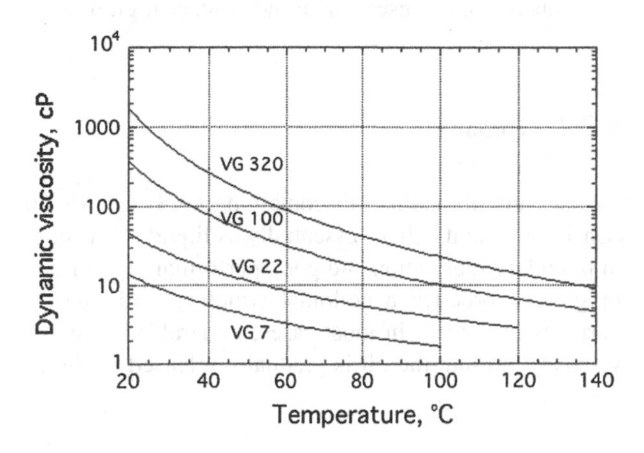


**Fig. 3.9** Dependence of dynamic viscosity on the temperature in the case of some industrial oils denominated using the ISO classification

temperatures ($T_1$ and $T_2$) is known, the oil viscosity at a temperature $T_x$ can be estimated using the Walther equation (also adopted in the ASTM D314 standard):

$$W_x = \frac{W_1 - W_2}{\log T_2 - \log T_1} \cdot (\log T_2 - \log T_x) + W_2 \tag{3.5}$$

where W = log log (ν + 0.7), ν is in $mm^2/s$ and T in K.

The viscosity of a lubricant depends also on pressure, especially if it is quite high as it may occur in non-conformal contacts. Several relationships have been proposed to express the dependence of viscosity on pressure. The simplest one is the exponential equation:

$$\eta = \eta_0 \exp(\alpha p) \tag{3.6}$$

where $\eta_0$ is the atmospheric viscosity and α is the so-called *Barus coefficient*. If not known, it can be estimated, in first approximation, by the following relationship [2]:

$$\alpha = (0.6 + 0.96 \cdot Log\eta_0) \cdot 10^{-8} \tag{3.7}$$

where α is in $m^2/N$ and $\eta_0$ in Pa s. The Barus coefficient typically ranges between 1 and 2 $10^{-8}$ $m^2/N$.

The International Organization for Standardization (ISO) classifies industrial oils on the base of viscosity ranges at 40 °C (some examples are shown in Fig. 3.9). Other classifications are used in practice, such as the SAE classification for automotive applications and the AGMA classification for gears. A useful parameter that accounts for the dependence of viscosity on temperature is the *viscosity index*, VI. It is determined in relation to two reference oils at which VI-values of 0 and 100 are assigned. In general, if oil is characterized by a low VI-value, its viscosity decreases greatly as temperature is increased; vice versa for oils with high VI-values. In Table 3.2 the characteristics of some mineral oils are listed.

The lubricant may be supplied in different ways: using baths, and then rings or chains to deliver the oil in the contact region; using circulating systems, in which oil is pumped from a reservoir in the contact region.

### 3.2.2 Greases

Greases are lubricating oils containing *thickeners* (5–20 %). They have a semi-fluid consistency, and behave essentially as liquid lubricants with excellent performances in boundary lubrication and good performances in fluid film lubrication. Thickeners in greases produce a colloidal structure with the dispersion of stabilized solid particles in the oil. In most cases, mineral oils are adopted with a well-controlled viscosity, so that the oil is gradually released in the region to be lubricated.

**Table 3.2** Dynamic viscosity and Barus coefficient for some industrial mineral oils (from Ref. [11])

| Lubricant | ISO VG | VI | Dynamic viscosity at atmospheric pressure (cP) | | | Barus modulus ($10^{-9}$ $m^2/N$) | | |
|---|---|---|---|---|---|---|---|---|
| | | | 30 °C | 60 °C | 100 °C | 30 °C | 60 °C | 100 °C |
| *High VI* | | | | | | | | |
| | 32 | 108 | 38 | 12.1 | 5.3 | | 18.4 | 13.4 |
| | 100 | 96 | 153 | 34 | 9.1 | 23.7 | 20.5 | 15.8 |
| | 150 | 96 | 250 | 50.5 | 12.6 | 25 | 21.3 | 17.6 |
| | 460 | 96 | 810 | 135 | 26.8 | 34 | 28 | 22 |
| *Medium VI* | | | | | | | | |
| | 15 | | 18.6 | 6.3 | 2.4 | 20 | 16 | 13 |
| | 32 | 68 | 45 | 12 | 3.9 | 28 | 20 | 16 |
| | 68 | 63 | 107 | 23.3 | 6.4 | 29.6 | 22.8 | 17.8 |
| | 75 | 84 | 122 | 26.3 | 7.3 | 27 | 21.6 | 17.5 |
| | 100 | 38 | 171 | 31 | 7.5 | 28 | 23 | 18 |
| *Low VI* | | | | | | | | |
| | 22 | −6 | 30.7 | 8.6 | 3.1 | 25.7 | 20.3 | 15.4 |
| | 100 | −7 | 165 | 30 | 6.8 | 33 | 23.8 | 16 |
| | 150 | 8 | 310 | 44.2 | 9.4 | 34.6 | 26.3 | 19.5 |
| | 1000 | | 2000 | 180 | 24 | 41.5 | 29.4 | 25 |

The thickeners are metallic fatty acid soaps or a clay mineral, and grease is usually indicated with reference to the type of thickener. Greases based on sodium, calcium and lithium are commonly used. Lithium greases are mostly used, since they can resist temperatures up to 200 °C. Anti-wear, anti-oxidation and EP additives, and solid lubricant powder (such as graphite or $MoS_2$) may be also added to improve the grease performances. Beside good lubricating properties, greases can remain long enough in place to protect the tribological system from external contaminants and corrosion. Greases are commonly used in rolling bearings.

## 3.3 Fluid Film Lubrication

In order to have fluid film lubrication, the lubricant in the meatus must be in pressure, and able to exert a thrust force that maintains the complete separation between the two surfaces ($\Lambda > 3$). Such force can be hydrostatic or hydrodynamic in nature [11, 13, 14]. In the case of *hydrostatic lubrication*, the lubricant is supplied and maintained under pressure by an external pump. As long as the pump feeds lubricant in the contact area, the separation between the surfaces may take place even in the absence of relative motion between the surfaces. This type of lubrication

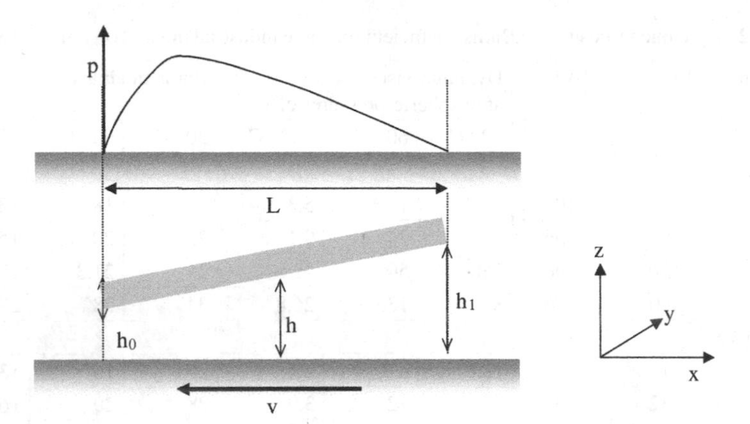


**Fig. 3.10** Principle of hydrodynamic lubrication

is used in special bearings (also with gaseous lubricants, such as air), if it is necessary to obtain very low values of the static friction coefficient.

In the case of *hydrodynamic lubrication*, the lubricant is able to exert a thrust force only if there is a relative motion between the surfaces in contact. The simplest tribological system to represent the condition of hydrodynamic lubrication is that shown in Fig. 3.10. It consists of an inclined upper plane and a horizontal lower plane moving with a velocity v. The lubricant adheres to both surfaces and is dragged into the contact region, which has a *converging shape in the direction of motion*. The entry section, with a height $h_1$, is greater than that the exit section, $h_0$. This configuration generates a pressure profile along the contact, which is responsible for the hydrodynamic thrust force.

In the case of incompressible fluids (i.e., typically in the case of liquids), the pressure evolution, p, in the meatus can be obtained by solving the Reynolds' equation:

$$\frac{d}{dx}\left(h^3\frac{dp}{dx}\right)+\frac{d}{dy}\left(h^3\frac{dp}{dy}\right)=6\cdot v\cdot\eta\cdot\frac{dh}{dx} \tag{3.8}$$

where h is the height of the meatus and η is the lubricant viscosity. By integrating, the equilibrium thrust force, $F_N$, and the tangential force, $F_T$, can be obtained [11]:

$$F_N=\frac{6\cdot\eta\cdot v\cdot L^2}{K^2\cdot h_0^2}\cdot\left(-\ln(K+1)+\frac{2K}{K+2}\right) \tag{3.9}$$

$$F_T=\frac{\eta\cdot v\cdot L\cdot b}{h_0}\cdot\left(\frac{6}{K+2}-\frac{4\ln(K+1)}{K}\right) \tag{3.10}$$

where b is the plane width and K is given by: $h_1/h_0-1$.

Note that both forces depend on the lubricant viscosity and on the relative velocity. If v tends to 0, no hydrodynamic force is exerted at all. Further note that $h_o$ is the minimum lubricant thickness. To really have fluid film lubrication, $\Lambda$ should be greater than 3, and therefore, from Eq. 3.1, $h_o$ should be greater than $3 \cdot \sqrt{R_{q1}^2 + R_{q2}^2}$, where $R_{q1}$ and $R_{q2}$ are the root-mean-square roughness values of the two surfaces. $h_o$ can be obtained from Eq. 3.9 since, in equilibrium, the thrust force corresponds to the normal applied force. Alternatively, the given relations can be also used to evaluate $\Lambda$ and therefore the effective lubrication regime. The hydrodynamic lubrication is present, for example, in thrust and journal bearings.

A special case of hydrodynamic lubrication is the *elasto-hydrodynamic lubrication* (also referred to as EHD), which occurs in the case of non-conformal contacts. In this case, the lubricant in the contact region interacts with the elastic deformations of the mating surfaces, and this gives rise to local variations in section producing a converging shape in the direction of motion. The resulting thrust force is thus connected with the elastic deformations, the lubricant viscosity, its dependence on pressure, and the relative motion. Figure 3.11 shows a schematization of the deformations in a non-conformal contact and the evolution of contact pressure. It can be noted that pressure evolution is modified with respect to the semi-elliptical Hertzian contact, with, in particular, the formation of an additional pressure spike at the contact exit. Such pressure spike may even exceed the Hertzian pressure even by 20–30 %, and this influences the performance of the tribological system since it induces a highly localized stress concentration at the surface [15].

The calculation of $h_{min}$ in the case of EHD lubrication is usually carried out with the relation proposed by Hamrock and Dowson [11, 13], which assumes that the lubricated surfaces are smooth:

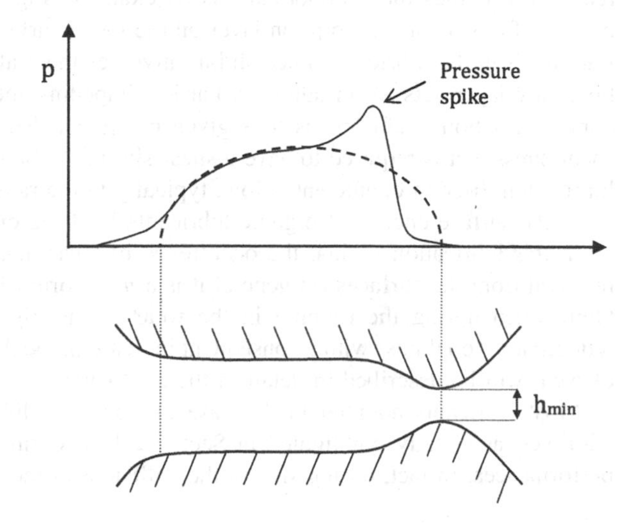


**Fig. 3.11** Hydrodynamic pressure distribution in an EHD contact (the *dashed line* represents the Hertzian pressure distribution)

$$h_{\min} = 3.63 \cdot R' \cdot \frac{G^{0.49} \cdot U^{0.68}}{W^{0.073}} \cdot \left(1 - \frac{0.61}{e^{0.73\,k}}\right) \tag{3.11}$$

where $G = \alpha E'$ (dimensionless materials parameter), $U = \eta_0 v'/E'R'$ (dimensionless speed parameter), $W = F_N/E'R'^2$ (dimensionless load parameter), $v' = (v_1 + v_2)/2$, being $v_1$ and $v_2$ the tangential velocities of the two surfaces ($v'$ is also called the rolling velocity), $\eta_0$ is the viscosity of the lubricant at ambient pressure and at the temperature of entry into the contact region, k is a parameter which is 1 for a point contact and infinity for a line contact (the other parameters in Eq. 3.11 were defined in Sect. 1.1).

To summarize, the ability of a liquid lubricant to produce fluid film lubrication is favoured by: high tangential speeds (this is the most important parameter; if they are excessive, however, the entrance of the lubricant in the meatus can be difficult); a low load; a high lubricant viscosity. Lubricant temperature at entry is also an important parameter. In fact, as such temperature is increased, both viscosity and Barus coefficient are decreased and the hydrodynamic thrust force (and thus $h_{min}$) is decreased. In most mechanical systems, the oil temperature at entry is maintained in the range between 40 and 60 °C.

## 3.4 Boundary Lubrication and Scuffing

When the load is high and/or the speed is low, $\Lambda$ may be very low (lower than 0.5), and the lubricant does not exert any supporting action. The surface asperities come into close contact and, at least in the case of metals, local plastic contacts take place. In this case, the role of the boundary layer, i.e., of the film that is formed on the surface to be lubricated by physical absorption, chemical adsorption or chemical reaction, becomes more important. As an example, Fig. 3.7 schematizes the formation of a boundary absorption layer on the metal surface, due to the presence of fatty acids in the lubricant. Since sliding involves the real area of contact ($A_r$), the boundary layer acts as an adhesion barrier, impeding the asperity metal-to-metal contact. Friction coefficient is thus given by: $\mu = \tau_m/H$, where $\tau_m$ is the average shear stress that is required to have a shear slip in the boundary layer. In boundary lubrication, friction coefficient is low, typically in the range between 0.1 and 0.15, since the surface energy of organic lubricants is of the order of 0.02 J/m$^2$.

In this lubrication regime the occurrence of wear, i.e., of progressive material removal from the surfaces (in general it is a *mild* form of wear), may be detected. Quite often during the running in the wear of the highest asperities leads to a reduction of roughness with a consequent increase in the $\Lambda$ factor. The phenomenon of wear will be described in detail in the next Chapters.

Most lubricants adopted in the case of boundary lubrication contain specific additives, as already highlighted in Sect. 3.2.1, to optimize the friction and wear performances. In fact, during sliding the collisions at the asperities can destroy the absorption layer. As a consequence, several dry asperity contacts are established and the local temperature rise may induce an avalanche effect, with the progressive desorption of the lubricant. This could have dramatic consequences, leading to *severe* wear (often called scoring, galling or seizure). The whole phenomenon is called *scuffing*, and is one of the most problematic failure modes in tribology.

If specific additives are present in the lubricant (such as anti-wear or EP additives), a stronger boundary layer is formed, often after a reaction with the metal surface (such layer is sometimes called *secondary boundary layer* [16] ). As already seen, anti-wear and EP additives contain atoms with good reactivity with the metal substrate, which promote formation of lubricating compounds. Such boundary layers have a complex structure and are made by a mixture of polymeric materials, wear debris and functional molecules. In this respect, surface temperature plays a very important role. In fact, the attainment of a critical flash temperature during sliding is required to activate the chemical reactions at the asperities, whose products form the protective boundary layer. Of course, if temperature is too low the reaction rate may be insufficient to quickly form a secondary boundary film. On the contrary, if it is too high the reaction rate may be too rapid, and chemical corrosion rather occurs, which in turn induces excessive wear [17].

The phenomenon of scuffing is usually investigated by carrying out lubricated tests in line or point contacts. In general, a point contact test is preferred since it allows the attainment of very high Hertzian pressures; in this respect, the *four-ball test*, described in Sect. 4.5.4, is very often employed to investigate the scuffing resistance of a lubricant. In such a test, the applied load is increased in a step-wise or continuous manner, in order to change the lubrication regime, i.e., to reduce the $\Lambda$-factor, and to increase the local pressure. At the same time, the friction coefficient is monitored as a function of time. Figure 3.12 schematizes a typical behaviour when using a lubricant without additives, with anti-wear additives, and with EP-additives [18]. At the beginning, friction coefficient is seen to increase since lubrication regime passes from mixed to boundary lubrication. At a critical load, known as the *scuffing load* (indicated with arrows in the graph) friction starts to dramatically increase. Scuffing is then seen to propagate, and when using oil lubricants without additives, friction coefficient rapidly rises up to values typical of dry sliding, at which severe wear occurs, characterized by large material removal and large increase in surface roughness. If the lubricant contains anti-wear or EP-additives, however, the scuffing load is increased and, most importantly, the propagation stage is made more difficult because of the formation of a secondary boundary layer, which is able to protect the surface and to allow a satisfactory lubrication regime. The data in Fig. 3.12 show that for the lubricant with anti-wear additives, the friction coefficient considerably increases prior to the formation of the secondary boundary layer. On the contrary, when using EP-additives, such a protective layer forms quite soon and friction coefficient does not rise much as the applied load is increased.

The prediction of scuffing in a specific tribological application would be very important, in particular when using lubricants without specific additives. However, it is a very difficult task, and the available models are strictly applicable to specific

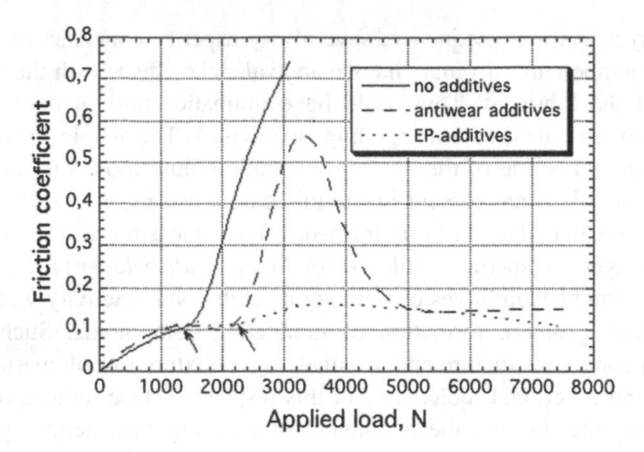


**Fig. 3.12** Evolution of friction coefficient with the applied load recorded in four-ball testing using chromium steel balls and oil lubricants without and with additives. The *arrows* indicate the scuffing loads (data from [18])

applications. As seen, scuffing occurs when the asperities of the mating surfaces collide, thus inducing the local removal of the lubricant. This may occur in boundary as well as in mixed lubrication regime. On such a basis, an *EHD breakdown criterion* may be used by setting that scuffing is triggered when Λ-factor becomes lower than a critical value and also lubricant viscosity is reduced below a critical value by frictional heating [19]. A considerable uncertainty, however, arises from the evaluation of Λ: $R_{q1}$ and $R_{q2}$ do not perfectly account for the actual asperity height distribution, and scuffing could initiate at the highest asperities where the collisions are more probable and severe; in addition, the equation commonly used for the calculation of $h_{min}$ (Eq. 3.11) is approximate since it is based on the assumption that the two surfaces in contact are smooth. *Thermally based models* were also proposed. A very common criterion was early suggested by Block: scuffing occurs when the contact flash temperature reaches a critical value, which is around 150 °C for most lubricants and depends on lubricant resistance to desorption [20]. To a first approximation, the flash temperature can be estimated by using the relations of Sect. 2.11; for the estimation of $T_s$, the friction coefficient for the upper limit of boundary lubrication could be used (μ around 0.1), whereas for the estimation of $T_f$, the friction coefficient for dry sliding could be used (μ around 0.7 for steels) since scuffing is initiated once the collisions have removed the lubricant from some asperities. Also in this case, however, there are a lot of uncertainties' that are not removed by using better temperature models. In fact, the desorption temperature depends on the chemical composition of lubricants, and it may also depend on pressure and other operational parameters.

The use of solid lubricants, such as soft metals, DLC or $MoS_2$ coatings, may considerably improve the performances of tribological systems operating in boundary lubrication conditions. In fact, when the boundary film fails, solid lubricant can carry the load, preventing the metal-to-metal contacts that produces

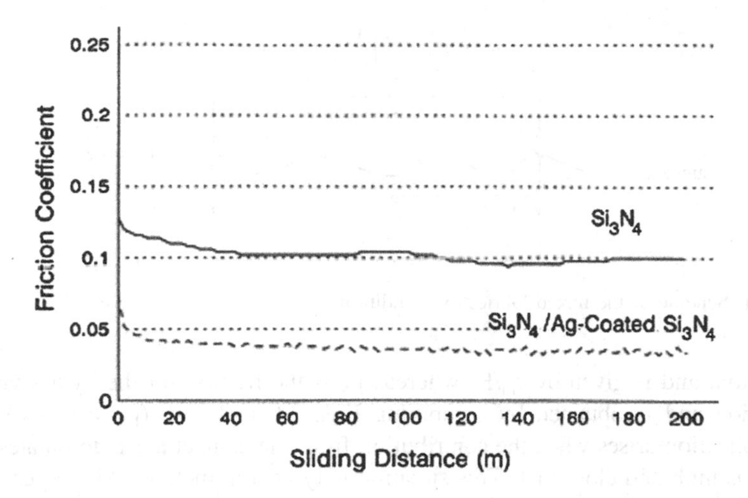


**Fig. 3.13** Friction evolution of uncoated and Ag-coated $Si_3N_4$ at 200 °C and in a synthetic oil [21]

wear and may lead to scuffing, therefore acting as a back-up lubricant [21, 22]. As an example, such an effect is often exploited in lubricated bearings that are coated with soft metals. In fact, during the run-in, the sliding speed is low and thus the hydrodynamic force exerted by the lubricant is also low making the system working in boundary-lubricated conditions. Beneficial synergistic effects can be also obtained by using a liquid lubricant in combination with a solid lubricant coating. As an example, Fig. 3.13 shows the friction evolution of an uncoated and an Ag-coated silicon nitride sample in a synthetic oil at 200 °C [21]. In this respect, for some applications the drastic reduction in S, P-base additives can be achieved thus favouring the successive oil recycling procedures.

Different systems work in boundary lubrication conditions, at least during part of their operating cycles. Examples include the piston ring and the cam/follower systems.

## 3.5 Mixed Lubrication

As schematized in Fig. 3.14, mixed lubrication occurs when both a hydrodynamic film and a boundary lubrication layer support the normal applied load, $F_N$. In this case, friction coefficient typically ranges between 0.07 and 0.1, and it may be evaluated by the following relation [23]:

$$\mu = \delta \cdot \mu_c + (1 - \delta) \cdot \mu_v \quad (3.12)$$

where δ is the contact area ratio, i.e., the ratio of the real contact area to the nominal area of contact, and $\mu_c$ and $\mu_v$ are the friction coefficients for the areas of contact and for the valley regions, respectively. $\mu_c$ is the friction coefficient for boundary

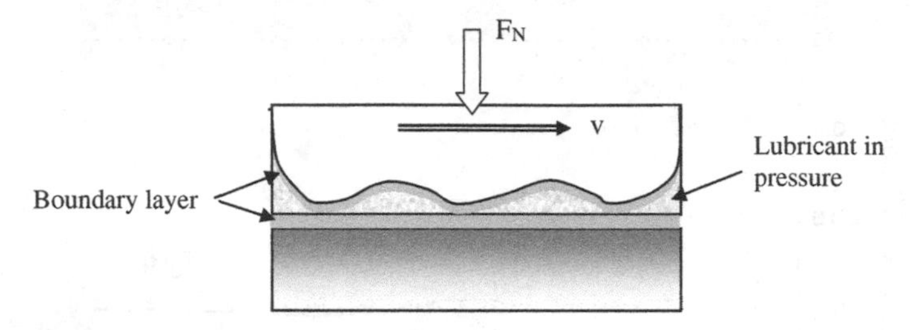


**Fig. 3.14** Scheme of the mixed lubrication condition

lubrication, and is given by $\tau_m/H$, whereas $\mu_v$ is the friction for the hydrodynamic lubrication and is obtained by combining Eqs. 3.9 and 3.10 ($\mu_v = F_T/F_N$) [24]. A complication arises when the contribution from the contact areas dominates, i.e., when $\delta$ is high and close to 1. This situation is typical in metalworking operations, and will be considered in the next paragraph.

$\Lambda \approx 1$ means that surface asperities are, on average, of the same height of the lubricant thickness. As a consequence, some collisions may occur during sliding between the highest asperities. In such conditions, local shear plastic deformations easily occur. They may lead to scuffing (see the previous paragraph), wear, or be the precursors for successive failures by contact fatigue, as shown in the next Chapter. Such surface localized plastic deformations are also known with the term *surface distress*.

However, a number of experimental investigations have shown that surface distress does not always occur even if the tribological system is operating at low $\Lambda$-values, well within the mixed lubrication regime. Two possible explanations have been proposed. The first one considers the so-called *micro-elasto-hydrodynamic lubrication* (micro-EHD): it is supposed that the lubricant interacts with the elastic deformations at the asperities, and this gives rise to local variations in section producing a converging shape in the direction of motion. Therefore, an EHD force is established between two mating asperities. A second explanation (probably correlated to the first one) considers the local increase in lubricant pressure that in turn induces an increase in viscosity: the lubricant becomes so thick to behave almost as a solid lubricant. It is clear that in such conditions no surface distress and also no scuffing may occur. No models are still available to predict the occurrence of surface distress (and scuffing) with adequate accuracy.

## 3.6 Lubricated Friction in Case of Large Plastic Deformations

Metals display particular behaviour when the normal load is very high and the real area of contact is close or equal to the nominal area of contact. As schematized in Fig. 3.15 [25], the friction force tends to be independent from the normal load, and

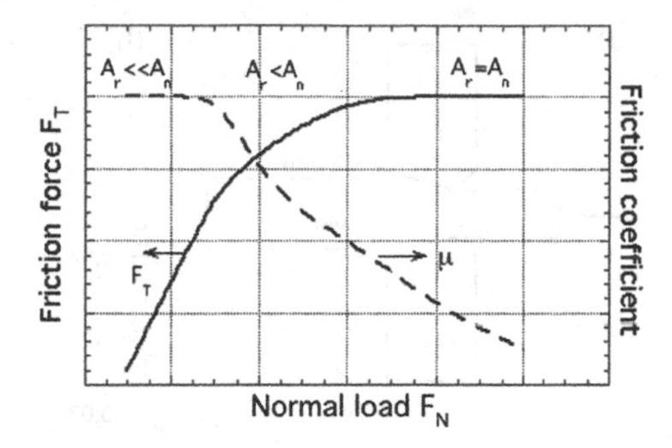


**Fig. 3.15** Schematic showing the dependence of the friction force to the normal load in metals

then friction coefficient tends to decrease with load. The situation $A_r \ll A_n$ is typical of mechanical contacts, such as those occurring in gears or in cams. The situation $A_r \approx A_n$ is typical of metalworking operations, typically conducted in lubricated conditions. In such operations, rigid (undeformable) tools plastically deform a work piece, to shape it as desired.

It is still not clear which lubricating mechanism is actually operating when δ, the contact area ratio, is close to 1 as a consequence of large plastic deformations in one of the contacting bodies. Certainly a boundary lubrication mechanism is operative, possibly favoured by the high local pressure. But other contributions may be present. The oil entrapped in the valleys could, for example, exert a hydrostatic support when pressed by the die, or it could squeeze in the contact regions exerting a hydrodynamic pressure (often called *plasto-hydrodynamic* lubrication). The trend displayed in Fig. 3.15 can be simply explained by only considering the role of boundary lubrication. Indicating with $\tau_m$ the average shear stress required to have shear slip in the boundary layer, the tangential force is $F_T = \tau_m A_n$ since $A_r \approx A_n$, and, consequently, $\mu = \tau_m A_n/F_N = \tau_m/p_0$ (where $p_0$ is the nominal pressure). This shows that when normal load is such that $A_r \approx A_n$, a further increase in load (or pressure) is accompanied by a decrease in friction.

An estimate of the lubrication regime when a work piece is in contact with a tool under large plastic deformations can be obtained by calculating the lubricant thickness, h, and then the Λ value, using the Wilson and Walowit's relation [26], obtained under the approximation of smooth surfaces. In the case of a *strip drawing* process (Fig. 3.16a), the relation is:

$$h = \frac{6 \cdot \eta_0 \cdot \alpha \cdot v_{entry}}{\Phi \cdot \left(1 - e^{-\alpha \cdot \sigma'_Y}\right)} \tag{3.13}$$

where $\eta_0$ is the oil viscosity at room pressure, α is the Barus coefficient, $v_{entry}$ is the entraining velocity, Φ is the entraining angle and $\sigma'_Y$ is the plane strain yield stress of the work piece. As an example, Fig. 3.16b shows the experimental values of friction coefficient as a function of h for the drawing of an aluminium alloy sheet with a thickness of 6 mm and a surface roughness of $R_q = 0.2$ μm [27]. Different

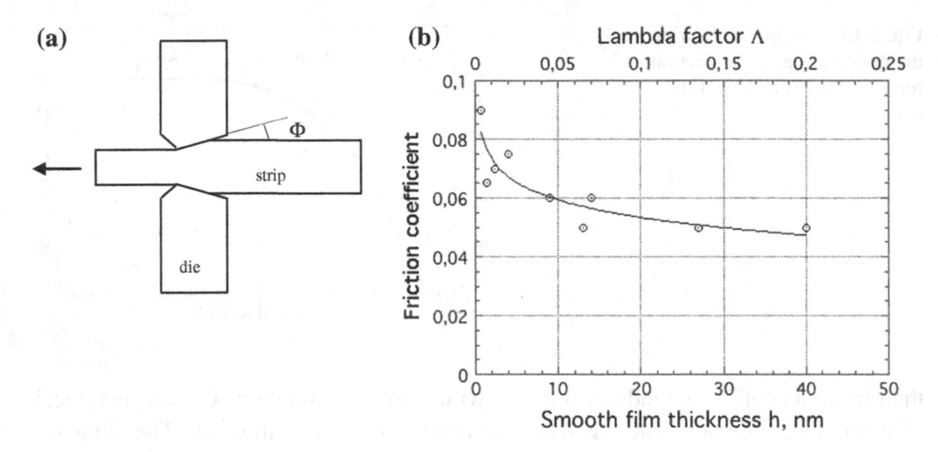


**Fig. 3.16** **a** Schematic of the strip drawing process; **b** friction coefficient versus smooth film thickness and Λ factor for strip drawing of an Al sheet (reduction between 0.1 and 0.15; lubricants: V68, M320, HVI 650) (data from [27])

half die-angles were employed to obtain different h-values. As expected, friction coefficient is seen to decrease with h. The obtained h-values are very small and this reinforces the idea of boundary lubrication. This is confirmed by the calculation of the corresponding Λ factors, which are all below 0.2, and by the observations of the surface topographies after drawing that highlighted the occurrence of remarkable surface distress phenomena. The comparison with the friction data reported in the previous paragraphs (see, for example, Fig. 3.1b), shows that the friction values in Fig. 3.16b are much lower than it would be expected for boundary lubrication conditions. This result is in agreement with the predictions of Fig. 3.15, since strip drawing was conducted under large plastic deformation of the work piece, and therefore under high loading conditions.

## 3.7 Lubricant Selection

The lubricants described in this chapter have been mainly developed to lubricate metallic surfaces and, in particular, ferrous alloys. The lubricating oils are also used in tribological systems made of ceramic materials. It is clear, however, that in this case the full potential of ceramics is not exploited, given the temperature limitations of oil lubricants. For use at high temperature, solid lubricants based on $MoS_2$ or $TiO_2$ should be selected. In the case of polymers, such a temperature limitation does not exist and the use of lubricants containing organic polar molecules is particularly effective and, therefore, recommended.

**Table 3.3** Comparison of the main properties of lubricants (modified from [28])

| Property | Solid lubricant | Oil | Grease |
|---|---|---|---|
| Hydrodynamic lubrication | Nil | Excellent | Fair |
| Boundary lubrication | Good to excellent | Poor to excellent (depends much on additives) | Good to excellent (depends much on additives) |
| Cooling | Nil | Very good | Poor |
| Ability to lubricate systems operating at high velocity | Very poor | Excellent | Poor |
| Ability to lubricate systems operating at low velocity | Good | Poor (excellent in hydrostatic lubrication) | Excellent |
| Ability to lubricate highly loaded systems | Good | Depends on additives | Fair |
| Protection against atmospheric corrosion | Poor to fair | Fair to excellent | Good to excellent |
| Temperature range | Good to excellent | Fair to excellent | Good |
| Cost | Fairly high | Low to very high | Fairly high to very high |
| Life determined by | Wear | Deterioration and contamination | Deterioration |

The choice of the optimum lubricant for each specific application is quite a difficult task and reference to the specialized literature should be made. In Table 3.3, the salient features of various types of lubricants are listed. They may be useful as a first guide in the correct selection of a lubricant.

## References

1. K.H. Czichos, K.H. Habig, *Tribologie Handbook* (Vieweg, Reibung und Verlschleiss, 1992)
2. I.M. Hutchings, *Tribology* (Edwald Arnold, London, 1992)
3. K. Miyoshi, *Solid Lubrication, Fundamentals and Applications* (Marcel Dekker, New York, 2001)
4. E. Rabinowicz, *Friction and Wear of Materials*, 2nd edn. (Wiley, New York, 1995)
5. T.W. Sharf, S.V. Prasad, Solid lubricants: a review. J. Mater. Sci. **48**, 511–531 (2013)
6. K. Holmberg, A. Matthews, *Coatings Tribology* (Elsevier, Amsterdam, 1994)
7. Y. Liu, A. Erdemir, E.I. Meletis, A study of the wear mechanism of diamond-like carbon films. Surf. Coat. Technol. **82**, 48–56 (1996)
8. K. Miyoshi, Lubrication by diamond and diamond-like carbon coatings. ASME J. Tribol. **120**, 379–382 (1998)
9. S. Rossi, F. Chini, G. Straffelini, P.L. Bonora, F. Moschini, A. Stampali, Corrosion protection properties of Nickel/PTFE, Phosphate/$MoS_2$ and bronze/PTFE coatings applied to improve the wear resistance of carbon steel. Surf. Coat. Technol. **173**, 235–242 (2003)
10. M.M. Khonsari, E.R. Booser, *Applied Tribology*, 2nd edn. (Wiley, New York, 2008)

11. G.W. Stachoviak, A.W. Batchelor, *Engineering Tribology*, 3rd edn. (Elsevier, Amsterdam, 2005)
12. A.M. Barnes, K.D. Bartle, V.R.A. Thibon, A review of zinc dialkyldithiophosphates (ZDDPs): characterisation and role in the lubricationg oil. Tribol. Int. **24**, 389–395 (2001)
13. B.J. Hamrock, *Fundamentals of Fluid Film Lubrication* (McGraw-Hill, New York, 1994)
14. E. Ciulli, *Elementi di meccanica* (Pisa University, Edizioni Plus, Pisa 2003)
15. L.D. Houpert, S. Ioannides, J. Kupers, The effect of the EHD pressure spike on rolling bearing fatigue. J. Tribol. 109 (1987)
16. A. Wachal, Analysis of boundary layer estimating criteria in lubricationg oils investigation. ZEM **3**, 325–332 (1983)
17. S.M. Hsu, *Boundary Lubrication of Materials* (MRS Bulletin, 1991) pp. 54–59
18. W. Tuszynski, M. Szczerek, Qualitative discrimination between API GL performance levels of manual transmission fluids by comparing their EP properties determined in a new four-ball scuffing test. Tribol. Int. **65**, 57–73 (2013)
19. W.F. Bowman, G.W. Stachowiak, A review of scuffing models. Tribol. Lett. **2**, 113–131 (1996)
20. S. Li, A. Kahraman, N. Anderson, L.D. Wedeven, A model to predict scuffing failures of a ball-on-disk contact. Tribol. Int. **60**, 233–245 (2013)
21. A. Erdemir, Review of engineered tribological interfaces for improved boundary lubrication. Tribol. Int. **38**, 249–256 (2005)
22. R. Michalczewski, W. Piekoszewski, W. Tuszynski, M. Szczerek, The problems of resistance to scuffing of heavily loaded lubricated friction joints with WC/C-coated parts. Ind. Lubr. Technol **66**, 434–442 (2015)
23. J.G. Lenard (ed.), *Metal Forming Science and Practice* (Elsevier, Amsterdam, 2002)
24. J. Xu, K. Kato, Formation of tribological layer of ceramics sliding in water and its role for low friction. Wear **245**, 61–75 (2000)
25. S. Kalpakjian, *Manufacturing Processes for Engineering Materials* (Addison-Wesley, Boston, 1984)
26. W.R.D. Wilson, J.A. Walowit, in An isothermal hydrodynamic lubrication theory for strip rolling with front and back tension. *Proceedings of Tribology Convention, I. Mech. Engrs*, Elsevier Science, London, 164–172 1972
27. H.R. Le, M.P.F. Sutcliffe, Measurements of friction in strip drawing under thin film lubrication. Tribol. Int. **35**, 123–128 (2002)
28. A.R. Lansdown, *Lubrication and Lubricant Selection. A practical Guide* (MEP, London, 1996)

# Chapter 4
# Wear Mechanisms

Wear is a damage of a surface in contact with another one, which results in the formation of *fragments* (or *debris*) that leave the tribological system. Wear may cause direct failure, may reduce tolerances and surface finish, or induce a surface damage that is responsible for the subsequent failure of the component (most often by fatigue).

The characteristics of the relative motion between two bodies in contact define the *wear processes*. Some examples are schematically shown in Fig. 4.1. If the bodies' slide one over the other, the resulting wear process is *sliding wear*. If they roll one over the other, the resulting wear process is *rolling wear*. A *rolling-sliding wear* is obtained if the two types of motion are superimposed. When a reciprocating sliding is present with very small displacement, the resulting wear process is called *fretting*. When one of the two bodies consists in one or more hard particles that abrade a softer surface, wear is called *abrasion by hard, granular material*. If a fluid carries such abrading particles, wear is called *erosion*.

Despite the high number of wear processes encountered in practice, the investigation of wear damage is facilitated by the observation that each wear process is determined by the action of a predominant *wear mechanisms*, and the wear mechanisms are only four [1, 2]:

(1) adhesive wear;
(2) tribo-oxidative wear;
(3) abrasive wear;
(4) wear by contact fatigue.

An understanding of the four wear mechanisms is crucial to properly control every wear process. This control can be done in the designing stage, when it is possible to recognize in advance the acting wear mechanism. This control can be also done subsequently, when there is the need to re-design a tribological system after a wear induced failure. To achieve this task, a proper *failure analysis* is required and this can be carried out only if the main wear mechanisms are correctly understood. In this chapter the four wear mechanisms will be described, while in the next chapter the salient features of the main wear processes will be outlined. At the



G. Straffelini, *Friction and Wear*, Springer Tracts in Mechanical Engineering,
DOI 10.1007/978-3-319-05894-8_4

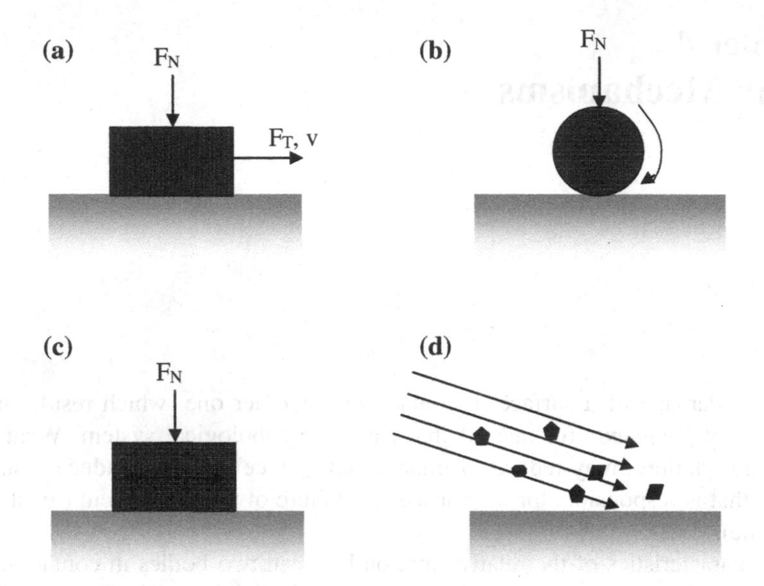


**Fig. 4.1** Examples of types of relative motion between bodies in contact and the related wear processes: **a** sliding wear; **b** rolling wear; **c** fretting wear; **d** hard particles transported by a fluid: erosive wear (erosion)

end of this chapter, the main wear testing procedures able to simulate the different wear mechanisms are also outlined.

## 4.1 Adhesive Wear

Adhesive wear takes place when the adhesion forces between the contacting asperities exert a predominant role in the formation of wear fragments. Historically, this mechanism is described by the *theory of Archard*, even if the current interpretation has been improved, thanks to the developments in the observation of the fragments and worn surfaces that have lead to a better understanding of the phenomena that are actually involved.

In the study of adhesive wear, it is useful to distinguish between wear of ductile materials (like most metals and polymers above their glass transition temperature) and wear of brittle materials (such as ceramics and polymers below their glass transition temperature).

### *4.1.1 Adhesive Wear of Ductile Materials*

In ductile materials, plastic junctions form at the contacting asperities during sliding (Fig. 1.11). As mentioned in Sect. 2.5, adhesion takes place at the junctions that in some cases may be more resistant than the bulk. As a consequence, the tangential displacement at some asperities may be due to fracture in the asperity bulk rather than by shearing at the interfaces. Such a fracture results in the formation of a loose wear fragment. As schematized in Fig. 4.2, the total *wear volume*, is given by: V = h $A_n$, where h is the *depth of wear*, and the *wear rate* is given by the ratio between the wear volume and the sliding distance, s: W = V/s. Since wear involves the contacting asperities, W is proportional to the contact area $A_r$:

$$W = K_{ad} \cdot A_r = K_{ad} \cdot \frac{F_N}{H} \tag{4.1}$$

having used Eq. 1.17 to express $A_r$. If the two materials in contact have different hardness values, H in Eq. 4.1 is that of the softer material, since it determines the extension of $A_r$ and produces the wear fragments. Following the Archard's view, the constant $K_{ad}$ (called the *wear coefficient for adhesive wear*) is representative of the fraction of junctions that give rise to the formation of a wear fragment, i.e., it provides the probability that a junction will form a wear fragment [3, 4].

However, it has been experimentally observed that during sliding different phenomena occur at the asperity contacts, and they have to be taken into account to properly understand the wear mechanism and, in particular, to clarify the meaning of the wear coefficient. First of all, the repeated plastic deformations at the asperities may induce local low-cycle fatigue damage or an accumulation of plastic deformation by ratcheting (see Sect. 2.1) [5, 6]. These processes involve quite extensive areas in the sub-surface contact regions and contribute to the material weakening. Hence, they lead to the formation of a wear fragment once a critical damage is attained. The intensity of the applied stresses is proportional to the local adhesion forces, and therefore to $W_{12}$. The resistance of the material to low cycle-fatigue/ratcheting is proportional, to a first approximation, to the material hardness, H. The adhesive wear coefficient, $K_{ad}$, can be thus considered proportional to the ratio $W_{12}$/H and therefore to friction coefficient (see Eq. 2.10a). The general validity of such observations is confirmed by the results reported in Fig. 3.6a and b, which refer to

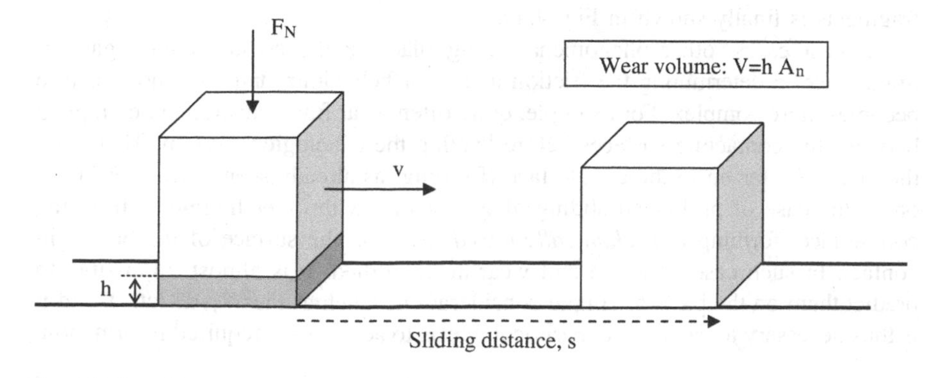


**Fig. 4.2** Definition of sliding distance and wear volume

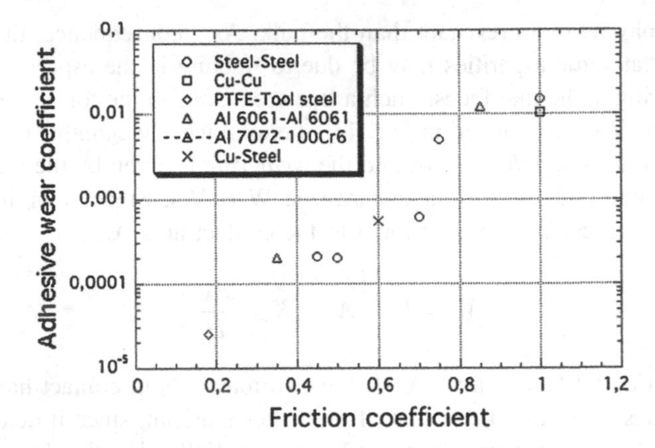


**Fig. 4.3** Relationship between experimental adhesive wear coefficient and friction coefficient in the case of different material pairs. The experimental data were obtained from tests conducted in order to have the sliding adhesive wear (in the case of different materials, the wear coefficient of the softest material is shown)

the wear behaviour of different solid lubricants. Additional results are displayed in Fig. 4.3 for different materials pairs (in each case, the tests were carried out in sliding condition chosen to obtain adhesive wear). Following the arguments outlined in Chaps. 1 and 2, the adhesive wear coefficient is then reduced as the tribological compatibility of the mating materials is decreased and their hardness is raised.

The typical morphology of the worn surfaces, the sub-surface damaged regions and the wear fragments are shown in Fig. 4.4. The reported observations refer to a titanium alloy (Ti-6Al-4V) after dry sliding against a steel counterface [7, 8]. The occurrence of large shear plastic deformation (by ratcheting/low cycle fatigue) at the asperities can be clearly appreciated in Fig. 4.4a. The cross section in Fig. 4.4b shows that plastic shearing involves the sub-surface regions too. This figure also shows the formation of a surface deformed scale that is about to leaving the tribological system, i.e., to *delaminate* as it is often said. The morphology of the wear fragments is finally shown in Fig. 4.4c.

In most cases, other phenomena taking place in the contact region gain in importance in determining the friction and wear behaviour, and the whole picture becomes more complex. For example, quite often wear fragments remain entrapped between the contacting surfaces before leaving the tribological system. Therefore, they can transfer on to the counterface (forming, as already seen, a transfer layer) or, in the case of prolonged sliding, they can mix with other fragments from the counterface, forming a *mechanically-mixed layer* on the surface of the bodies in contact. In such cases, friction and wear are modified. It is almost impossible to predict them on the basis of simple considerations (such as the $W_{12}/H$ ratio), and it is thus necessary to carry out experimental tests to achieve the required information.

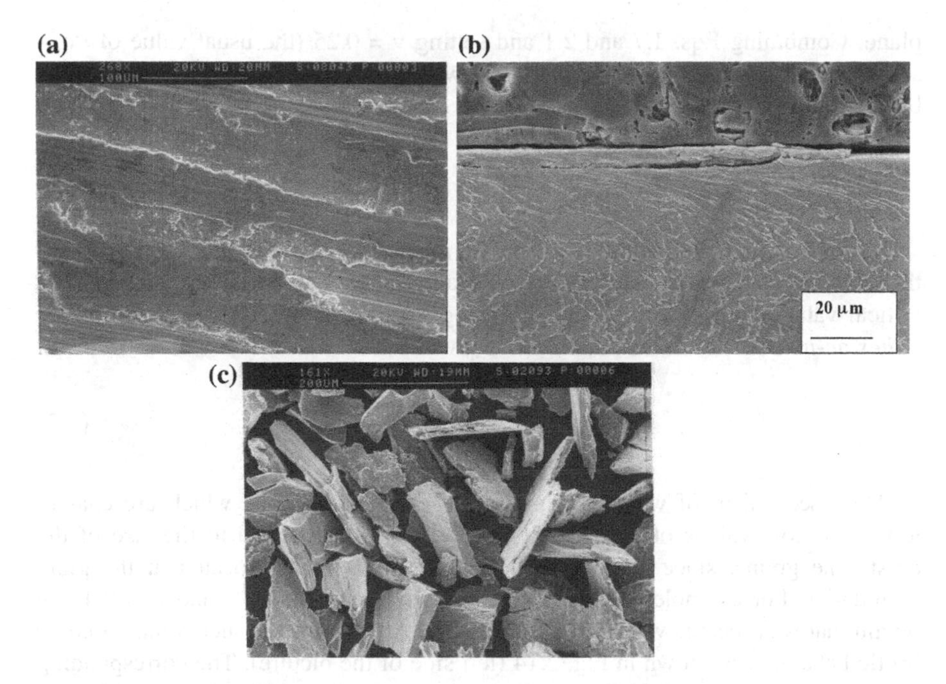


**Fig. 4.4** Ti-6Al-4V sliding against a steel counterface. **a** Worn surface showing large shear plastic deformations at the asperities; **b** plastic shearing in the sub-surface region; **c** scale morphology of the metallic fragments [7, 8]

Although the theory of Archard states that adhesive wear is restricted to the body with the lowest hardness, the wear of the tribological system is in general given by the sum of the wear volumes of the two bodies in contact. If one body is much softer than the other, it contributes alone to wear and the hardest body in contact may also increase in weight (and volume) after sliding because of transfer phenomena. But if the two bodies have comparable hardness, both contribute to the total wear volume. In this case, Eq. 4.1 can be used to determine the wear volume of each body. Indicating with $K_{ad1}$ and $K_{ad2}$ the wear coefficient of material 1, the hardest one, and material 2, the softest one, respectively, from a number of experimental data Rabinowicz obtained that $K_{ad1} = 1/3\ K_{ad2}$ [9].

### *4.1.2 Adhesive Wear of Brittle Solids*

A special case of adhesive wear occurs in brittle contacts (Sect. 1.1.4). The adhesive interaction at the contacting asperities, in fact, induces the appearance of a surface tensile stress during sliding, as described in Sect. 2.1. Such a stress can induce the formation of a wear fragment by brittle contact, if the material has sufficiently low fracture toughness [10]. Consider, for example, the contact between a sphere and a plane. Combining Eqs. 1.7 and 2.1 and setting $\nu = 0.25$ (the usual value of Poisson's ratio in the case of ceramic materials, which are typical materials that can give brittle contact), we get:

$$\sigma_t \cong \frac{p_{max}}{6}(1 + 10\mu) \tag{4.2}$$

If a micro-crack of length c is present on the surface of the plane (see Fig. 1.6c), the surface tensile stress, given by Eq. 4.2, can produce brittle fracture if it reaches a critical value given by Eq. 1.10. Combining Eqs. 4.2 and 1.10, the condition for *adhesion-induced brittle contact wear* is given by:

$$p_{max} \geq \frac{5.36 \cdot K_{Ic}}{\sqrt{\pi c} \cdot (1 + 10\mu)} \tag{4.3}$$

This mechanism of wear is common in ceramic materials, which are characterized by low values of $K_{Ic}$. In this case, c is closely related to the size of the crystalline grains, since defects, such as pores, are mainly located at the grain boundaries. For example, setting $c = 10$ μm, $K_{Ic} = 2$ MPa $m^{1/2}$ and $\mu = 0.4$, we obtain that wear occurs when $p_{max} \approx 380$ MPa. An example of such a macroscopic brittle behaviour is shown in Fig. 2.14 (left side of the picture). The corresponding values of wear rate and friction coefficient are quite large. The fragments have a blocky or plate-like shape on the scale of the grain size.

If the applied pressure is lower than the critical value given by Eq. 4.3, massive wear by brittle contact (and fragmentation) does not occur. Fragmentation is rather confined at the asperities, and the fragments are compacted to form surface scales able to support the applied load (Fig. 2.14, right side). As a consequence, both wear rate and friction coefficient are relatively low. Wear debris are finer than the average grain size [11].

Equation 4.1 is often used to express the adhesive wear behaviour of brittle solids too. The experimental $K_{ad}$-values typically range between $10^{-8}$ and $10^{-2}$. The highest values are attained when the applied stress is above the critical value.

## 4.2 Tribo-Oxidative Wear

Tribo-oxidative wear is due to the interaction of the surfaces with an environment containing oxygen. Tribo-oxidative wear is thus given by a combination of oxidative and mechanical actions at the contacting asperities. In general, it is accompanied by the formation of a surface oxide scale, which avoids the metal-to-metal contact at the asperities and may act as a sort of solid lubricant, thus reducing friction and wear. There are different situations that may lead to tribo-oxidative wear. First of all, it is convenient to distinguish between *tribo-oxidative wear at high temperatures*, and *tribo-oxidative wear at low sliding speed* [12].

### 4.2.1 Tribo-Oxidative Wear at High Temperatures

Tribo-oxidative wear at high temperatures may occur in two situations:

(1) at high sliding speeds, greater than 1 m/s in steels;
(2) when the contacting materials are exposed to high temperatures.

In both cases, surface temperature is sufficient to promote the *direct oxidation* of the asperities.

Consider first the direct tribo-oxidation at high sliding speeds. The model here presented is mainly due to the work by Quinn [13]. Oxidation is regarded to be activated by the contact flash temperature, $T_f$, and involves the contacting asperities. In the case of steels, $T_f$ has to be greater than about 700 °C to trigger this type of wear (and the average surface temperature is greater than 300–400 °C). The oxide grows at the asperity tips and spalls off once a critical thickness, $Z_c$, is reached (about 10 μm in steel materials). The oxide breaking thus produces wear fragments and generates a fresh surface that can oxidize again, thus continuing the process. As the oxidation involves the contacting asperities, the wear rate, W, is directly dependent on the contact area, $A_r$. W can be expressed with the following equation [14]:

$$W = \frac{V}{s} = \frac{Z_c \cdot A_r}{v \cdot t_c} \tag{4.4}$$

where v is the sliding speed and $t_c$ is the time required to reach the critical oxide thickness. In most cases, including the case of steels, oxidation follows a parabolic kinetics and then: $\Delta m^2 = k\, t$, where Δm is the mass increase per unit area due to the oxygen taken up to form the oxide, and k is the rate constant:

$$k = A \cdot \exp\left[-\frac{Q}{RT_f}\right] \tag{4.5}$$

where A is the Arrhenius constant, Q is the activation energy for oxidation (in steels: $A \approx 10^6$ kg/m$^4$s and $Q \approx 138$ kJ/mol [14]), R is the gas constant and $T_f$ is the flash temperature. Δm is connected with the stoichiometry of the oxide that is formed. In the case of steels, it may be assumed that $Fe_3O_4$ is formed. Then, if a volume $\Delta V_{Fe}$ of iron is oxidized per unit area, $\Delta m = 2/3\ \Delta V_{Fe}\ \rho_{Fe}(M_{O2}/M_{Fe})$, where $\rho_{Fe}$ is the density of iron, $M_{O2}$ is the molecular weight of oxygen and $M_{Fe}$ that of iron. Neglecting the volume expansion that occurs in oxidation, the oxide thickness, Z, is equal to the thickness of iron from which it originates, and thus equal to $\Delta V_{Fe}$. As a consequence: $Z^2 = C^2\, k\, t$, where $C = (3M_{Fe})/(2M_{O2}\ \rho_{Fe}) = 3.4 \times 10^{-4}$ m$^3$/kg. Since $t_c$ is given by: $Z_c^2/C^2\, k$, using Eq. 4.4 and considering $A_r = F_N/H$, it is finally obtained:

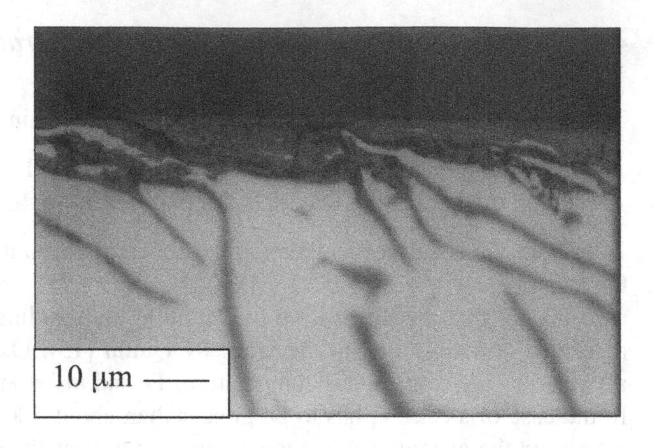


**Fig. 4.5** Cross section of tribo-oxidized cast iron after sliding against a breaking pad [15]

$$W = \frac{C \cdot k}{v \cdot Z_c} \frac{F_N}{H} = K_{ox} \frac{F_N}{H} \quad (4.6)$$

Equation 4.6 is similar to the relation 4.1, and the constant $K_{ox}$ is called the *wear coefficient for tribo-oxidative wear.* $K_{ox}$ is typically lower than $10^{-5}$ and strongly depends on the flash temperature.

As an example, Fig. 4.5 shows the cross section of a pearlitic cast iron after sliding against a braking pad [15]. The test simulated a severe braking condition, and the average contact temperature was greater than 500 °C. It is observed that direct oxidation took place in the surface and subsurface regions, with also some oxide penetration at the boundaries between the graphite lamellae and the cast iron matrix.

From a practical viewpoint, it is important to clarify the role exerted by the sliding parameters, such as normal load, sliding speed, and ambient temperature, on the tribo-oxidative wear behaviour of materials.

- $F_N$. As shown by Eq. 4.6, as $F_N$ is increased, W is increased too. But if $F_N$ overtakes a critical value, the counterface may penetrate the oxide layer and destroy it by brittle fracture. In this way a metal-to-metal contact is established, and thus adhesive wear may occur.
- v. If sliding velocity is decreased and it is lower than a critical value (about 1 m/s for steels), the flash temperature is too low to trigger this wear mechanism. But if it is too high, greater than about 10 m/s for steels, *severe oxidation* occurs because of the high surface heating. The oxide becomes thick and plastic, and it flows on the metal insulating it and protecting it from wear.
- $T_O$. If $T_O$ is increased, $K_{ox}$ is also increased because of the increase in flash temperature. However, an excessive temperature rise may cause a softening of the substrate that becomes unable to conveniently sustain the protective oxide layer. A transition to adhesive wear may thus occur, with a consequent increase in the wear rate.

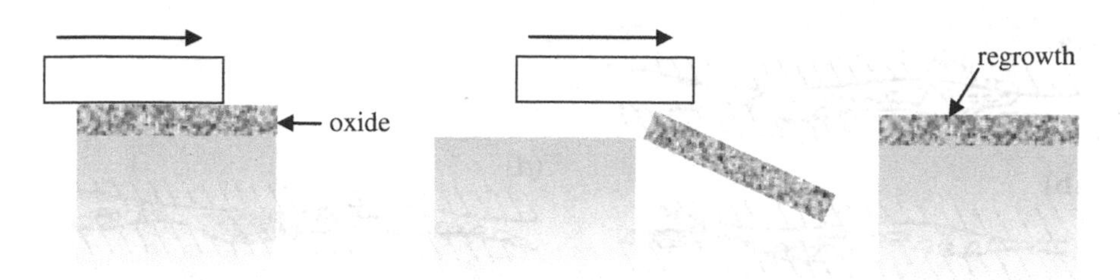


**Fig. 4.6** Schematic of the oxidation-scrap-reoxidation mechanism, which occurs when ambient temperature is high

A particular situation takes place in case of *repeated sliding*. A typical example is given by the hot-rolling rolls (and hot working tools, in general). The surfaces undergo oxidation during the contact with the hot strip and, most of all, a general oxidation during the *out-of-contact* periods, due to environmental oxygen and also the vapour produced by cooling water. During the subsequent sliding contact the oxide scale can be then removed (partially or totally) and fresh metal is exposed to the environment for a reoxidation. This wear mechanism is called *oxidation-scrape-reoxidation*, and a schematic is shown in Fig. 4.6 [16]. The wear rate depends on the oxide growth kinetic, and therefore on the surface temperature reached during the in contact and out-of-contact periods (in between $T_s$ and $T_0$). If the ambient temperature is quite high, the oxide formed during the out-of-contact periods can be sufficiently thick and can be only partially removed during the subsequent contact. In such a case, this type of tribo-oxidative wear is rather similar to the wear encountered at high sliding speeds (with severe oxidation), and it is therefore quite mild in nature.

### *4.2.2 Tribo-Oxidative Wear at Low Sliding Speed*

Experience shows that tribo-oxidative wear, with evidence of oxide particles in the wear fragments, can take place even at low sliding speeds, i.e., when the flash temperature is not sufficient to trigger the direct oxidation of the asperity tips. It is particularly important in the case of reciprocating sliding, when wear fragments can be easily retained between the surfaces in contact. This type of tribo-oxidative wear has been extensively investigated by Stott [16]. The proposed model is shown in Fig. 4.7. Wear proceeds through the following steps:

(a) At the contacting asperities, metallic fragments are generated by adhesive wear. Some may leave the tribological system and some may remain trapped between the mating surfaces;
(b) Such fragments are strain-hardened, fractured, oxidized (oxidation is activated by the very high surface area and the high density of surface defects) and agglomerated;

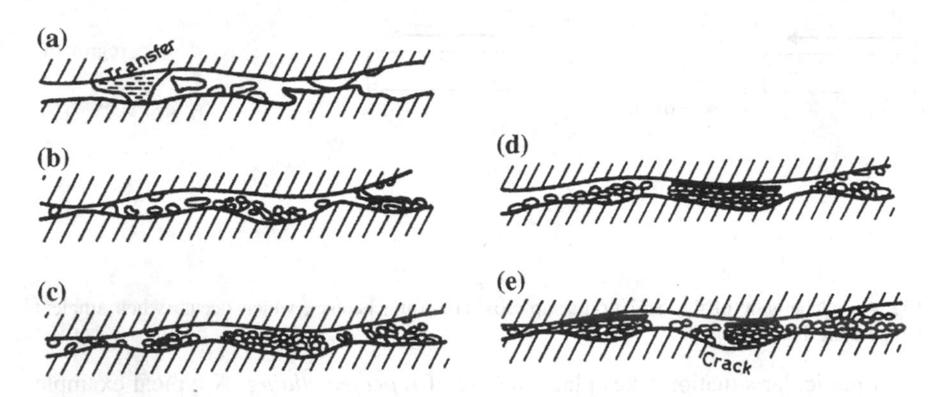


**Fig. 4.7** Tribo-oxidative wear at low sliding speed: schematic representation of the Stott model (see text for details) [16]

(c) If the load and sliding conditions are intense, a tribological layer made of compacted scales is formed;
(d) If contact temperature is sufficiently high, the scales sinter and form a very protective *glaze* layer on the top;
(e) The possible brittle fracture of the scales (orthogonally to the sliding direction) leads to the generation of fragments that may remain in the contact region or leave the tribological system.

The intensity of wear is thus given by the attainment of a dynamic equilibrium between steps (a), (d) and (e). Depending on step (a), wear is often mixed, with the coexistence of metallic fragments (formed by adhesion), and oxidized fragments, formed at step (e). Equation 4.6 can be used to model the wear behaviour and $K_{ox}$ is determined by experimental testing (such as pin-on-disc tests). In general, differently from the tribo-oxidative wear at high temperatures, wear rate decreases as surface temperature is increased, since temperature determine the conditions for a better formation of protective oxide scales.

An example of the typical morphology of the worn surfaces and the wear fragments is shown in Fig. 4.8. The pictures refer to a Ti-6Al-4 V alloy after dry sliding against a steel counterface [7]. On the worn surface (Fig. 4.8a), the presence of typical scales of compacted oxides (tribological layer) can be clearly observed. In some areas, such scales are also fragmented by brittle fracture, perpendicularly to the sliding direction. The wear fragments (Fig. 4.8b), are composed of very small equiaxed particles with a sub-micrometric size, which are oxide formed by the fragmentation of the compacted scales.

The role of ambient temperature in the tribo-oxidative wear at low sliding speed has been evidenced by Stott in a like-on-like reciprocating sliding test on a nickel alloy (with 20 % chromium), at temperatures between room temperature and 600 °C [16]. Figure 4.9 shows the results of the tests carried out at 15 N (6 h of sliding). Most of the oxide was NiO, and wear decreased with temperature since at

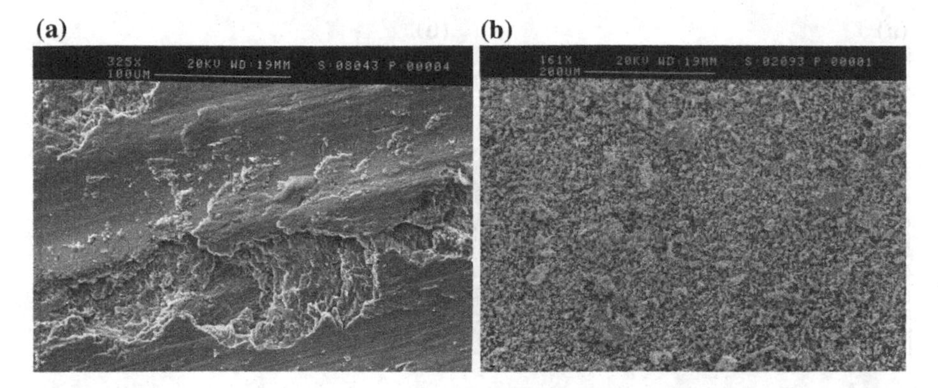


**Fig. 4.8** Ti-6Al-4V sliding against a steel counterface. **a** Worn surface showing compacted oxide scales fractured perpendicularly to the sliding direction; **b** morphology of the wear fragments [7]

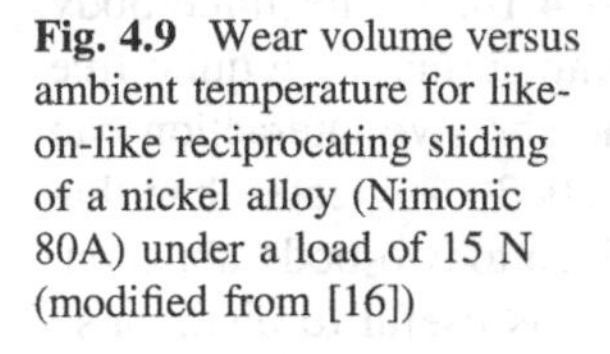

**Fig. 4.9** Wear volume versus ambient temperature for like-on-like reciprocating sliding of a nickel alloy (Nimonic 80A) under a load of 15 N (modified from [16])

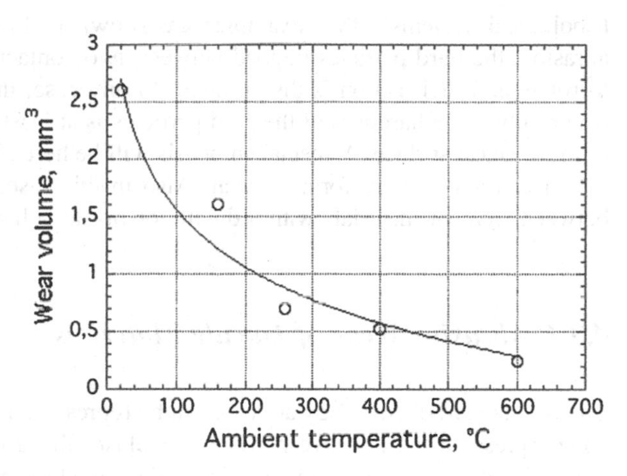


higher temperatures glaze layers developed on top of the compacted debris, leading to a more protective surface.

## 4.3 Abrasive Wear

As introduced in Sect. 2.8, there are two types of abrasive interaction: two-body and three-body abrasion. In the first case, a hard particle or a hard protuberance would plastically penetrate a softer counterface and groove it. Abrasive hard particles may be embedded in the material microstructure (such as in ceramic-reinforced composites, in steels or cast irons containing hard carbides, or in grinding wheels where the particles are held together by a specific bonding system), or may come from the surrounding environment (typical examples are the sand particles that contaminate

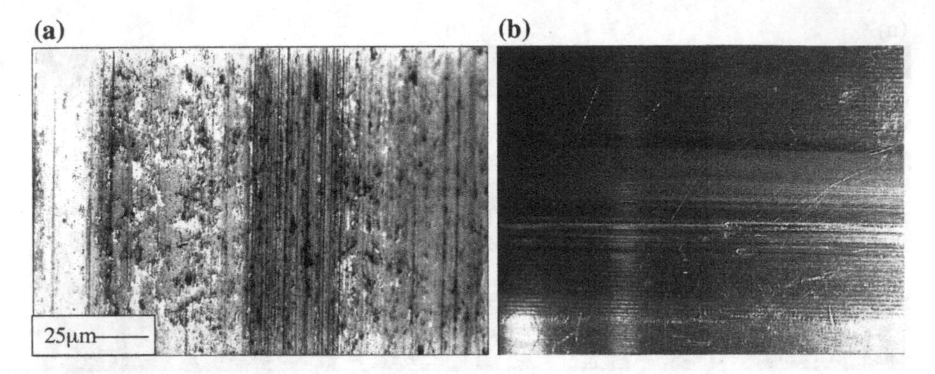


**Fig. 4.10** **a** Cast iron disc abraded by the hard ingredients of a breaking pad (*vertical grooves*); **b** bronze surface abraded by contaminating sand particles (*horizontal grooves*)

tribological systems). Two examples are shown in Fig. 4.10. In the three-body abrasion, the hard particles trapped between two contacting surfaces are quite free to rotate and their action is thus limited. In any case, the abrasive interaction can occur only if the hardness of the hard particles is at least 20–30 % greater than that of the weakest surface. A special emphasis will be here given to two-body abrasion, which is a very severe form of wear. Also in this case, it is useful to distinguish between wear of materials with a ductile or brittle behaviour.

## 4.3.1 Abrasive Wear of Ductile Materials

The mechanism of two-body abrasioncan be represented as shown in Fig. 2.25. The cone represents the abrasive particle that plastically grooves the weakest surface during its movement. If all the plastically deformed material is removed, wear is by *microcutting* and it is maximum. If all the plastically deformed material flows to the sides of the groove, wear is by *microploughing* and it is zero, even if the surface can be severely damaged. A schematic of these two limiting cases is shown in Fig. 4.11.

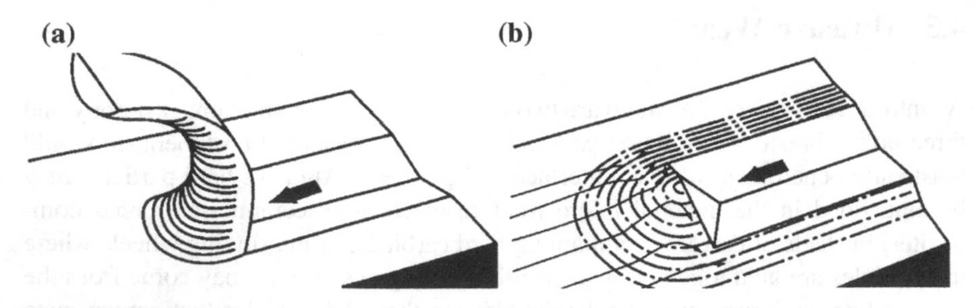


**Fig. 4.11** Abrasive grooving by microcutting (**a**) and microploughing (**b**) (modified from [17])

Consider first the wear by microcutting. From the schematisation of Fig. 2.25 it is easily obtained that wear rate, W, is given by: $\frac{A_p \cdot s}{s} = A_p$, where $A_p$ is the groove section generated by the cone in the direction of sliding ($A_p = r^2 tg\Theta$) and s is the sliding distance. At equilibrium: $F_N = p_Y \pi r^2/2 = H\pi r^2/2$ (only the front half of the moving cone supports the load), and thus [3, 4]:

$$W = \frac{2tg\Theta}{\pi} \cdot \frac{F_N}{H} \tag{4.7}$$

In general, however, microcutting and microploughing work together, and Eq. 4.7 has to be rearranged in the following way:

$$W = K_{abr} \cdot \frac{F_N}{H} \tag{4.8}$$

where $K_{abr}$ is the *wear coefficient of abrasive wear*, given by:

$$K_{abr} = \Phi \cdot \frac{2tg\Theta}{\pi} \tag{4.9}$$

where $\Phi$ varies between 1 and 0. When $\Phi = 1$, wear is ideally by microcutting only. In this case $K_{abr} = \mu_{abr}$ (see Eq. 2.14) and the wear coefficient is very high. On the other hand, when $\Phi = 0$, wear is ideally by microploughing and $K_{abr} = 0$.

$K_{abr}$ thus depends on the attack angle $\Theta$ and on the coefficient $\Phi$, which are also interrelated. $\Theta$ depends on several factors including:

(1) the angularity of the particles (that is, if they are more or less rounded, and this depends on the type of the particles);
(2) the size of the particles;
(3) the hardness of the particles in relation to the hardness of the material being abraded;
(4) the replacement, or not, of the abrasive particles;
(5) the presence of any lubrication.

Obviously, the higher the angularity of the particles, the higher $\Theta$. The attack angle also increases with increasing particle size, given that large particles may shatter during the contact and produce abrasive particles with sharp edges and thus high values of $\Theta$. For example, it has been found that SiC particles in abrasive paper, with an average diameter of 16.3 μm, display an average attack angle of about 5°. If the average diameter increases to 57.3 μm, $\Theta$ increases to about 10°, and if the average diameter is 125 μm, $\Theta$ becomes approximately 15° [18]. In general, a plateau is reached around 150 μm, and $\Theta$ does not further increase if the particles size overtakes this limiting value. If, moreover, the antagonist were sufficiently hard, a rounding at the particles edges would most likely occur, thus reducing $\Theta$. A particularly intense decrease in $\Theta$ is observed in the case of sliding without renewal of the abrasive particles: the transfer of the abraded debris on the abrasive particles and even the detachment of the particles from the surface where they are embedded, can considerably reduce Θ. Conversely, in the presence of lubrication, flowing water included, the lubricant can remove the wear fragments of the abraded material, avoiding their accumulation that would block the abrasive action of the particles. In addition, lubricant reduces friction between the hard particle and the abraded surface, and these favours wear by microcutting [2].

The constant Φ mainly depends on the geometry of the abrasive particles (it increases as the attack angle is increased) and on the plastic properties of the worn material, in particular its ductility. The latter point may be interpreted considering that every abrasive interaction is characterized by a very intense plastic deformation, which can be also accumulated by ratcheting if the interaction is repeated. Such deformation can lead to the formation of a wear fragment by ductile fracture (even in the case of microploughing), if the ductility of the material is relatively low. In general, ductility is inversely proportional to hardness, and hence a dependence of Φ on the hardness of the abraded material has to be expected. In Fig. 4.12 such experimental dependence is shown for different metals and for high and low values of Θ. In the more general case of low Θ values, Φ is little dependent from hardness (and it is quite low, lower than 0.3) as long as hardness is lower than about 250 kg/mm$^2$. Therefore, in this region the *specific wear rate*, $W/F_N$, decreases as hardness is increased. But in the region 250–400 kg/mm$^2$, Φ strongly increases as hardness is increased. This means that the specific wear rate is almost independent from hardness (and it could also decrease as hardness is increased). If hardness is high, greater than, say, 400 kg/mm$^2$, Φ is, again, little dependent from hardness (and it is quite high, greater than 0.8). Also in this region, the specific wear rate is expected to decrease as hardness is increased further. The behaviour here described is typically found in many tribological systems and will considered in more detail in Sect. 5.4 [2].

In the case of three-body abrasion, the coefficient $K_{abr}$ is much lower than in two-body abrasion, and it typically varies between $10^{-3}$ and $10^{-4}$. This is due to the

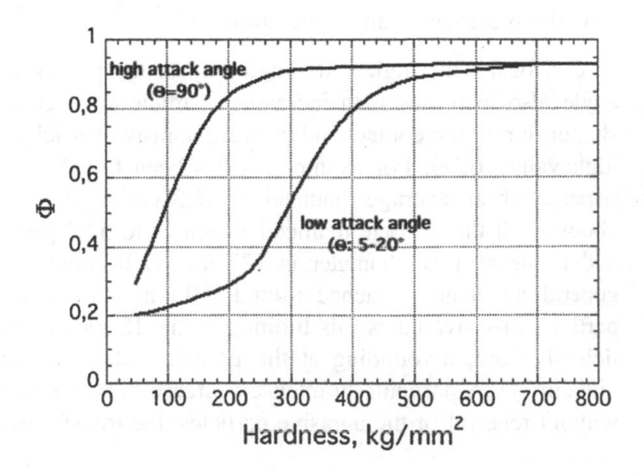


**Fig. 4.12** Experimental dependence of parameter Φ (Eq. 4.9) with the hardness of (ductile) materials for high and low attack angles (data from [2, 18])

fact that the particles can roll between the bodies in contact (Fig. 2.24b), dissipating energy and also rounding their edges.

### 4.3.2 Abrasive Wear of Brittle Materials

In brittle materials, with low fracture toughness ($K_{IC}$), the abrasive interaction can produce a wear fragment by the Lawn and Swain mechanism, described in the Sect. 1.1.4 (see Fig. 1.7). In fact, the hard and angular particles may exert an indentation on the brittle surface over which they are sliding. The subsequent formation and propagation of lateral cracks (almost parallel to the surface of the abraded material) can lead to the formation of fragments of wear by spallation. If the hard particles are very rounded, they may induce a brittle fracture of the counterface following the Hertzian brittle contact (Fig. 1.6c).

Various models have been proposed to assess the wear rate in the case of abrasive wear by brittle fracture. In most models, W results directly proportional to the load applied by each abrasive particle, and inversely proportional to the material's hardness and fracture toughness. The model proposed by Evans and Marshall [19] is based on the Lawn and Swain mechanism and states that:

$$W = \alpha_3 \cdot \frac{F_N^{9/8}}{K_{Ic}^{1/2} \cdot H^{5/8}} \cdot \left(\frac{E}{H}\right)^{4/5} \tag{4.10}$$

where $\alpha_3$ is an experimental constant. As an example, Fig. 4.13 shows the abrasive wear resistance (1/W per unit time) as a function of the so-called *abrasion parameter*, given by $H^{5/8} \cdot K_{IC}^{1/2}$. The experimental relationship is almost linear, in agreement with Eq. 4.10 (consider that the ratio E/H does not vary greatly among brittle materials). Such equation, however, should be used with caution. In fact, it

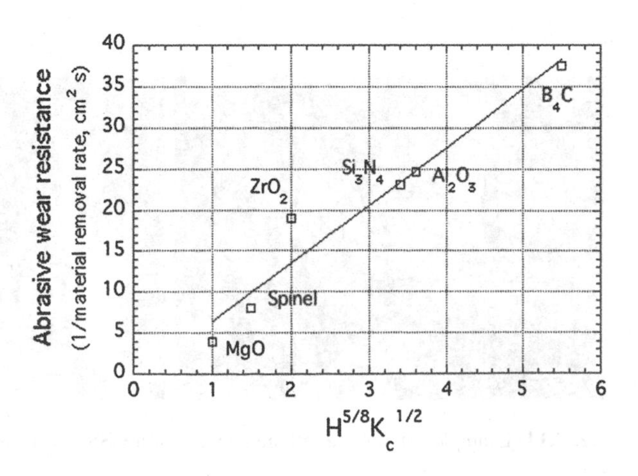


**Fig. 4.13** Experimental relationship between the abrasive resistance and the abrasive parameter for different ceramic materials (modified from [19])

does not properly take into account the role of microstructure. In particular, weak grain boundaries could anticipate brittle fracture by favouring a grain boundary separation followed by grain ejection.

## 4.4 Wear by Contact Fatigue

The wear mechanisms considered so far are progressive in nature: in each case the material removal starts from the beginning of the contact and it continues almost linearly with time. In addition, the contact phenomena are always characterized by large plastic deformations. On the other hand, wear by contact fatigue is a typical *fatigue failure*: with the application of cyclic loading, a crack is nucleated and then it propagates up to the final fracture. This means that a wear fragment is produced after some cycles that correspond to the *fatigue life* of the loaded part. In addition, in most cases and depending on the intensity of the applied load, the overall damaging process takes place under small-scale plastic deformation, and the worn region appears macroscopically free from large plastic deformations. An example of surface fatigue damage in a bearing steel is shown in Fig. 4.14 [20].

In most cases, wear by contact fatigue occurs in non-conformal contacts, when at least one of the two bodies *rolls* over the other. In such cases, this wear mechanism is also called *rolling contact fatigue* (RCF). Wear involves the contact regions and is induced by the cyclic contact Hertzian stresses. The wear mechanism is rather complex, since it is influenced by numerous factors: contact stresses, lubrication regime (Λ-factor), sliding, material's properties (e.g. mechanical strength, fracture toughness, microstructural cleanliness), and residual stresses [21]. The phenomena involved in this wear mechanism are simply presented with reference to the operative lubrication regimes. Further details will be given in Sect. 5.3.

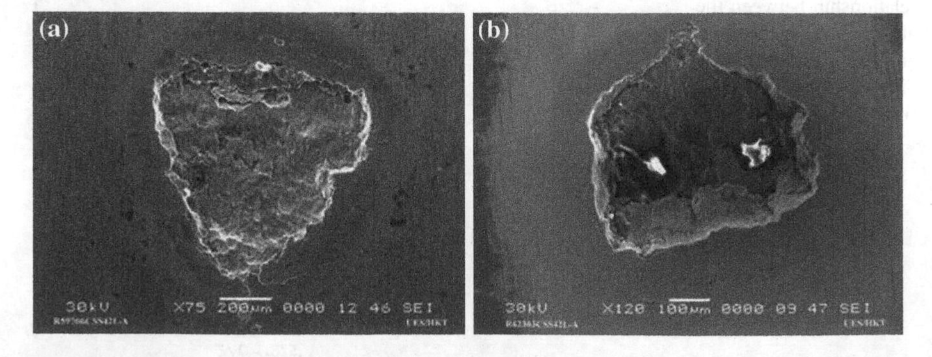


**Fig. 4.14** Examples of surface fatigue damage in a CSS 42L bearing steel [20]

### 4.4.1 Contact Fatigue Under Fluid Film Lubrication

Under fluid film lubrication ($\Lambda > 3$), the asperities of the surfaces in contact do not touch and friction coefficient is very low, typically lower than 0.05. Assumed that the Hertz theory is applicable also in the case of cyclic loading, at each contact the surfaces are submitted to the stress distribution reported in Sect 1.1. The maximum shear stress, $\tau_{Max}$, is located at a certain depth, $z_m$ (see Table 1.1), and the sub-surface shear stress $\tau_{yz}$, which is parallel to the surface (see Fig. 1.3c), changes direction during each contact (its maximum amplitude is $0.5p_{max}$).

Fatigue failure occurs by crack nucleation, crack propagation, and fracture. The sequence of phenomena leading to failure is shown in Fig. 4.15. Due to the cyclic loading, a subsurface crack nucleates after a number, $N_n$, of cycles. Crack nucleation typically occurs at areas with high stress concentration, i.e., non-metallic inclusions (oxides, sulphides), precipitates (carbides, nitrides) or pre-existing flaws. Alternatively, it may also occur at soft spots in the microstructure. Quite often, the nucleation stage is characterized by a local microstructural alteration, with dislocations build-up or transformation of retained austenite (this is quite typical in heat-treated steels), which induce a local hardening. Eventually, crack nucleation may be indirectly induced by carbide dissolution or other phenomena that are associated with local softening. Nucleation occurs at a depth close to $z_m$, where a critical combination of applied stress and microstructural defect is achieved (Fig. 4.15a). From a mechanical viewpoint, nucleation generally occurs after an elastic shake-down has been reached [22]. The number of cycles for crack nucleation, $N_n$, decreases as the contact load is increased, and therefore the subsurface stress, is

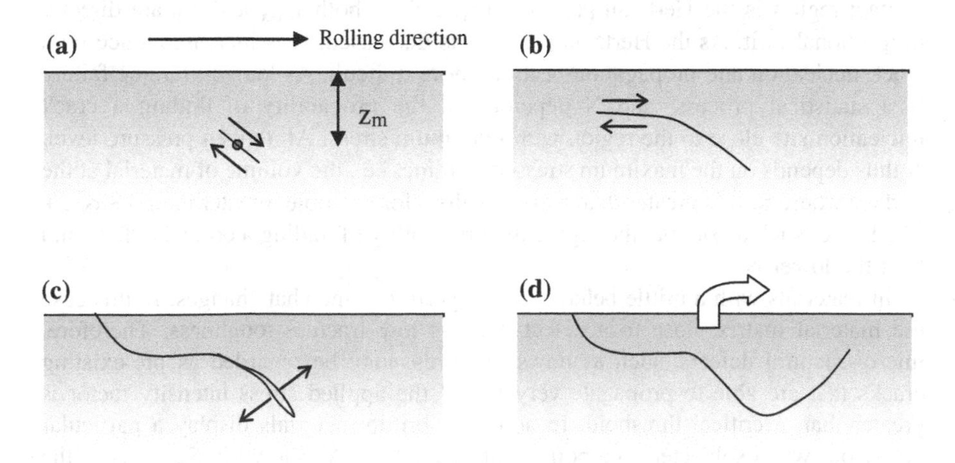


**Fig. 4.15** Contact fatigue damage under fluid film lubrication. **a** Nucleation of a sub-surface crack at a depth close to $z_m$; **b** crack propagation towards the surface and then parallel to the surface; **c** crack branching to the surface (by ligament collapse) and oil pumping effect; **d** final formation of a large fragment by spalling

increased. If the elastic shakedown is overwhelmed, cracks can nucleate more easily by low-cycle fatigue.

After nucleation, the subsurface cracks propagate driven by the contact stresses. In general, the propagation is a mixed mode II (shear mode, which is predominant) and mode I (tensile mode) propagation. In a simplified view, it can be assumed that cracks initially propagate along the maximum shear stress forming an angle close to 45° with the surface. After some propagation, the cracks change direction and propagate parallel to the surface, driven by the $\tau_{yz}$ shear stress. Subsequently they propagate until the ligament between their tip and the surface fracture by plastic collapse [23]. These latter two events are schematized in Fig. 4.15b, c. From this moment, the lubricating fluid is entrapped in the crack, and exerts a *pumping effect*, which increases the driving force for crack propagation. The crack may then reach the surface also on the other side, thus producing a wear fragment by *spalling* (Fig. 4.15d). The wear fragment results quite large, typically greater than 100 μm in size.

The total number of cycles to failure, N, provides the fatigue life and is given by: $N = N_n + N_p$, where $N_p$ is the number of cycles for crack propagation.

- $N_n$ mainly depends on the microstructural features promoting crack nucleation. $N_n$ thus increases as their density and size is decreased [24].
- $N_p$ depends on microstructural defects that may offer energetically favourable paths for crack propagation and, most importantly, on microstructural strength and fracture toughness. In ductile materials, such as in most metals, as strength, i.e. hardness, is increased, the crack propagation rate is decreased, since the material strength in the plastic region at the crack tip is larger [25].

The parameter that is usually employed to quantify the stress intensity in the contact region is the Hertzian pressure, $p_{max}$, since both $\tau_{max}$ and $\tau_{xy}$ are directly proportional to it. As the Hertzian pressure is decreased, N is increased since both crack nucleation and propagation become more difficult. As known, fatigue failure is a statistical process, and N depends on the probability of finding a crack nucleation site close to the region with maximum stress. At a given pressure level, N thus depends on the maximum stressed volume, i.e., the volume of material at the surface where $\tau_{max}$ is greater than a given value (for example, greater than 0.8 $\tau_{Max}$). The higher such a volume, the higher the probability of finding a critical defect, and thus the lower N.

In materials with a brittle behaviour, the picture somewhat changes. In this case the material matrix close to a defect possess low fracture toughness. Therefore, microstructural defects, such as flaws or pores, may be regarded as pre-existing cracks that are able to propagate very fast if the applied stress intensity factor is greater than a critical threshold. In addition, brittle materials display a particular behaviour when subjected to a point contact loading. As shown in Sect. 1.1.4, the surface radial tensile stress (Fig. 1.3a) may lead to brittle contact with the formation of C-cracks. The presence of a cyclic surface stress may favour such a crack nucleation, or cause the fatigue propagation of pre-existing cracks. As an example, Fig. 4.16 shows the crack network formed on the surface of a $Si_3N_4$ disc after a

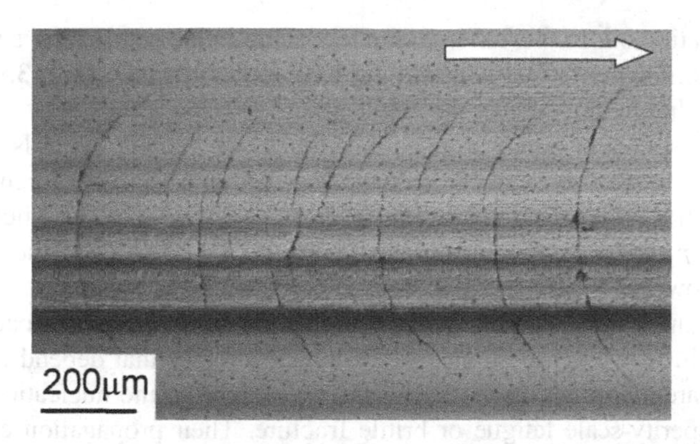


**Fig. 4.16** Crack network on the surface of a $Si_3N_4$ disc after a rolling-sliding test period of 10 h (the *arrow* indicates the imposed sliding direction) [26]. Semi circular cracks are formed on the rear of the contact because of the additional contribution to the surface tensile stress given by friction (see Fig. 2.2)

rolling-sliding contact of 10 h against a hardened steel (contact pressure = 4.55 GPa; lubricant: emulsion of 5 % oil in water; recorded friction coefficient: 0.085) [26].

### *4.4.2 Contact Fatigue in Mixed and Boundary Lubrication Regime*

The fatigue performance in fluid film lubrication can be considered as an upper limit, since it refers to a condition of optimum lubrication. A decrease in Λ induces a decrease in the contact fatigue resistance. If lubrication is mixed, a number of asperities are brought into repeated contact. As a consequence, crack nucleation at the surface is made easier, since the associated plastic deformation (that can be also characterized by asperity-scale ratcheting) may easily induce the formation of surface micro-cracks. They usually form a shallow angle with the surface because of some sliding at the asperity contacts that may be present even in the case of pure rolling if a torque is transmitted through the contact. Note that some micro-slip between two mating surfaces is almost always present even in case of free rolling, as shown in Sect. 2.10.

Geometrical discontinuities, such as grinding marks, are preferential sites for surface micro-crack nucleation. Cracks may then propagate towards the interior of the material if the local stress intensity factor exceeds the relevant threshold. Two main factors contribute to the stress intensity factor. The first one is represented by the contact stress. The presence of friction (which increases as Λ-factor is decreased) induces an increase in the local stress, and the Hertzian profile is also shifted to the surface (Fig. 2.3). The second one is due to the pumping effect exerted by the lubricating oil. In this case, such effect starts immediately after surface crack nucleation. In non-conformal contacts the EHL pressure spike (Fig. 3.9) exerts an additional hydraulic pressure at the crack tip.

In most cases, after some propagation the cracks branch towards the surface because of the instability of shear propagation. A contact fatigue fragment is thus formed, which is relatively small, of the order of 10 μm in size. This phenomenon is often called *pitting* and it may anticipate failure by spalling if the Λ-factor is sufficiently low.

In brittle and hard materials, surface nucleation by plastic deformation is generally difficult. However, the high surface tensile stresses that depend on $p_{max}$ and friction and are amplified by the EHL effect may lead to the nucleation of micro-cracks by asperity-scale fatigue or brittle fracture. Their propagation can produce severe damage by pitting.

## 4.5 Wear Testing

The coefficient of friction and all wear responses under the action of the different wear mechanisms, are not intrinsic materials properties, since they depend on the *tribological system*, i.e., on the mating materials, the type of contact, the surface characteristics of the mating bodies (including the possible presence of lubrication), and so on. Therefore, it is clear that for the determination of the realistic tribological behaviour of a given system, *field tests* are required. The tests should be then carried out on the actual system in service. Such tests, however, are expensive, long-term, complex and the results are often difficult to interpret, because it is very difficult to make out the influence of individual variables. For these reasons, it is preferred, in most cases, to perform laboratory tests in simpler configurations, such as:

(1) *bench tests* on the system of interest;
(2) accelerated *tests on single components* isolated from the real system;
(3) *simplified tests* that simulate the acting wear mechanism.

In passing from field tests, bench tests, accelerated component tests, and simplified tests, the relevant costs and test durations become generally lower. In addition, the test parameters are better controlled (specific standardized procedures have been also developed for many laboratory tests), and the obtained results can thus be more easily interpreted. Simplified tests can be conducted with adequate statistical confidence, and with the possibility of comparing the performance of different materials, even with literature data that refer to tests carried out under nominally identical or similar conditions. However, the simplified laboratory tests must be selected and performed with special care, and data have to be interpreted with caution, in order to transfer safely the obtained information to the real tribological system. A correct approach is based on the preliminary identification of the dominating wear mechanism responsible for the damage in operation, and on the ability to carry out laboratory tests that wear the material specimens with the same mechanism.

A critical aspect of simplified laboratory tests regards the control of contact temperature. In fact, the adopted test specimens are often small in size in comparison to the real components they intend to simulate. Therefore, they may easily reach contact temperatures much different than those found in the real components (often higher but not always), even if the same tribological parameters (such as contact load and sliding velocity) are adopted. The different surface temperatures may induce a different wear mechanism (e.g. tribo-oxidation in place of adhesion). Its monitor and control (but adopting, for example, specific forced cooling systems, such as jets of compressed air) is thus paramount.

The data obtained from the laboratory tests may have a direct use in designing real tribological systems and configurations, like for instance, the sliding wear coefficients for dry sliding bearings, or the S-N curves for gears. In most cases, however, they are intended to produce a *ranking* between several candidate materials for a given application and relevant selection criteria. Consider that in many situations there is an uncertainty in using new materials, materials produced with a different process route or by different manufacturers, materials obtained by special coating techniques, and so on.

The available testing machines (*tribometers*) are very numerous and in most cases the test procedures are not governed by specific standards. This paragraph describes the main laboratory tribometers for basic simplified tests, which are able to simulate the different wear mechanisms. Table 4.1 lists the tribometers under consideration, together with the wear process and wear mechanism they are able to simulate. At the end of the paragraph, a short indication on the most common techniques for the examination of the wear products (wear debris and tracks) is given. In fact, in order to check for the validity of the executed tests, it is necessary to verify that the wear mechanism evidenced by the test sample is the same of that

**Table 4.1** Simplified tribological tests with the related wear process and wear mechanisms that they are able to simulate

| Test machine | Simulated wear process | Simulated wear mechanisms |
|---|---|---|
| Pin-on-disc | Sliding wear | Adhesive wear<br>Tribo-oxidative wear |
| Block-on-ring | Sliding wear | Adhesive wear<br>Tribo-oxidative wear |
| Disc-on-disc | Rolling-sliding wear | Contact fatigue<br>Adhesion/tribo-oxidation |
| 4 Balls | Rolling wear | Contact fatigue |
| Pin abrasion test (PAT) | High-stress abrasion | Abrasive wear |
| Dry-sand, rubber-wheel wear test (DSRW) | Low-stress abrasion | Abrasive wear |

met in the real components. The same examination techniques are used for the failure analysis of the real components, a necessary step to establish which kind of test has to be performed.

### *4.5.1 Pin-on-Disc*

This test setup is schematically shown in Fig. 4.17. The test configuration is constituted by a stationary pin with a cylindrical shape and a diameter of few millimetres, which is pressed against a rotating disc. The contact can be conformal or non-conformal. In the latter case, a sphere usually substitutes the pin. In the case of conformal contact, the edges of the contact are often rounded, in order to avoid (especially in lubricated tests) disturbing uncontrolled effects, due to stress concentration (see Fig. 1.4).

The control of the contact temperature is usually attained by placing one or two thermocouples in the pin, at a certain distance from the contact surface. The contact temperature is then evaluated using the relations introduced in Sect. 2.11. In this regard it should be noted that each region on the wear track on the disc gets in contact with the pin once per each revolution. If the disc is not able to dissipate the heat with sufficient efficiency, its temperature may rise continuously up to an equilibrium value. The contact temperature may thus become much higher than expected.

The depth of wear may be continuously recorded using a linear displacement transducer. In this case the measurement may also account for the wear contribution of the disc. Of course, the displacement of the transducer may be affected by thermal expansion of the pin due the frictional heating and this contribution has to be accounted for. Alternatively, wear may be measured by weighing the pin before and after each test (or at regular test intervals), and converting the mass loss to the volume loss using the density of the worn material. The wear of the disc is usually low. If necessary, it can be determined by obtaining the wear track profile, as measured with a profilometer. It is then possible to obtain a wear curve similar to

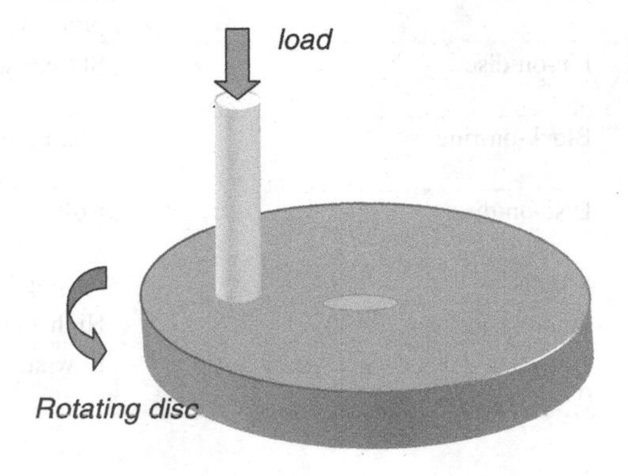


**Fig. 4.17** Schematic of the pin-on-disc test

those shown in Fig. 5.1. With reference to the steady state stage, the wear rate (W), the specific wear coefficient ($K_a$) or the wear coefficient (K) can be evaluated quite easily. In the testing arrangement shown in Fig. 4.17, wear debris remain in the contact region. This may be important when simulating low-sliding speed tribo-oxidative wear. On the contrary, if the plane of the disc is vertical, all the debris falls away from the contact region. The test configuration has to be selected with respect to the wear mechanism to be simulated, and to the real contact conditions that need to be reproduced.

During each test it is also possible to follow the evolution of the friction coefficient. This is achieved by recording the tangential force required to restrain the pin. Such a measurement is very useful since, inter alia, it affords the possibility to detect the presence of transitions in the mechanism of wear that are typically evidenced by transitions in the friction coefficient. All the test procedures have been standardized in the ASTM G99-95 norm.

## *4.5.2 Block-on-Ring*

In this test a stationary block is pressed against the outer surface of a rotating ring. As shown in Fig. 4.18, the contact can be conformal or non-conformal (at least at the beginning of the test). This test is very similar to the pin-on-disc test. It is usually preferred when there is the need to simulate sliding conformal contacts with large values of the nominal area of contact, and allowing the debris to be free to fall away the contact region. This type of test is also used to investigate lubricated wear, including the phenomenon of scuffing, in a non-conformal configuration, since it allows for the obtainment of high contact pressures.

In the case of conformal contact, a quite long running in stage is required to eliminate some unavoidable misalignments between the shaped block and the ring. Wear can be quantified by weighing the block and the ring before and after each

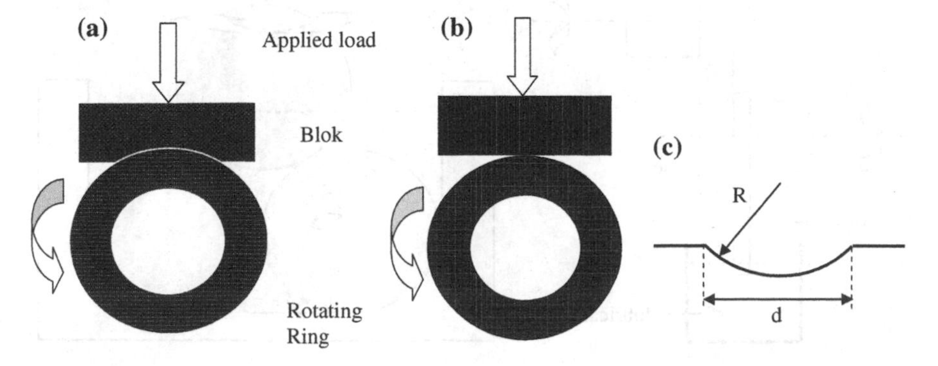


**Fig. 4.18** Schematic of the block-on-ring testing rig. **a** Conformal contact; **b** non-conformal contact; **c** block wear volume determination using Eq. 4.11

test. In the case of non-conformal contact, the wear volume, V, is usually assessed by measuring the size of the wear track. The following relation can be used (see Fig. 4.18c for the meaning of the symbols):

$$V \cong \frac{d^3}{12R} L \tag{4.11}$$

The procedures to be followed for this type of test are given in ASTM G77-93.

### 4.5.3 Disc-on-Disc

In this test (also called *twin disc test*), two discs are in contact along a generatrix, realizing a non-conformal line contact. Figure 4.19a shows a schematic of the test setup. By changing the rotation velocity, rolling-sliding tests can be carried out at different percentages of sliding. The tests can be run in dry or lubricating conditions. In this case, lubricant can be taken to the contact region by means of a chain, which drags it from a container positioned below the specimens (Amsler-type tribometer; in this test, the discs have a diameter usually between 40 and 50 mm, and a height of 10 mm). The lubricant temperature is controlled and maintained on a fixed set value.

This type of test is mainly used to simulate wear by contact fatigue. By varying the test conditions (specimen geometry, surface roughness, applied loads and speeds), tests at different values of the lambda factor as well as of the Hertzian pressure can be conducted. By using crowned specimens, a point contact is established and very large contact pressures can be attained. The onset of fatigue

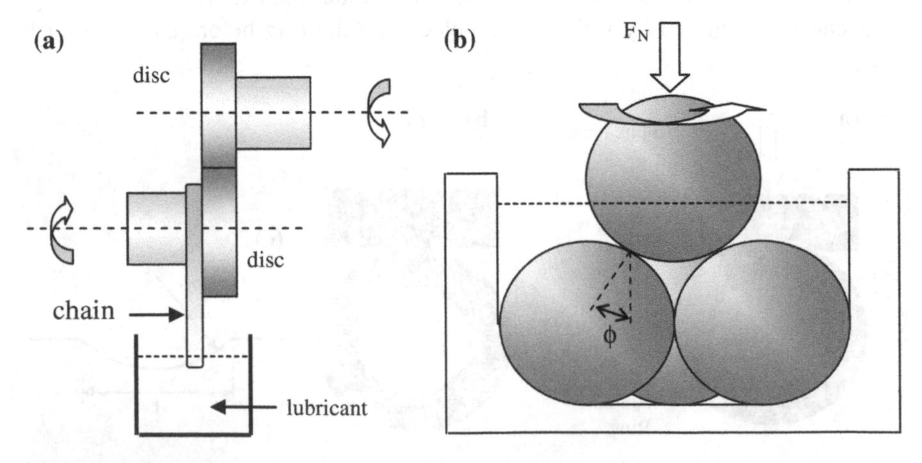


**Fig. 4.19** Schematic side view of the **a** disc-on-disc testing rig (Amsler type), and **b** the 4-Ball tribometer

damage is usually determined in correspondence of a transition in the recorded friction coefficient, or by the appearance of noise or vibrations. In fact, if the wear fragments remain trapped between the bodies in contact, they induce an increase in the coefficient of friction and also induce vibrations in the system. The surface damage can be inferred by examining the surface state of the samples at regular intervals. Also this tribological test is widely used for studying the resistance of materials and lubricants to scuffing.

Different test configurations have been developed for reducing the running time. One example is the *four-roller test*, in which a cylindrical specimen is loaded by three counter discs (120° apart), giving three load applications for revolution.

### *4.5.4 Four-Ball Tribometer*

The test setup is schematically shown in Fig. 4.19b. It realizes a point contact. The top rotating ball is kept in contact against the three stationary lower balls immersed in the lubricant. The load, $F_N$, is applied to the top ball and the load acting on each ball is then $F_N/3 \cos\phi$. For each rotation of the top ball produces three contacts with the lower balls that are typically made of an AISI 52100 steel. The eventual wear volumes of the balls are determined by measuring the wear scars on their surfaces. This type of test is mainly used to study the properties of lubricants (including their scuffing resistance). The standard test procedure is described in ASTM D 4172–94.

### *4.5.5 Dry-Sand, Rubber-Wheel Wear Test (DSRW)*

This type of test is used to evaluate the tribological behaviour of materials under low-stress abrasion conditions. The test setup is schematically shown in Fig. 4.20. A rotating rubber-rimmed wheel slides against the surface of a specimen in the presence of abrasive particles, which are fed by gravity from a hopper. The

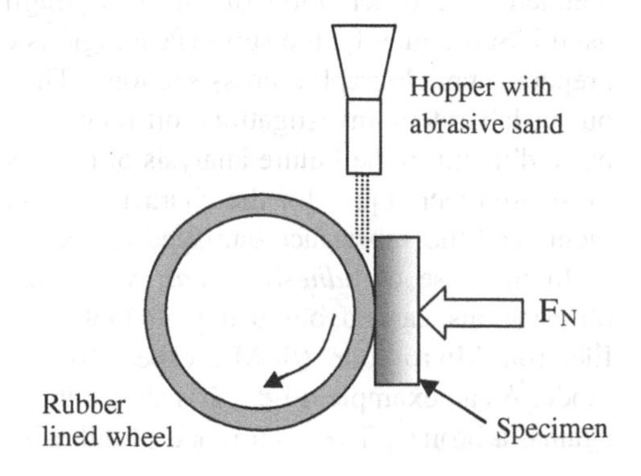


**Fig. 4.20** Scheme of the DSRW test setup

procedure is described by the ASTM G65 standard. Sand particles (Ottawa sand, Fig. 5.17b) with a size of about 200 μm are used. In the procedure B, the applied load is 130 N and the sliding distance is 1430 m. Wear is determined by weighing the sample before and after the test. This allows obtaining the relevant wear rates and wear coefficients. As in the block-on-ring test, wear induces a continuous increase in the nominal area of contact.

The DSRW test can be modified to investigate the influence of specific parameters that may play an important role under particular tribological conditions. For example, the tests can be carried out using an abrasive slurry (ASTM B611) or in a corrosive environment.

### 4.5.6 Pin Abrasion Wear Test (PAT)

This type of test is suitable for the study of the high-stress abrasive behaviour of materials. The geometric configuration is similar to that shown in Fig. 4.17. A cylindrical pin (the sample to be studied) slides against an abrasive paper containing ceramic particles, typically alumina or silicon carbides. In order to avoid any interference between the abrading particles and wear debris, a spiral track is realized by moving the specimen towards the centre of the disc. Alternatively, a *pin on abrasive drum* test is used. In this test, a rotating cylinder is covered with an abrasive paper, and a pin is pressed against it whilst moving along a generatrix, in order to be continuously in contact with fresh abrasives. The evolution of wear is determined by periodically interrupting the test and weighing the pin.

### 4.5.7 Examination of the Wear Products

In order to understand the acting wear mechanism in a given tribological system, with the aim of validating the laboratory tests or in the failure analysis of real components, the observation of the *worn surfaces* (or *wear tracks*) is recommended. The observation of the wear fragments, if available, can be also very useful. Subsequently, the subsurface regions can be possibly observed, on carefully prepared metallographic cross-sections. The latter operation can be easily carried out in laboratory investigations on relatively small specimens, whereas it may be more difficult in the failure analysis of real components. Table 4.2 lists some of the most used techniques for the characterization of the worn surfaces, the wear fragments and the subsurface damaged regions.

In the case of *adhesive wear*, very useful information is obtained from the observations carried out using an Optical Microscope (OM), or in a Scanning Electron Microscope (SEM), especially in the Back-Scattered Electron (BSE) mode. As an example, Fig. 2.20a shows the wear surface of a steel after dry sliding against a bronze. The occurrence of adhesive wear is clearly demonstrated by the

**Table 4.2** Most used techniques to characterize the wear damage

| Worn surfaces (or, wear tracks) | Visual inspection | Optical microscopy | Electronic microscopy (SEM) equipped with micro-analysis (EDS) | Special techniques (such as XPS) |
|---|---|---|---|---|
| Wear fragments (o, debris) | Visual inspection | Optical microscopy | Electronic microscopy (SEM) equipped with micro-analysis (EDS) | X-ray diffraction (XRD). Transmission electron microscopy (TEM). Special techniques |
| Sub-surface regions | Optical microscopy | Microhardness profiles | Special techniques (such as SIMS) | |

presence of transferred fragments that have a plate-like shape. The occurrence of transfer is also well evidenced by SEM observations in BSE mode or using the EDXS analysis (see, for example, Fig. 2.29).

Visual inspection and OM observations are very fruitful also for detecting wear by *tribo-oxidation*. As an example, Fig. 4.21 shows the OM planar view of the surface of a steel that underwent tribo-oxidative wear at low sliding speed. The presence of dark scales of compacted oxides can be clearly appreciated. In some cases such dark scales or fragments are simply detected by naked eye.

The presence of grooves on the wear surface of specimens or components is a clear indication of *abrasive wear*. Such grooves can be detected by OM, as shown in Fig. 4.10b. Such an operation can be easy accomplished when the grooves are all aligned along the same direction. It is much more difficult when they are produced by particles moving in different directions.

Figure 4.14 shows a steel surface damaged by *contact fatigue*, observed in a SEM. This technique allows observing surfaces with a high depth of focus, and it is

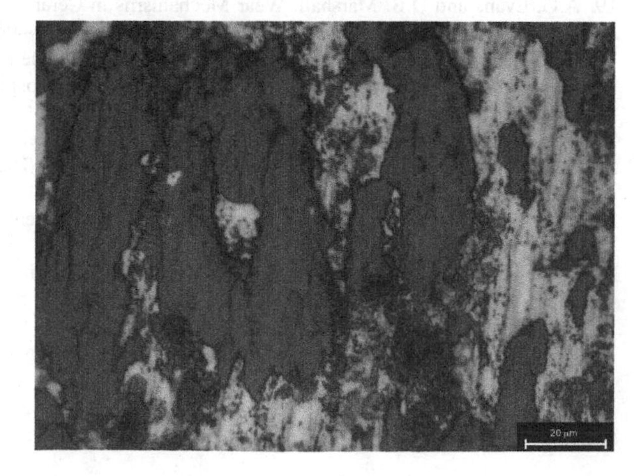


**Fig. 4.21** OM observation of the wear track of a steel after tribo-oxidative wear [27]

therefore very useful in detecting spalled layers and pits produced by contact fatigue. An evidence of the occurrence of this kind of damage is also given by the limited, if any, presence of plastic deformations in the damaged areas. Indeed, in all other cases of wear damage, extensive plastic deformation is present.

## References

1. K.H. Czichos, K.H. Habig, *Handbook Tribologie* (Vieweg, Reibung und Verlschleiss, 1992)
2. K.H. Zum Gahr, *Microstructure and Wear of Materials*, (Elsevier, New york, 1987)
3. E. Rabinowicz, *Friction and Wear Materials*, 2nd edn. (Wiley, New york, 1995)
4. I.M. Hutchings, *Tribology*, ed. by Edwald Arnold (London, 1992)
5. D.A. Rigney, L.H. Chen, M.G.S. Naylor, A.R. Rosenfield, Wear processes in sliding systems. Wear **100**, 199–219 (1984)
6. B.S. Hockenbull, E.M. Kopalinsky, P.L.B. Oxley, An investigation on the role of low cycle fatigue in producing surface damage in sliding metallic friction. Wear **148**, 135–146 (1991)
7. G. Straffelini, A. Molinari, Dry sliding wear of Ti-6Al-4V alloy as influenced by the counterface and sliding conditions. Wear **236**, 328–338 (1999)
8. G. Straffelini, A. Molinari, Mild sliding wear of Fe-0.2 %C, Ti-6 %Al-4 %V and Al-7072: a comparative study. Tribol. Lett. **41**, 227–238 (2011)
9. M.B. Peterson and W.O. Winer (eds) *Wear control handbook*, ASME (1981)
10. H.S. Kong, M.F. Ashby, Wear mechanisms in brittle solids. Acta Metall. Mater. **40**, 2907–2920 (1992)
11. K. Adachi, K. Kato, N. Chen, Wear map of ceramics. Wear **203–204**, 291–301 (1997)
12. G.W. Stachowiak and A.W. Batchelor *Engineering Tribology*, 3rd edn. (Elsevier, Amsterdam, 2005)
13. T.F.J. Quinn, Oxidational wear. Wear **18**, 413–419 (1971)
14. S.C. Lim, M.F. Ashby, Wear mechanism maps. Acta Mater. **35**, 1–24 (1987)
15. G. Straffelini, L. Maines, The relationship between wear of semimetallic friction materials and pearlitic cast iron in dry sliding. Wear **307**, 75–80 (2014)
16. F.H. Stott, The role of oxidation in the wear of alloys. Tribol. Int. **31**, 61–71 (1998)
17. D.A. Rigney (ed.) *Fundamentals of friction and wear of materials*, ASM (1981)
18. T. Hisakado, T. Tanaka, H. Suda, Effect of abrasive particle size on fraction of debris removed from plowing volume in abrasive wear. Wear **236**, 24–33 (1999)
19. A.G. Evans and D.B. Marshall, Wear Mechanisms in Ceramics, in Fundamentals of Friction and Wear of Materials, ed. by D.A. Rigney, ASM, pp. 439–452
20. H.K. Trivedi, N.H. Foster, L. Rosado, Rolling contact fatigue evaluation of advanced bearing steels with and without the oil anti-wear additive tricresyl phosphate, Tribol. Lett. **41**, 597–506 (2010)
21. S. Suresh, *Fatigue of materials*, 2nd edn. (Cambridge university press, Cambridge, 1998)
22. G. Donzella, M. Faccoli, A. Ghidini, A. Mazzù, R. Roberti, The competitive role of wear and RCF in a rail steel. Eng. Fract. Mech. **72**, 287–308 (2005)
23. Y. Ding, R. Jones, B.T. Kuhnell, Elastic-plastic finite element analysis of spall formation in gears. Wear **197**, 197–205 (1996)
24. D. Nelias, M.L. Dumont, F. Champiot, A. Vincent, D. Girodin, R. Fougeres, L. Flamand, Role of inclusions, surface roughness and operating conditions on rolling contact fatigue. J. Tribol.: ASME Trans **121**, 240–251 (1999)
25. N.A. Fleck, K.J. Kang, M.F. Ashby, The cyclic properties of engineering materials. Acta Metall. Mater. **42**, 365–381 (1994)

26. M. Hartelt, H. Riesch-Oppermann, I. Khader, O. Kraft, Probabilistic lifetime prediction for ceramic components in rolling applications. J. Eur. Ceram. Soc. **32**, 2073–2085 (2012)
27. G. Straffelini, M. Pellizzari, L. Maines, Effect of sliding speed and contact pressure on the oxidative wear of austempered ductile iron. Wear **270**, 714–719 (2011)

# Chapter 5
# Wear Processes

In the study of the wear failures, it is usual to consider the *wear processes* defined with reference to the type and geometry of relative motion between two mating surfaces (Fig. 4.1). In general, each wear process is due to one (or more) wear mechanisms. Table 5.1 lists some examples of tribological systems with the main wear mechanism. In the next paragraphs, the characteristics of the main wear processes will be outlined, including the methods to control them. The role of materials will be also indicated, although more detailed information on the materials' selection and surface engineering in tribology will be given in the next two chapters.

## 5.1 Sliding Wear

The mechanisms that determine the surface damage in sliding systems are *adhesion* and *tribo-oxidation*. If the tribological system is characterized by the presence of hard particles (much harder than the two surfaces in contact), abrasive wear may also occur. In this case wear rate is generally higher than when adhesion or tribo-oxidation are the only acting mechanisms. For this reason, the wear process changes into *abrasive wear by hard particles*, and will be considered separately in Sect. 5.4.

### 5.1.1 Wear Curves

During sliding between two bodies in contact (Fig. 4.2), the wear volume, V, increases with distance of sliding, s, with the typical trends schematically shown in Fig. 5.1a. We can distinguish three stages [1, 2]:

(1) Running in (run-in or break-in). During this stage (quite short in dry sliding), the wear rate, W, which is given by $\Delta V/\Delta s$, is usually very high. In fact, the mating surfaces are wearing out to remove the asperity peaks and to



G. Straffelini, *Friction and Wear*, Springer Tracts in Mechanical Engineering,
DOI 10.1007/978-3-319-05894-8_5

**Table 5.1** Main wear processes with the corresponding predominant wear mechanism and some relevant tribological systems

| Wear process | Main wear mechanism | Examples of relevant tribological systems |
|---|---|---|
| Sliding | Adhesion Tribo-oxidation (abrasion) | Sliding bearings; cam and tappets; guides; seals; piston rings; disc-pad braking system; forming dies; cutting tools |
| Fretting | Adhesion (at run-in) Tribo-oxidation | Riveted joints, bolted flanges, shrink fits, all in the presence of vibrations; reciprocating arms; electrical contacts; wire ropes |
| Rolling | Contact fatigue | Rolling bearings; cam and follower |
| Rolling-sliding | Contact fatigue Adhesion (tribo-oxidation in dry conditions) | Gears; wheel-rail system |
| Impact | Contact fatigue Adhesion | Equipment's with sliding and percussive systems |
| Abrasion by hard, granular materials | Abrasion | Ore processing machinery; tillage tools; sliding systems with hard particles in between |
| Erosion | Abrasion | Slurry pipelines; centrifugal pumps for slurry; turbine blades; nozzles for sand blasters |
| Cavitation | Contact fatigue | Turbine blades; centrifugal pumps; ships' propellers; high-speed lubricated sliding bearings |

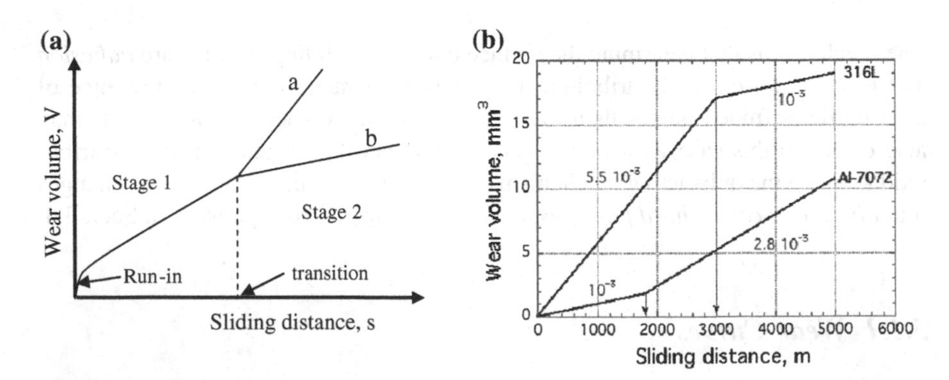


**Fig. 5.1** **a** Scheme of typical sliding wear curves; **b** sliding wear curves for an aluminium alloy and 304L stainless steel (pin on disc test; average pressure: 1 MPa; sliding speed: 0.8 m/s) (data from [3, 5])

compensate for possible misalignments. Moreover they get cleaned, totally or partially, from the contaminants.

(2) Stage 1. After run-in, a *steady state stage* is entered. Wear rate is usually lower than during run-in. Very often, this is the main stage of the tribological process

and it lasts up to the end of the component life. It is typically controlled by adhesion, tribo-oxidation or a combination of the two.

(3) Stage 2. In some cases, a wear *transition* may occur after a certain sliding distance. Wear rate may increase (curve 'a' in Fig. 5.1a) or it may decrease (curve 'b'). Such transitions are caused by a change in the wear mechanisms, often accompanied by a change in the friction coefficient.

Two examples of wear curves with sliding distance transitions are shown in Fig. 5.1b. The first curve pertains to an aluminium alloy, dry sliding against a heat-treated steel counterface (pin-on-disc test; applied pressure 1 MPa; sliding speed 0.8 m/s) [3]. After the run-in stage (very short and not shown in the figure), Stage 1 starts. It is characterised by a low wear rate: $10^{-3}$ $mm^3/m$. After 1800 m of sliding, a transition is observed with an increase in the wear rate to $2.8 \times 10^{-3}$ $mm^3/m$. The observation of the wear debris and tracks revealed that in Stage 1 wear was by tribo-oxidation, whereas in Stage 2 it was by adhesion. The transition can be explained using a criterion based on the attainment of a *critical flash temperature*: if $T_f$ becomes close to about 0.4–0.5 $T_M$ (where $T_M$ is the melting point of the alloy in Kelvin), it may trigger a noticeable thermal softening at the contacting asperities [4]. As a consequence, the material is no longer able to support the oxide layer, and wear changes from tribo-oxidation to adhesion. From the record of the contact temperature during sliding a flash temperature of about 130 °C at the transition was estimated. This is sufficient to induce thermal softening [4]. The second wear curve in Fig. 5.1b pertains to an AISI 304L stainless steel dry sliding against a heat treated steel, in a testing configuration very similar to that used for the former aluminium alloy [5]. In this case, however, Stage 1 was characterised by quite a high wear rate, and after the transition, occurring after 3000 m of sliding, wear rate decreased. Wear was found to pass from adhesion to tribo-oxidation. During Stage 1 a noticeable subsurface strain hardening was detected; the frictional heating was not able, in this case, to soften the material, given its relatively higher melting point and, consequently, temperature resistance. This Stage can be thus regarded as an incubation period required for the growth of a surface oxide layer and for the hardening of the subsurface material, which supports better the aforementioned oxide layer [6].

### 5.1.2 Mild and Severe Wear

As seen in Sects. 4.1 and 4.2, wear rate, W, for adhesive as well as tribo-oxidative wear can be expressed by the following generalised relationship:

$$W = K \cdot \frac{F_N}{H} = K_a \cdot F_N \tag{5.1}$$

where K is the wear coefficient and $K_a$ is the *specific wear coefficient* (units: $m^2/N$). Both coefficients depend on the active wear mechanism and can be obtained from the experimental wear curves, like those shown in Fig. 5.1.

In the case of metals sliding in dry conditions, wear by tribo-oxidation is typically *mild* in nature, whereas wear by adhesion is *severe*. Considering a large number of experimental data, it may be stated that mild wear is attained when:

$$W < 5 \times 10^{-3}\ mm^3/m$$
$$K < 10^{-5}$$
$$K_a < 10^{-4}\ m^2/N$$

For larger W-, K- and $K_a$-values, wear is severe. The boundary between mild and severe wear is not sharp, and an intermediate region with mixed transitional behaviour, as shown in Fig. 5.1b, may be present. As an example, Fig. 5.2 shows the wear coefficient and the surface temperature as a function of sliding speed, for a steel containing 0.21 % C (H: 180 $kg/mm^2$) dry sliding at room temperature against an AISI 52100 bearing steel (H: 850 $kg/mm^2$) [3]. The composition of the fragments and the fractions of the different phases were obtained by XRD analysis and using the Rietveld method. The results for the tests at 0.2, 0.6 and 1 m/s are listed in Fig. 5.2. As sliding speed is increased, i.e., as the contact temperature is raised, the contribution of the tribo-oxidative wear increases (the oxides are 90 % of the fragments at 1 m/s), and this explains the decrease in wear rate that approaches the mild wear regime. The fraction of magnetite in the oxides also increases with sliding velocity, and this also explains the recorded trend in friction coefficient that decreased from 0.65 at 0.2 m/s to 0.51 at 1 m/s, in agreement with the observations reported in Sect. 2.4.

A number of investigations have been carried out to determine the load and sliding speed ranges that result in mild wear, which is regarded as an acceptable form of wear in many applications, whereas severe wear cannot be generally tolerated. For this purpose, an approach based on the so-called *wear maps* is extremely fruitful [7]. Sliding tests are carried out at different loads and sliding speeds, and the relevant wear rates are reported on a graph having the speed and the load (or the nominal pressure) on the x- and y-axes respectively. The regions with mild wear are hence identified following the definitions given above. In addition, by means of the analysis of wear debris and tracks, the relevant wear mechanisms are recognized and indicated in the different regions of the map. The experimental investigations are often completed with the record of the steady-state contact temperatures, which may be included in the map to help in the explanation and discussion of the results. As an example, Fig. 5.3 shows the wear map for the Al 7010 alloy, dry sliding against an AISI 32100 steel (in a pin on disc testing rig) [8]. It can be seen that mild wear by tribo-oxidation is prevailing at low nominal pressures and sliding speeds. A boundary line is shown, separating the regions of mild from severe wear (by adhesion). A second boundary line is shown: the *seizure* line. Above it, wear is still by adhesion but particularly intense, with massive transfer of metallic debris onto

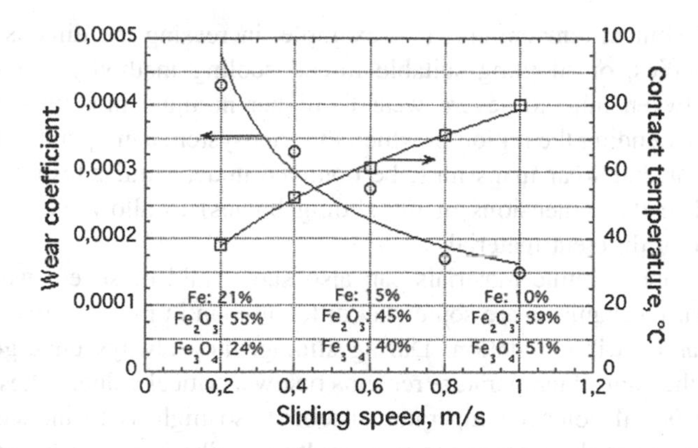


**Fig. 5.2** Dry sling wear of a 0.2 % C steel against bearing steel. Wear coefficient and contact temperature versus sliding speed. The composition of the collected fragments is also indicated (data from [3])

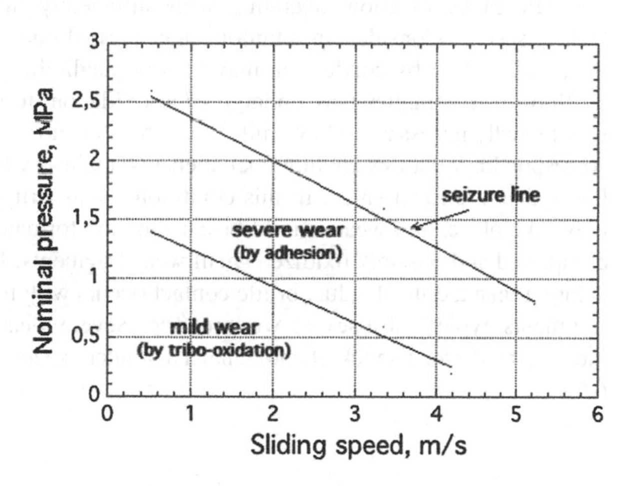


**Fig. 5.3** Wear mechanism map for the Al 7010 alloy dry sliding against an AISI 32100 steel counterface (modified from [8])

the counterface. The measurements of contact temperature showed that seizure was induced by the attainment of a critical surface temperature of 115 °C. In general, seizure is reached when contact temperature becomes so high as to induce intense material softening. The transition between mild and severe wear was also occurring at a critical contact temperature: 60 °C in this case. As shown in the previous paragraph for the Al 7072 alloy, this critical contact temperature is associated to a critical flash temperature, able to induce asperity-scale softening.

All metals feature sliding wear maps similar to that shown in Fig. 5.3, with a transition from mild tribo-oxidative wear to adhesive severe wear at the attainment of a critical surface temperature. This means that wear maps strongly depend on the geometry of the tribological system, which determines the contact temperature field, as shown in Sect. 2.11. As a consequence, all engineering solutions aimed at reducing the contact temperatures (for example: increasing the dimensions of the contacting bodies, or adopting suitable forced cooling methods) would shift the boundary between mild to severe wear to higher nominal pressures and sliding speeds, thus extending the region in which the tribosystem can operate safely [9]. It is further clear that wear maps must be obtained under similar testing conditions (with regard to the dimensions of the sliding bodies) to allow meaningful comparisons among different materials.

Polymeric and ceramic materials can also show mild or severe wear, but the involved wear mechanisms are somewhat different than in metals. Polymers do not undergo wear by tribo-oxidation. During sliding, they always undergo adhesive wear. But if the contact temperature remains below a critical value, adhesive wear is mild. Otherwise, if contact temperature becomes so high as to induce extensive thermal softening at the surface (or even melting), adhesive wear becomes severe with large material transfer, and cannot be accepted. Since the involved critical temperatures can be comparatively low (lower, for instance, than 100 °C), an external forced cooling system can be very important to keep the temperatures in the right range to allow operating with sufficiently high contact pressures and sliding speeds. Consider, in addition, that thermal conductivities of polymers are low, and cooling by conduction may thus be negligible.

Non-oxide ceramics may undergo tribo-oxidation during sliding. However, this is not strictly necessary to have mild wear. As shown in Sect. 4.1.2, mild wear (and, correspondingly, a low friction coefficient) is attained when the applied pressure is lower than a critical value. In this condition, an asperity-scale brittle contact only may take place, and wear is mild because of the formation of protective scales of compacted and possibly oxidized small wear fragments. But if the applied pressure is larger than a critical value, brittle contact occurs with the formation of quite large fragments, typical of a severe wear regime. Severe wear in ceramics may also be induced by thermal shock phenomena. This latter aspect will be considered in Sect. 6.5.1.

### 5.1.3 Mild Wear of Materials

Table 5.2 shows the typical ranges of the specific wear coefficient in the case of mild wear for different materials, all dry sliding against a high strength steel counterface. Data were obtained from a number of investigations that were carried using similar testing conditions (i.e., pin on disc or block on disc testing rigs, with nominal contact areas of the order of 30 $mm^2$). The analysis of the data allows us to highlight some points that are worth to be remarked:

(1) In steels (but this applies also for other metals), $K_a$ decreases markedly with increasing hardness. In fact, as hardness increases, the ability of the underlying metal to support the oxide layer also increases.

**Table 5.2** Typical ranges for hardness and specific wear coefficient for mild wear, in case of different materials dry sliding against high-strength steels (data taken from a number of literature investigations)

| Material | Typical hardness ($kg/mm^2$) | Specific wear rate $K_a$ ($m^2/N$) |
|---|---|---|
| High strength steels | 300–600 | $\approx 5 \times 10^{-14}$ |
| Tool steels | 600–850 | $10^{-15}$–$10^{-14}$ |
| Nitrided steels | 900 (at the surface) | $\approx 5 \times 10^{-15}$ |
| Grey cast iron (with pearlitic or martensitic matrix) | 300–500 | $3 \times 10^{-16}$–$5 \times 10^{-14}$ |
| Bronzes | 300–400 | $10^{-15}$–$10^{-14}$ |
| Al alloys | 100–250 | $\approx 10^{-14}$ |
| Hard Chromium (coating) | 900–1000 | $\approx 5 \times 10^{-16}$ |
| WC-Co | 1000–1600 | $\approx 5 \times 10^{-16}$ |
| Ni-P/SiC (coating) | 800–900 | $\approx 5 \times 10^{-16}$ |
| Polymers | 10–100 | $10^{-15}$–$10^{-14}$ |
| Reinforced polymers | – | $2 \times 10^{-16}$–$10^{-15}$ |
| Ceramics | 2000–3000 | $10^{-16}$–$10^{-15}$ |

(2) In grey cast iron, low values of $K_a$ are recorded because of the presence graphite nodules or flakes in the microstructure. After a run-in phase, they emerge at the surface and exert a solid lubricant action, which reduces friction and wear.
(3) Bronzes also display low values of $K_a$ thanks to their good hardness, their low compatibility versus ferrous alloys (see Fig. 1.13) and their high thermal conductivity that facilitates the removal of frictional heat, thus reducing contact temperatures.
(4) Aluminium alloys may provide relatively low $K_a$-values but their hardness is quite low and decreases considerably as contact temperature is increased. In these alloys, the transition to adhesive wear is thus quite easy.
(5) Some metallic coatings, such as hard chrome or nickel reinforced with ceramic particles, are characterised by very high hardness values. Therefore, they display quite low $K_a$-values.
(6) Polymers, especially polymeric composites, are characterised by low $K_a$-values when sliding against steels or other metals. This is due to their relatively low work of adhesion. However, their performance is strongly limited by temperature rises.
(7) Ceramic materials (often employed as coatings), display low values of $K_a$. Ceramics are generally hard, and they are able to maintain their hardness at high temperatures. They undergo mild wear as long as contact pressure is lower than a critical value, and they are not submitted to thermal shocks.

### *5.1.4 The PV Limit*

As already highlighted, tribological systems working in dry sliding should operate in the mild wear region. For many design purposes, Eq. 5.1 can be rewritten with reference to the depth of wear, h (see Fig. 4.1 for its definition):

$$h = K_a \cdot p_0 \cdot s \tag{5.2}$$

or, better, to the *linear wear rate*, $\dot{h}$ (units: m/s):

$$\dot{h} = K_a \cdot p_0 \cdot v \tag{5.3}$$

In different tribological systems, a limit is set to $\dot{h}$ in order to achieve an acceptable operating life. If we indicate with $\dot{h}_{al}$ the *allowable linear wear rate* (which is the maximum linear wear divided by the desired lifetime), Eq. 5.3 can be recast in the following useful form:

$$p_0 \cdot v = \frac{\dot{h}_{al}}{K_a} = PV_{limit} \tag{5.4}$$

The so-called $PV_{limit}$ is therefore given by the ratio between the allowable linear wear rate for a specific tribosystem and the specific wear coefficient provided by the materials pair, operating in the mild wear regime. Consider, for example, the case of *dry bearings*. The rotating shaft is generally made by heat-treated steel. Therefore, the materials for dry bearings have to guarantee low friction and relatively low wear rate when sliding against steel. When using thermoplastics bearings, a linear wear rate of about $5 \times 10^{-4}$ mm/h is typically requested [10]. Nylon 6.6 filled with 20 % PTFE displays a low friction coefficient and a specific wear coefficient of $2.4 \times 10^{-16}$ $m^2$/N. Therefore, from Eq. 5.4 it is obtained that $PV_{limit}$ = 0.58 MN/(m s). In the design stage, it is thus necessary to verify that the product between the nominal pressure on the bearing and the tangential shaft speed is below this limit. Additionally, it is also required to verify that pressure and speed are both lower than a maximum allowable value. In Table 5.3, typical characteristics and operating limits for a number of materials for dry bearings are listed.

All values of the friction coefficient and the specific wear rate listed in Table 5.3 were obtained by sliding tests at room temperature. They are valid as long as the contact temperature is lower than a critical value, above which wear changes from mild to severe. As seen in Sect. 2.11, contact temperature depends on several aspects, including the friction coefficient, the load and sliding speed, the geometry of the tribosystem (in particular, its dimensions and cooling capacity) and the ambient temperature. For sliding bearings, the average surface temperature can be easily evaluated using Eq. 2.29. Note that the allowable maximum temperatures listed in Table 5.3 for materials operating in dry conditions, refer to the temperature above which extensive mechanical deterioration occurs. When the ambient

**Table 5.3** Characteristics and operating limits for a number of materials used for dry bearings (taken from [10] and other sources in the literature; the sintered bronze impregnated with oil operate under boundary or mixed lubrication)

| Material | Friction coefficient (against steel) | Specific wear rate against steel, $K_a$ ($m^2/N$) | Maximum pressure (MPa) | Maximum sliding speed (m/s) | Maximum temperature (°C) |
|---|---|---|---|---|---|
| Nylon 6,6 (filled with PTFE) | 0.2 | $2.4 \times 10^{-16}$ | 10 | – | 200 |
| PTFE | 0.03–0.15 | $4 \times 10^{-13}$ | 3.4 | 0.3 | 250 |
| Filled PTFE (15 % glass fibre) | 0.1 | $1.4 \times 10^{-16}$ | 17 | 5 | 250 |
| Acetal resin (filled with PTFE) | 0.07–0.1 | $4.9 \times 10^{-16}$ | – | – | 110 |
| Carbon-graphite | 0.06–0.15 | $1.4 \times 10^{-15}$ | 4.1 | 13 | 400 |
| Sintered bronze (filled with oil) | 0.05–0.15 | $4 \times 10^{-17}$ | 6.1 | 28 | 80 |

temperature approaches these values, the contact temperatures are much higher and the severe wear regime is most likely entered. Correspondingly, the calculated $PV_{limit}$ should be decreased by 70 % as ambient temperature approaches such maximum values [10].

### *5.1.5 The Effect of Lubrication*

The presence of a lubricant film between two mating surfaces greatly reduces friction and wear. As a matter of fact, the lubricant prevents direct contact between the asperities, reducing the average shear stress at the junctions.

In fluid film lubrication, no wear is thus recorded because of the absence of any contact between the asperities. However, in case of mixed or, specially, boundary lubrication, adhesive wear may take place where metal-to-metal contacts are established. Since lubricant greatly reduces the availability of oxygen in the contact areas, wear by tribo-oxidation is somewhat difficult, although its contribution cannot be excluded. Wear in lubricated sliding is really difficult to model. A promising approach is provided by a refinement of the model presented in Sect. 3.5, where parameter δ, i.e. the contact area ratio, has been introduced. δ represents the fraction of real area of contact at which boundary lubrication occurs. In the present model, the parameter α′ is introduced, which represents the fraction of boundary film that is defective, i.e., the fraction of real area of contact that is metallic [11]

($\alpha'$ is often called the *fractional film defect*, and it is smaller than $\delta$). Equation 4.1 can thus be rewritten for lubricated sliding wear:

$$W = \alpha' \cdot K_{ad} \cdot \frac{F_N}{H} = K_{lub} \cdot \frac{F_N}{H} \quad (5.5)$$

where $K_{lub}$ is the *coefficient for adhesive wear in lubricated conditions.*

Therefore, $K_{lub}$ decreases as the $\Lambda$ factor is increased and also depends on the lubricant quality. In mixed lubrication, $K_{lub}$ typically ranges from $10^{-10}$ to $10^{-6}$, while in boundary lubrication it typically ranges between $10^{-6}$ and $10^{-5}$ [12]. In case of boundary lubrication, the following relation can be used as a first approximation:

$$K_{lub} = x\, K_{ad}$$

where:

$x = 10^{-1}$ for poor lubrication ($\alpha'$ is quite high; a poor lubricant, such as water, is used);

$x = 10^{-2}$ for average lubrication ($\alpha'$ is low; a common mineral oil is used);

$x = 10^{-3}$ for excellent lubrication ($\alpha'$ is very low; lubricants with EP additives are employed)

The evaluation of $\alpha'$ is rather difficult. In general, $\alpha'$ decreases as [10]:

- the lubricant heat of adsorption is increased (if the lubricant adsorption on the metal surface is strong, the desorption during sliding can be more difficult);
- the surface contact temperature is decreased (desorption is due to the attainment of a critical contact temperature; in addition, temperature also limits the mechanical performances of the secondary boundary layerthat is formed when EP additives are used);
- the sliding speed is decreased (in this way a shorter contact time at the asperities is allowed).

Two important aspects have to be considered when dealing with lubricated sliding wear: *running in*, and *scuffing* (i.e., the failure of boundary lubrication). Figure 5.4 shows the wear evolution of a heat-treated Ni-Cr-Mo steel (H: 420 kg/mm$^2$) tested in a pin-on-disc and in a like-on-like configuration under lubricated conditions [13]. The applied load was 200 N ($p_0 = 6.3$ MPa) and the sliding speed 2 m/s. A quite long running in stage (see Fig. 5.1a) was recorded, characterised by rapid wear, followed by a steady-state stage. The calculated $K_{lub}$ during the steady-state stage is equal to $1.3 \times 10^{-8}$. This value is representative of mixed lubrication as confirmed by the recorded friction coefficient that was quite low, around 0.025. During running in, $K_{lub}$ was between one and two order of magnitudes larger than during steady state. Adhesive wear was mainly caused by the elimination of initial misalignments between the mating surfaces and the highest peaks in the roughness profiles. The correct execution of the run-in stage is thus of paramount importance

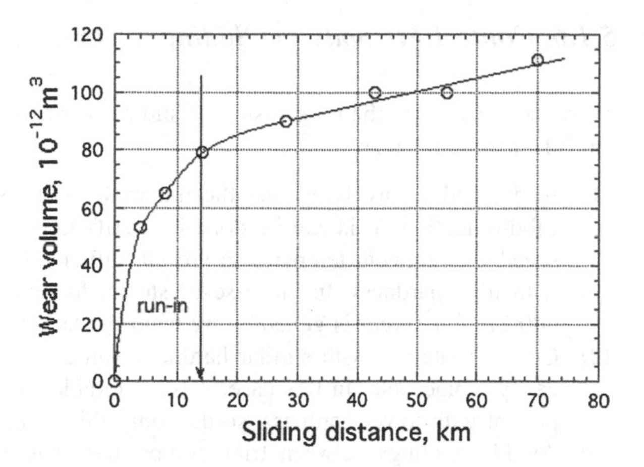


**Fig. 5.4** Wear evolution in a Ni-Cr-Mo steel during sliding at 2 m/s speed and 200 N applied load in lubricated conditions (modified from [13])

in lubricated tribological systems, since it will increase the $\Lambda$ factor and decrease the subsequent risk of scuffing. This fact introduces the necessity of a careful operation at the beginning of service of lubricated machinery (like gears, bearings and engines). A commonly adopted strategy is to operate with mild loading conditions in freshly assembled surfaces, and gradually increase the loads to the design levels [11]. In some cases, small abrasive particles are introduced between the surfaces to help surface polishing, and therefore attain a very low roughness after running in (such particles have to be washed out before entering service). Most operations are still based on experience, and there is a continuous industrial need to optimise the run-in procedures, including a reduction in the run-in times.

If the operating conditions are severe, *lubricant failure* by scuffing can occur during lubricated sliding in boundary conditions. The phenomenon of scuffing has been treated in Sect. 3.4. As seen, there is no universally accepted procedure to predict the conditions leading to scuffing. The most common approach is to evaluate the average surface temperature, and make sure that it is less than the critical value, which is around 150 °C for mineral oils. Following the arguments highlighted in the previous paragraph, a design procedure based on the $PV_{limit}$ concept can be possibly adopted. In order to minimise the risk of scuffing, it is useful to select a heat resistant lubricant or to adopt lubricants with EP additives, and/or to modify the materials at the surface, by using surface treatments that reduce friction and/or are able to act as oil reservoirs (examples include chromium plating, molybdenum coating and gas nitriding, see also Chap. 7). It may be also useful to favour a high value of the $\Lambda$ factor (by properly optimising the surface microgeometry and the running in stage); a low lubricant temperature (by properly cooling the lubricant); a high thermal conductivity of the mating materials (which favours the achievement of low contact temperatures).

### *5.1.6 Control Methods for Sliding Wear*

To reduce or control the progression of sliding wear, it may be useful to consider the following guidelines:

(1) In dry sliding involving metallic materials, make sure that wear is mild by tribo-oxidation. This can be done by verifying that contact pressure, sliding speed and ambient temperature are all sufficiently low, and by using alloys with high hardness. In the case of steels, for example, if hardness exceeds 600 kg/mm$^2$, wear is generally by tribo-oxidation.
(2) Couple materials with similar hardness values, except when a component is easily replaceable. In this case, it is preferable to have the replaceable component with lower hardness, so that only this component is worn out.
(3) Avoid couplings between tribo-compatible materials (see Sect. 1.4). When possible, prefer metal/polymer or metal/ceramic couplings (by adopting, if required, suitable surface treatments, see Chap. 7). When using polymers, verify that surface temperature does not exceed a critical value, in order to avoid transition to severe wear and thermal distortions as well. When using filled polymers, verify that the antagonist has a hardness of at least 30 % greater than that of the reinforcing fibres (glass fibres, with a hardness of about 500 kg/mm$^2$, are widely employed) to avoid abrasive wear (alternatively, design the production process of the polymer composite with the aim of realizing a fibre-free skin on its surface).
(4) If possible, adopt lubrication or improve its quality, aiming at high $\Lambda$ factors since for mixed and boundary lubrication this reduces the risk for scuffing. If necessary, use lubricants with EP additives, or combine the use of normal mineral oils with the employment of solid lubricants (Sect. 3.1).
(5) Use proper self-lubricating materials (PTFE-based materials, for example), if a low friction coefficient is required and fluid lubrication cannot be adopted. The selection should be based on the $PV_{limit}$ concept; if necessary, design a proper forced cooling system to avoid transition to severe wear.
(6) Ensure that abrasive particles do not contaminate the tribological system. Possibly use appropriate filtration systems and protective tools (note that grease is a protective lubricant).
(7) Reconsider the design of the tribological system. For example, reduce the applied load (note that overloading can also damage the adsorbed layer of the lubricant in the case of boundary lubrication) and, when possible, prefer rolling (or rolling-sliding) contact to pure sliding.
(8) When sliding against steel, materials that give the best behaviour are (in decreasing order of performance): ceramic materials; filled polymers; bronzes; cast irons; martensitic steels.

## 5.2 Fretting Wear

Fretting damage involves contacting surfaces that are subjected to small amplitude oscillatory movement, with displacements typically between 1 and 100 μm. In most cases (see Table 5.1), such relative displacements are induced by vibrations imposed to mechanical systems that normally are at rest. Contact cyclic stresses may nucleate surface cracks by contact fatigue. Such cracks may thus be responsible for subsequent fatigue failure if further bulk stresses are superimposed to one or both components in contact. This type of damage is also called *fretting fatigue*. The wear mechanisms acting during fretting wear are tribo-oxidation and contact fatigue.

Three main *fretting regimes* can be recognized. They mainly depend on the amplitude of the oscillation [14, 15]. In the following, they will be briefly described with reference to the schematization of Fig. 5.5 that refers to a cylinder on flat contact.

*Stick regime*. If the amplitude is very low (roughly less than 2–4 μm), the junctions between the asperities in contact do not undergo sliding and the relative motion between the two bodies in contact is simply due to elastic deformations at the junctions. In this case, the surface damage is absolutely negligible. This condition occurs when $F_T \ll \mu_s F_N$, where $\mu_s$ is the coefficient of static friction (Fig. 5.5a).

*Partial slip regime*. If the amplitude of the oscillation is larger (typically between 5 and 20 μm), the peripheral areas of the contact undergo micro slip, because of the lower Hertzian pressure with respect to the centre of the contact. As shown in Fig. 5.5b, a stick region is present at the centre of the contact, whereas at the periphery a micro slip region is attained. Indicating with c the half-width of stick

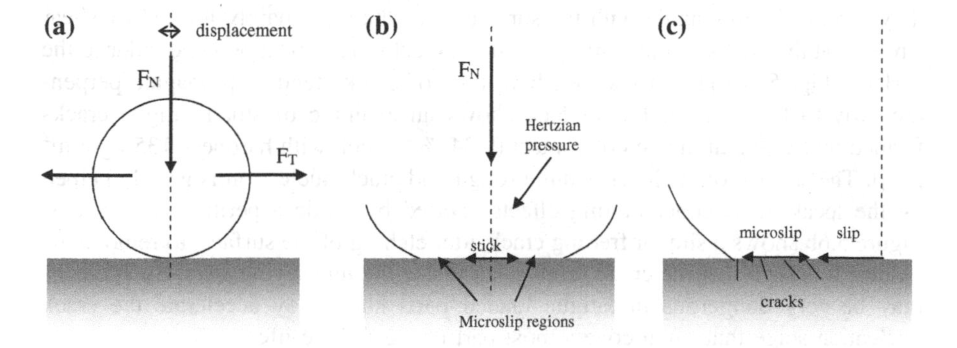


**Fig. 5.5** **a** Cylinder on flat contact; **b** partial slip regime characterised by the presence of a central stick region and two peripheral micro slip regions; **c** formation of surface microcracks due to oscillatory motion

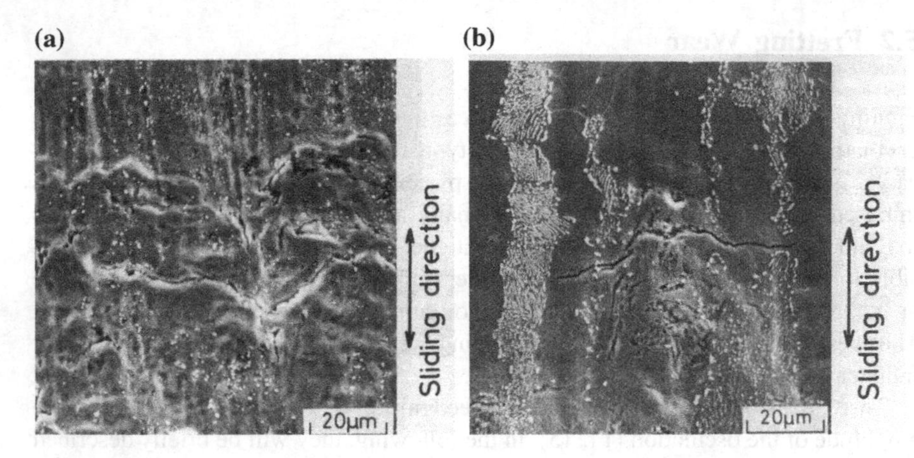


**Fig. 5.6** **a** Small surface fretting cracks at the end of the contact; **b** appearance of a fretting cracks after etching [18]

region and with a the half-width of the Hertzian contact (a can be calculated using the relation reported in Table 1.1), the ratio c/a is given by [16]:

$$\frac{c}{a} = \sqrt{1 - \frac{F_T}{\mu_s \cdot F_N}} \tag{5.6}$$

for example, if $F_T$ is 0.5 $F_N$, then c/a = 0.4 if $\mu_s = 0.6$. In the micro slip regions, wear by tribo-oxidation may occur, facilitated by the alternating motion as seen in Sect. 4.2. Wear rate is rather low (around $5 \times 10^{-16}$ $m^2/N$ in steels), but the repeated contact shear and tensile stresses due to friction (Fig. 2.2) may induce the formation of surface microcracks. At the boundary between the stick and micro slip regions, they form a shallow angle with the surface since they are mainly formed by shear stresses; at the edges of the contact, the microcracks are almost perpendicular to the surface (Fig. 5.5c) [17]. However, both type of cracks tend to propagate perpendicularly to the surface. Figure 5.6a shows an example of small fatigue cracks formed by fretting at the end of contact (0.34 % C steel, with hardness 135 $kg/mm^2$ [18]). The surface oxide layer is quite rough and crack nucleation is greatly helped by the local stress concentrating effects exerted by oxide asperities and valleys. Figure 5.6b shows a similar fretting crack after etching of the surface, to remove the oxides. It is seen that the crack propagated inside the underlying steel. Such cracks may be very dangerous in fatigue loaded parts since they accelerate the crack nucleation stage that often covers most part of the fatigue life.

*Gross slip regime*. If the oscillation amplitude is greater than approximately 20 μm, the whole area of nominal contact undergoes sliding. This occurs when $F_T$ is close to $\mu_s$ $F_N$ and c ≈ a. Wear rate increases with the oscillation amplitude, until it reaches typical steady-state values for sliding wear (by tribo-oxidation) when the amplitude is greater than, say, 300 μm. In this case, wear may be tolerated or it may give rise to failures. A failure example is provided by the loss of interference in flanged connections. At the same time, the formation of a layer of compacted oxides prevents the formation of surface cracks, which could originate fatigue cracks.

In summary, the conditions to have:

- absence of damage (stick regime)
- surface cracking (partial slip regime)
- tribo-oxidative wear (gross slip regime)

mainly depend on the oscillation amplitude, the tribological system that determines the coefficient of friction, the applied load and the ambient temperature.

As far as the friction coefficient is concerned, it is evident that the higher it is the larger is the possibility for surface cracking. In fact, as friction increases the contact stresses also increase, and the transition to the gross slip regime becomes more difficult. But if the applied load ($F_N$) is particularly high, it might be more convenient to have a high friction coefficient, which can prevent the transition from the stick to the partial stick regime.

High frequency of oscillation can prevent proper formation of a protective oxide layer thus favouring the formation of surface cracks. If the system geometry facilitates the ejection of wear fragments from the contact region, the formation of the protective layer is further prevented or slowed down.

Ambient temperature plays an important role in systems operating in space applications or in power plants. It has been experimentally observed that an increase in ambient temperature above a critical value may decrease by even five times the intensity of fretting wear. In steels, for example, such a critical temperature is around 400 °C [19]. As described in Sect. 4.2.1, above this temperature a tribo-oxidative wear by direct oxidation is achieved, with the formation of a thick and protective oxide layer. Even titanium alloys, widely used in the chemical industry, show an increase in fretting resistance as temperature rises. Titanium alloys, as will be discussed in Sect. 6.4, display difficulty in establishing a mild wear regime by tribo-oxidation despite their high reactivity towards oxygen. For this reason, as temperature is increased over room temperature, fretting damage is initially increased. However, as ambient temperature overtakes about 500 °C, fretting damage begins to decline and then become very low, due to the formation of a thick oxide layer by direct oxidation. Note that ambient temperature must not be too high if an adequate fatigue resistance is required; in fact, if temperature increases too much the material's strength is reduced and this impairs its bulk mechanical performances.

Different approaches can be used to avoid or to control surface damage by fretting, although fretting is a very complex process with numerous interdependent variables that are very difficult to consider together in the design stage.

(1) Reduce the contact stresses. This can be accomplished by lowering the contact (clamping) load and/or by reducing friction. The use of solid lubricant coatings typically lowers friction. In most circumstances liquid lubricants, including greases, cannot be used, since they would be squeezed out of the contact.

Steel ropes that form a close system able to retain grease or other lubricants constitute an exception.

(2) Reduce excessive vibrations. When possible, a vibration analysis should be carried out in the design stage and allowable vibration limits should be assessed.

(3) Proper materials selection. In this respect, it is worth considering that materials with high fatigue notch sensitivity also display a high tendency to form surface cracks by fretting [19]. In addition, surface treatments that induce *compressive residual stresses*, like shot peening, are also reported to have beneficial effects on fretting fatigue since they prevent crack initiation and propagation.

## 5.3 Rolling—Sliding Wear

Rolling-sliding wear typically occurs when two surfaces are in repeated contact. As an example, Fig. 5.7a shows two rolling cylinders that are in contact along a generatrix. The cylinders are counter rotating, and the tangential speeds are indicated with $v_1$ and $v_2$. Surface elements of both cylinders undergo repeated contact when the cylinders are rotating. Pure rolling is attained when $v_1 = v_2$. If $v_1 \neq v_2$, sliding occurs and the *sliding speed*, $v_s$, is defined by: $v_1 - v_2$. The *rolling speed*, $v_r$, is defined by: $(v_1 + v_2)/2$ (in pure rolling, $v_r = v_1 = v_2$). In rolling-sliding contacts, the contribution of sliding is then quantified by the ratio between $v_s$ and $v_r$:

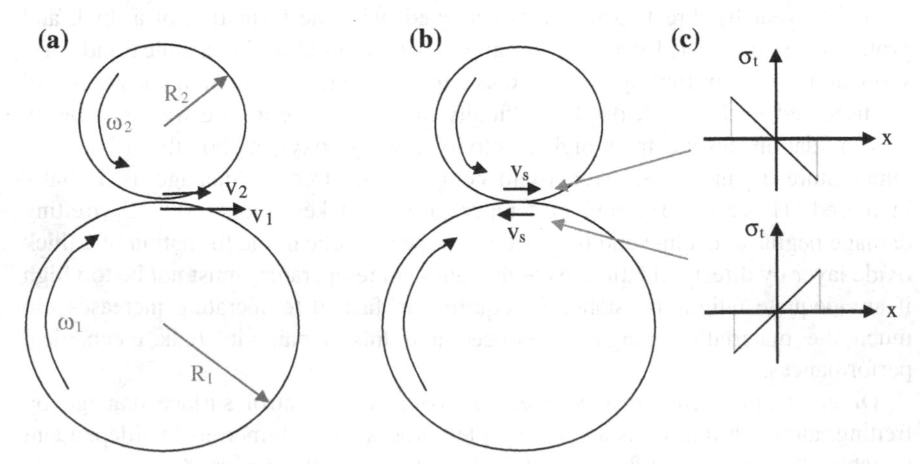


**Fig. 5.7** Rolling cylinders in contact along a generatrix. The cylinders are counter rotating and the tangential velocity of the *upper* cylinder is lower than that of the *lower* cylinder (this may occur, for example, in tractive rolling when the *lower* cylinder is the driver roller and the *upper* one is the driven roller)

$$\% \, sliding = 100 \cdot \frac{|v_1 - v_2|}{1/2 \cdot (v_1 + v_2)} \tag{5.7}$$

Conditions *close* to pure rolling occur in mechanical components such as rolling bearings or roller-cam systems (however, as shown in Fig. 2.31, some micro slip in the contact region is always present in case of tractive rolling). Typical examples with rolling-sliding conditions arc the teeth to teeth contact in gears, and the wheel-rail system. Figure 5.7b shows the direction of the sliding velocity for the two mating cylinders. In the presence of friction, which depends on the lubrication regime and possibly on the mating materials, the contact stresses are modified as illustrated in Sect. 2.1. In particular, Fig. 5.7c schematically shows the contact tensile/compressive stress fields (indicated with $\sigma'_y$ in Fig. 2.2). In some conditions, such stresses play an important role in determining the rolling-sliding performance of the mating bodies.

In rolling sliding–wear processes, wear mechanisms are contact fatigue and/or adhesion/tribo-oxidation. This depends on many parameters that are quite often interrelated. The most important one is the lubrication regime. Generally speaking, in the case of fluid or mixed lubrication, the main wear mechanism is contact fatigue, and wear resistance is generally provided by the achievement of an excessive damage by spalling or pitting (if the latter is particularly intense). The occurrence of these failures modes typically brings about noise and vibrations, which cannot be tolerated for the correct operation of the tribological system. In addition, noise and vibration are a clear warning that damage has started, and a prompt to stop the system to avoid more dramatic failures. In the case of boundary lubrication or dry contact, sliding damage is prevailing, and wear may be by adhesion or tribo-oxidation.

### 5.3.1 S-N Curves and the Role of Material

The rolling–sliding wear behaviour is generally investigated using the *disc-on-disc tribometer* (line contact) or the *4-ball tribometer* (point contact), described in Sect. 4.5. To obtain the S-N curves, different tests are carried out at different load levels (expressed by the Hertzian pressure, $p_{max}$) and the number of cycles, N, to reach fatigue damage is recorded. If the experimental data are plotted in a log-log graph, the interpolating curve is typically a straight line, as schematized in Fig. 5.8. N is thus given by the following exponential relationship:

$$N = \left(\frac{K}{p_{max}}\right)^n \tag{5.8}$$

where K and n are two constants. K depends on the material strength and on the lubrication regime, whereas n mainly depends on the lubrication regime, and

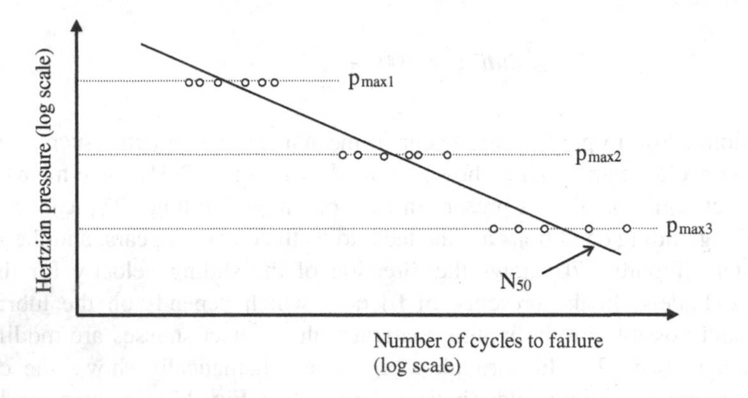


**Fig. 5.8** Schematisation of S-N curve for contact fatigue. Examples of repeated test results are displayed for three stress levels

increases with Λ factor [20]. In some cases, the analysis is complicated by the dependence of n on Hertzian pressure. For example, in high quality steels, such as some bearing steels, n is found to increase as $p_{max}$ is decreased. If Eq. 5.8 is used to fit the mean values at each pressure level, N is thus indicated with $N_{50}$, and provides the number of cycles with a failure (or survival) probability of 50 %. Equation 5.8 can be also used to fit the data that give a different (typically lower) failure probability, for example $N_{10}$.

In designing for fatigue resistance, it would be useful to know the so-called *fatigue limit* that in the present case is the Hertzian pressure below which an infinite fatigue life is attained. Nonetheless, many Authors believe that fatigue failure is reached under cyclic contact stresses even if the Hertzian pressure is very small. In general, it is thus preferred to use the concept of *endurance limit*, $p_{end}$ which is the contact pressure that guarantees a fatigue life of at least $10^8$ (or $5 \times 10^7$) cycles which, for practical purposes, can be considered as infinite [21]. Equation 5.8 is usually rewritten in the following way:

$$N_{50} = 10^8 \cdot \left(\frac{p_{end}}{p_{max}}\right)^n \tag{5.9}$$

In the case of materials with a ductile behaviour, the endurance limit scales with the yield strength (or the hardness, H) roughly in a linear manner [22]:

$$p_{end} = a \cdot H \tag{5.10}$$

where a is a material constant. In fact, as hardness is increased, the dislocation motion required to form the plastic regions necessary for crack nucleation and propagation is made more difficult. If we simply consider a line contact, we obtain that the Hertzian pressure, $p_{max}$, is proportional to $\sqrt{F_N \cdot E}$. Therefore, for a given geometry and a given applied load, the contact fatigue life increases with the ratio $H^2/E$, which can be considered as a simple *contact fatigue index* for ductile

**Table 5.4** Ratio $H^2/E$ for some engineering materials

| Material | Hardness (MPa) | Young's modulus (GPa) | $H^2/E$ (MPa) |
|---|---|---|---|
| Low-C steel | 2000 | 207 | 19.3 |
| Heat-treated steel | 6000 | 207 | 237 |
| Carburized steel | 8000 | 207 | 310 |
| Pearlitic grey cast iron | 2100 | 180 | 24.5 |
| P-bronzes | 1200 | 110 | 13 |
| Cu-Be alloys | 3000 | 110 | 82 |
| Hardened Al alloys | 1200 | 75 | 19 |
| PMMA | 300 | 8.6 | 10.4 |
| Nylon | 120 | 3 | 4.8 |
| Alumina | 14,000 | 300 | 650 |
| Si nitride | 18,000 | 310 | 1045 |

materials. In Table 5.4, representative values of such an index are listed for different materials.

The most used materials in rolling–sliding applications are metals and, in particular, steels. Steels all have similar Young's moduli and their rolling-sliding resistance increases with hardness. The best performances are then expected by *carburised steels* (or carbonitrided steels), which will be further considered in the next two chapters. Carburising is a surface treatment that noticeably increases surface hardness and relevant performances of steels if the fracture toughness is correspondingly not excessively reduced, and/or the size of microstructural defects is not too large. In addition, carburising introduces a beneficial surface layer, featuring *residual compressive stresses* (of the order of 200–300 MPa). It is assumed that such a stress field partially compensate the applied Hertzian stresses thus reducing the effective contact loads. The effective maximum shear stress, $\tau'_{max}(z)$, is thus given by [21]:

$$\tau'_{max}(z) = \left| \tau_{max}(z) + \frac{\sigma_{rs}(z)}{2} \right| \tag{5.11}$$

where $\sigma_r$ is the residual stress that is generally assumed to be equibiassial (i.e., having the same value in the x-y plane) and dependent only on z (see Fig. 1.2 for the definition of the coordinates; note that Eq. 5.11 should be modified to account for the relaxation of the residual stresses during cyclic loading). Table 5.5 reports some experimental data obtained for different alloys in a line contact and using mineral oil as a lubricant [23]. The tests were carried out in almost pure rolling (0 % sliding), and with 9 % sliding. It should be noted that the reported data were not obtained recently. After the introduction of new production technologies, for example the secondary metallurgy, the alloys produced today are characterised by definitely better microstructural cleanliness and, thereby, mechanical properties.

**Table 5.5** Experimental $p_{end}$ values for different alloys rolling sliding against a tool steel with H = 62 HRc (data from [23])

| Material | Pure rolling | 9 % sliding |
|---|---|---|
| Low-C steel (H: 140 kg/mm$^2$) | 615 | 535 |
| Carburized 1020 steel | 1770 | 1610 |
| 4340 induction hardened steel (H: 50–58 HRc) | 1800 | 1500 |
| Grey cast iron (H: 240 kg/mm$^2$) | 650 | 600 |
| Nodular cast iron (H: 260 k/mm$^2$) | 700 | 650 |
| Phosphorus bronze (H: 70 kg/mm$^2$) | 400 | 250 |

Ceramic materials generally feature very high hardness values (retained also at high temperature) and display very high values of the $H^2/E$ index. They typically display contact fatigue lives that are longer than those of steels. However, their performance may be limited by their low inherent fracture toughness, and the microstructural defects may be regarded as pre-existing cracks that are able to propagate if the applied stress intensity factor is greater than a critical threshold. In case of contact fatigue with subsurface nucleation, we may refer to shear propagation parallel to the surface, and then obtain the following relation:

$$\Delta K_{II} = Y \cdot p_{max} \cdot \sqrt{\pi c} \tag{5.12}$$

where $\Delta K_{II}$ is the applied *stress intensity factor*, Y is a non-dimensional geometric factor of the order of 0.5 (it depends on the z/a value, where z is the distance from the contact surface and a is the half width of the contact region, see Sect. 1.1.1), and c is the length of the defect. Setting $p_{max} = p_{end}$, $\Delta K_{II} = \Delta K_{IIth}$ that is the mode II *stress intensity threshold* (if $\Delta K_{II} < \Delta K_{IIth}$, no propagation occurs). $\Delta K_{IIth}$ is roughly proportional to $K_{Ic}$ [22]. From Eq. 5.12, it is then obtained:

$$p_{end} = C \cdot \frac{K_{Ic}}{\sqrt{c}} \tag{5.13}$$

where C is a system constant. Equation 5.13 shows that in materials with brittle behaviour the fatigue endurance is increased if the defect size (i.e., the maximum size, in the case of a size distribution) is reduced, and the fracture toughness is increased. In general, the S-N curve of brittle materials is very flat. In fact, if $p_{max}$ is increased over $p_{end}$, N decreases very rapidly since crack propagation is very fast.

In order to exploit at best the mechanical properties of ceramic materials, components with very small defects should be produced. Today, ceramics like silicon nitride ($Si_3N_4$) are increasingly used in rolling elements of bearings, in particular when additional properties, such as the electrical insulation, are required, thus justifying higher production cost. But the presence of surface cracks, which might form during the polishing process or following the subsequent handling with unavoidable collisions, may jeopardise the reliability of the mechanical parts, and the presence of unexpected overloads may then have serious consequences [24]. All

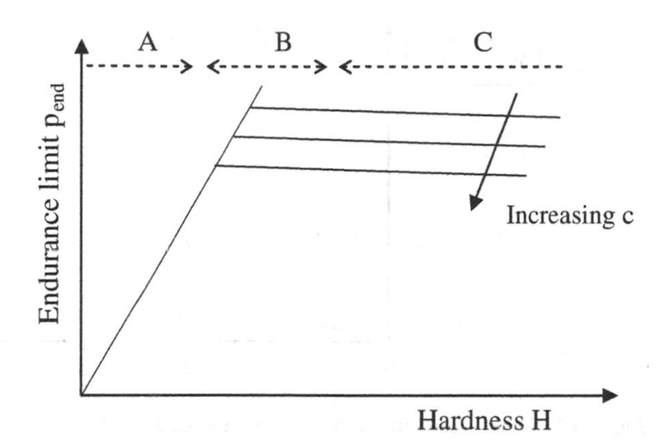


**Fig. 5.9** Schematic diagram showing the correlation between the endurance limit and materials' hardness

this claims for the necessity of adopting a probabilistic design approach when dealing with ceramic components that are subjected to contact fatigue loading.

It is quite interesting to consider Eqs. 5.10 and 5.13 together, as shown in the schematization of Fig. 5.9. Generally speaking, ductile materials possess relatively low hardness and high fracture toughness. Thus, they fall in region A. On the other hand, brittle materials possess high hardness and relatively low fracture toughness, which may also depend on their hardness as well. They fall in region C, and their endurance limit depends much on defects. In the intermediate region, B, materials possess intermediate values of both hardness and fracture toughness, and they may behave in a ductile or brittle manner depending on the presence and size of the defects. This is the case, for example, of carburized steels. They may achieve very high hardness values after heat treatment but may show a lower endurance limit than expected, because of the presence of defects (typically inclusions or precipitates).

Finally, Table 5.4 shows that polymers typically display limited performances as concerns contact fatigue (even if they display $H^2/E$ values similar to those of bronzes). Such materials are consequently used in mild loading conditions, against themselves or metals, typically when additional advantages are attained. For example, they are often used in mild loaded gears working in dry conditions, when low noise or high resistance to a specific environment are required. As already highlighted, polymers are also very sensitive to temperature rises, which may also occur during a cyclic contact loading due to their viscoelastic behaviour.

### *5.3.2 The Influence of Lubrication Regime*

Figure 5.10 schematises the role of the lubrication regime (i.e., of the $\Lambda$ factor) on the rolling-sliding wear. Fatigue life is seen to increase with $\Lambda$ reaching a plateau for $\Lambda \approx 3$, when fluid film lubrication is attained. A further increase in $\Lambda$ does not

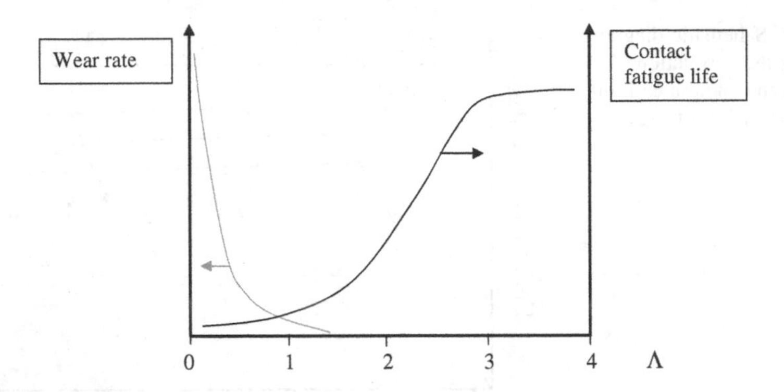


**Fig. 5.10** Scheme showing rolling-sliding damage mechanisms as a function of Λ factor

lead to an appreciable increase in the fatigue life. If Λ is lower than about 1.5, adhesive wear becomes more important, and the two wear mechanisms may have a competitive role (wear is important also in pure rolling because of the presence of micro slip). In this case, a simplified approach is typically adopted, and fatigue life is firstly assessed. After checking that it is sufficiently long for the given application, wear is considered, and the component wear life for an allowable depth of wear is then verified. However, it is possible that the two mechanisms interact in a negative or positive way. For instance, wear may produce a surface damage that favours the surface nucleation of fatigue cracks. Alternatively, wear may remove, partially or entirely, the surface cracks nucleated by contact fatigue. The subsequent crack propagation is thus delayed and the overall rolling-sliding resistance is increased [25]. Different parameters must be balanced in order to take advantage of wear in limiting contact fatigue damage, and this perspective is limited to a restricted number of systems.

Most of the experimental investigations carried out so far to assess the role of Λ on the contact fatigue of materials have been focussed on steels. The researches reported by Niemann are well known [26]. The Author proposed the following relationships for various types of steels under pure rolling: $p_{end} = 3$ H for a line contact (H is in kg/mm$^2$ and $p_{end}$ in MPa), and $p_{end} = 5.25$ H for a point contact (the difference can be clearly attributed to the statistical effects described in Sect. 4.4.1). If $\Lambda > 3$, Eq. 5.9 can be then used, with $n > 16$. Table 5.6 reports simplified relations for the evaluation of $p_{end}$ as a function of the lubrication regime, and the

**Table 5.6** Typical $p_{end}$ and n values for steels and in line contact for $N_{50}$ (from different literature sources)

| | $p_{end}$ (MPa) | n range |
|---|---|---|
| Rolling-sliding with boundary lubrication ($\Lambda < 0.5$, $\mu > 0.1$) | 1.7 H | <8 |
| Rolling-sliding with mixed lubrication | 2.25 H (or: 2.76 H −70) | 8–16 |
| Rolling-sliding with fluid film lubrication ($\Lambda > 3$, $\mu < 0.07$) | 3 H | >16 |

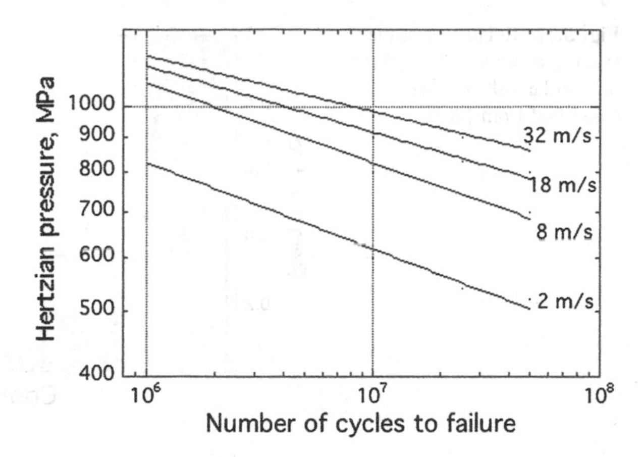


**Fig. 5.11** Influence of pitch line speed on the contact fatigue behaviour of a 53MnSi4 steel (modified from [26])

relevant n-ranges. It has to be considered that fluid film lubrication is certainly desirable but not always possible. It cannot be reached in conditions characterised by low speed, high load and high ambient temperature. In addition, low $\Lambda$ factors are usually attained during the starting and stopping phases of rolling-sliding components, when steady-state EHD conditions are not yet attained.

In the mixed and boundary lubrication regime, cracks nucleate at the surface, favoured by sliding. Correspondingly, the oil pumping effect (Fig. 4.15) gains in importance. As a consequence, the contact fatigue life decreases. As an example, Fig. 5.11 shows the S-N curves for a 53MnSi4 steel [26]. The tests were carried out on gears (line contact) at different pitch line speeds. It may be assumed that at 32 m/s the contact conditions were in between fluid film and mixed lubrication. As the pitch line speed is decreased, the thrust force exerted by the lubricant is reduced (see Sect. 3.3), and the $\Lambda$-factor is in turn decreased, entering a mixed lubrication regime. This produces a decrease in the contact fatigue resistance.

In case of mixed and boundary lubrication regime, propagating cracks were observed to be more frequent on the driven cylinder (such as the upper cylinder in Fig. 5.7). In fact, the surface of such cylinder enters the contact with a tensile stress that opens cracks allowing lubrication to pour into them. This strongly favours the oil pumping effect. It has been further shown that lubricant reduces friction between the crack surfaces, and this enhances crack propagation rate [27]. Of course, all this does not mean that lubrication should be avoided, since it strongly reduces the intensity of the asperity contacts, therefore greatly reducing the surface damage.

Several factors can deteriorate the lubricating conditions. One of these is certainly the *contamination* of the lubricant by foreign particles. Such contaminants can be mineral particles, like sand particles, or metallic particles, like cutting chips, grinding dust or wear debris. Such contaminants may form grooves that act as nucleation sites for further surface fatigue cracks, and may also reduce the adsorption capacity of the lubricant. Contaminants are particularly damaging when their dimensions are of the same order as the lubricant thickness. Figure 5.12 shows the relative reduction in lifetime in a ball bearing due to different contaminants [28].

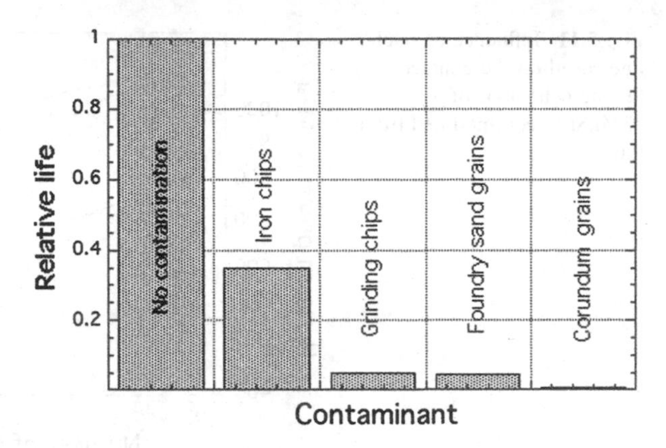


**Fig. 5.12** Relative lifetime reduction due to contaminants in angular ball bearing (modified from [28])

Hard abrasive particles are particularly harmful. Lubricant cleanliness is therefore a clear precondition for an adequate rolling-sliding life.

## *5.3.3 The Influence of Sliding*

Another factor playing an important role in determining the lifetime of a component in rolling-sliding is the fraction of sliding between the mating surfaces (Eq. 5.7). In the case of mixed or boundary lubrication, sliding may favour the shear plastic deformation at the contacting asperities, thus amplifying the possibility of surface crack nucleation. Generally speaking, poor lubrication (low Λ-factor) and high sliding greatly contribute to contact fatigue failure. Furthermore, sliding may induce a temperature rise of the lubricant that, in turn, produces a reduction in the lubricant viscosity, and then in Λ. Such a process may also lead to scuffing.

The experimental data in Table 5.5 show the influence of 9 % sliding in reducing the endurance limit with respect to pure rolling for different metals. The reduction is of the order of 10 %. Figure 5.13 shows the influence of sliding on the endurance limit of a 100Cr6 through hardened bearing steel (50 % probability of survival) [29]. It is seen that the reduction in endurance limit continues up to 24 % sliding (with a reduction of 35 % in the endurance limit).

As schematised in Fig. 5.10, a low Λ factor induces wear (by adhesion), whose intensity depends clearly on the amount of sliding. Sliding wear can then be calculated using Eq. 5.5. With reference to the system made by two rolling cylinders (Fig. 5.7), the evolution of the depth of wear, h, can be easily obtained:

$$h = \frac{K_{\text{lub}}}{H} \cdot \frac{F_N}{L} \cdot \frac{v_s}{v_t} \cdot N \tag{5.14}$$

where L is the length of the contact line, $v_t$ is the tangential velocity ($v_1$ or $v_2$) and N is the number of cycles (the other parameters were already defined). As an example,

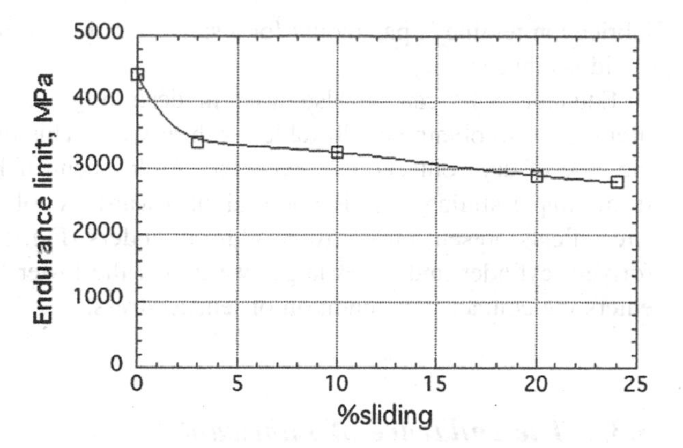


**Fig. 5.13** Influence of % sliding on the endurance limit of a 100Cr6 bearing steel (line contact; lubricant ISO VG 100 + 4 % S–P additive; 50 % failure probability) (modified from [29])

Fig. 5.14 shows the wear evolution of two rolling-sliding cylinders made with a 18NiCrMo5 carbonitrided steel (surface H: 700 kg/mm$^2$; $R_q$ = 0.88 μm) and an AISI M2 steel (H: 860 kg/mm$^2$; $R_q$ = 0.11 μm). The tests was carried out with $v_s$ = 1.27 m/s, $F_N$ = 50 N, L = 10 mm and using a mineral oil with VI = 150 as a lubricant. The two discs behave in a different way. The carbonitrided steel shows quite a long run-in stage, characterised by a high initial wear rate that progressively decreases reaching a steady state after about 5 × 10$^4$ cycles. During running in, the recorded Λ factor was about 0.13 and then typical of boundary lubrication. Using Eq. 5.14 it is obtained that $K_{lub}$ is 7.7 × 10$^{-6}$ in the first testing interval. $K_{lub}$ then decreases to 5.7 × 10$^{-7}$ at steady state, where a mixed lubrication regime is entered because of the reduction in the surface roughness. The hardest and smoother M2 disc displays negligible wear at the beginning of the test. A steady state wear, with an intensity similar to that displayed by the counterface disc in its steady state regime, is attained after about 2 × 10$^4$ cycles. Such behaviour is very similar to that described in Sect. 5.1.5, and is typical of lubricated systems (operating under sliding as well as rolling-sliding conditions). As mentioned in Sect. 5.1.5, the correct execution of the run-in stage with the transition from boundary to mixed

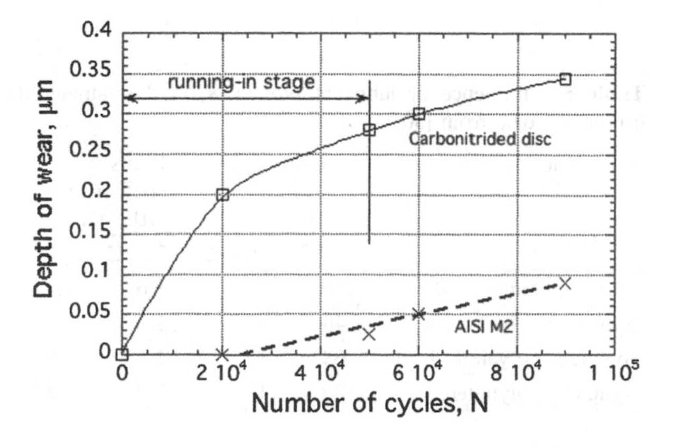


**Fig. 5.14** Wear evolution as a function of the number of cycles in a lubricated disc on disc test (see text for details)

lubrication regime is paramount for a safe operation of the system (in particular, to avoid scuffing).

Equation 5.14 can be also used in designing, i.e., in the evaluation of the lifetime, N, to obtain an allowable depth of wear. This relation can be also used in the case of dry contacts. However, the employment of K (or $K_a$) values obtained from simple sliding tests is not straightforward. A role is played by the distinct stress fields present on the two rolling cylinders (Fig. 5.7). In general, the upper (driven) cylinder undergoes larger wear than the lower (driver) one, as its surface enters the contact in a condition of tensile stress.

### *5.3.4 The Influence of Lubricant*

The lubricant properties play a critical role in determining the rolling-sliding wear resistance of a mechanical coupling. The data in Tables 5.5 and 5.6 refer to tests carried out using basic mineral oils. However, since the resistance to surface fatigue increases with the $\Lambda$ factor, it is clear that the lubricants characterized by high viscosity and high Barus coefficient (especially at the working temperature) are those that guarantee the best performance. As an example, Table 5.7 shows the results of an experiment conducted with a 4-ball tribometer, using an AISI 52100 (100Cr6) bearing steel [30]. The tests were carried out with different mineral and synthetic lubricants. The table shows the main properties of the lubricants, along with the results of the contact fatigue tests, in particular the number of cycles required to reach spalling at the Hertzian pressure of 8700 MPa (which is quite a high pressure). As expected, the fatigue life increases with increasing both the viscosity and the Barus coefficient of the adopted lubricant.

As already mentioned, specific additives are often added to the base oil to improve its performance. Indeed, the use of specific additives is very effective in the case of boundary or mixed lubrication, while in the case of fluid lubrication such operation is not useful. Due to the high pressure adopted for the tests of Table 5.7, a mixed lubrication regime was mostly attained. As a consequence, the Authors

**Table 5.7** Influence of lubricant properties on the fatigue life of AISI 52100 balls (4-ball tribometer, data from [30])

| Lubricant | Viscosity (cSt) | | Barus coefficient ($10^{-8}$ $Pa^{-1}$) | | $N_{50}$ for $p_{max}$ = 8700 MPa |
|---|---|---|---|---|---|
| | 40 °C | 100 °C | 23 °C | 80 °C | |
| Paraffinic mineral oil | 120 | 13 | 1.6 | 1.1 | $4 \times 10^5$ |
| Synthetic polyol ester | 46 | 9 | 1.1 | 0.7 | $6 \times 10^4$ |
| Synthetic polyalkil glycol | 68 | 12 | 1.55 | 0.9 | $2.5 \times 10^5$ |
| Synthetic polybutene | 130 | 13 | 2.3 | 1.45 | $6 \times 10^5$ |

observed that the use of synthetic oil with EP additive induced a longer contact fatigue life than without such an additive.

### 5.3.5 Control Methods for Rolling-Sliding Wear

The following guidelines can be useful for reducing or controlling the rolling-sliding wear of mechanical components:

(1) Reconsider the system design, verifying the possibility of reducing the contact pressure.
(2) Use materials with high hardness. Possibly use high-strength and high quality bearing steels (such as the well-known AISI 52200, or 100Cr6, steel in the hardened state); when using case hardened steels, verify that the effective thickness of the surface treated layer (see Chap. 7) is sufficiently high for the design requirements. As a rule of thumb make sure that it is greater than 2 $z_m$ (Table 1.1). Ceramic materials, such as silicon nitride, can be used in applications where high performance products are required.
(3) Use high-quality homogeneous materials, with a very low content of defects, such as inclusions or flaws.
(4) Improve the lubrication regime by increasing the $\Lambda$ factor, that is, by reducing the surface roughness of the mating surfaces and/or by increasing the thickness of the lubricant meatus (for example, by reducing the operating temperature of the lubricant or by using a lubricant with high VI index).
(5) If the $\Lambda$ factor is greater than 3, the lubrication is fluid and the rolling-sliding wear resistance is maximum. In many applications, however, it is also acceptable $\Lambda$ less than 3 and a mixed lubrication regime is attained in service (for example in gears). In this case, it is mandatory to avoid contamination by foreign particles (use of filters or barrier systems).
(6) Reduce the sliding component; this also reduces the risk of scuffing (if $\Lambda$ is less than 1).
(7) In case of boundary or mixed lubrication regime, it may be convenient to use cast iron or copper alloys against high-strength steels. Solid lubricants (typically in the form of coatings on at least one of the mating surfaces) or lubricants with additives are also recommended in these systems. In these lubrication conditions it is also necessary to check for the wear evolution in order to verify that the depth of wear is lower than an allowable value at the end of the expected component life.
(8) In the absence of lubrication, follow the guidelines proposed in Sect. 5.1.6.

## 5.4 Abrasive Wear by Hard, Granular Material

The abrasive wear by hard, granular material is generally a severe damaging process, and the relevant wear mechanism is abrasion. The hard, granular material that is responsible for the abrasive interaction can be extraneous to the tribological system, or be part of it. The particulate matter present in the atmosphere is an example of foreign particles that can contaminate a tribological system. Air typically contains 0.01 mg/m$^3$ of dust, with a maximum size between 10 and 20 μm. A large fraction of this dust is composed of silica (with a hardness of around 1000 kg/mm$^2$), which can exert an abrasive interaction on many materials, including steels. Several lubricated systems in automotive engines (like bearings, gears and the piston-ring system) may undergo severe failures if contaminated with these hard particles. In fact, the trapped particles may exert an abrasive action or prevent the correct formation of the lubricant meatus. Other foreign particles may be dirt and sand (sometimes also made of silica), which may contaminate several mechanical parts, including pumping systems or water transport ducts.

Tribological systems in which the abrasive particles are an integral part of the whole system are, for example, the crushing or grinding mills, the blades in earthmoving machines, the conveyors of minerals, the moulds in powder metallurgy, and the screws for the extrusion of polymeric composites (two examples are shown in Fig. 5.15).

In general, abrasive wear by hard, granular material increases as the following parameters increase:

- Ratio of the hardness of the abrasive material to that of the abraded surface;
- Amount of abrasive material involved;
- Size of the abrasive particles (up to a maximum of about 150 μm);
- Angularity of the abrasive particles;
- Sliding amount between the particles and the worn surface.

As an example, Fig. 5.16 schematically shows the wear rate of different materials as a function of the hardness of various mineral granules [20]. It is shown that wear

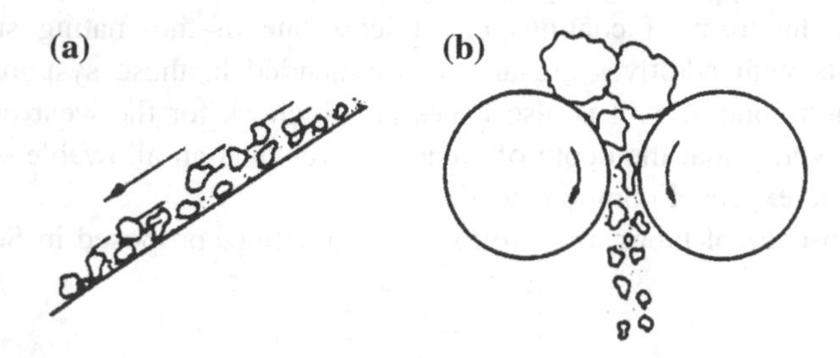


**Fig. 5.15** Examples of tribological systems that are typically damaged by abrasive wear by hard, granular materials. **a** Conveyor of granular materials (low-stress abrasion); **b** crushing mill of mineral ores (high-stress abrasion) (modified from [20])

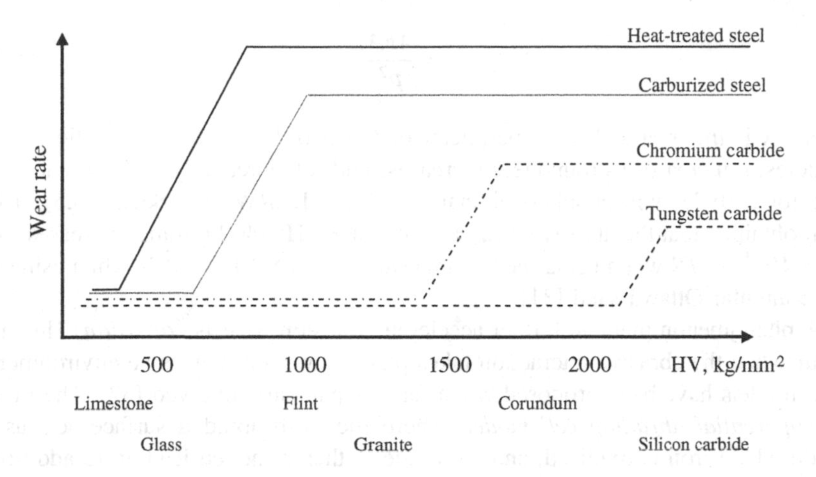


**Fig. 5.16** Schematization showing wear rate of various materials as a function of the hardness of abrading minerals (modified from [20])

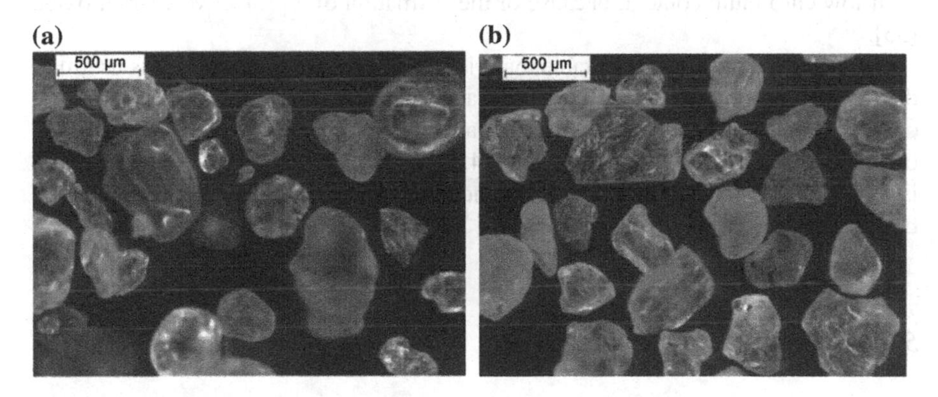


**Fig. 5.17** Stereo microscope pictures of **a** Gambia sand (κ = 0.62), and **b** Ottawa sand (κ = 0.6) [31]

rate starts to increase when the hardness of the abrasive particle is sufficiently high. In addition, wear rate is seen to decrease with the hardness of the abraded counterface.

Figure 5.17 shows stereomicroscope pictures of two types of sand. By using an image analyser interfaced with an optical microscope, it is possible to evaluate the average size of the particles and also their *shape factor*, which is a simple evaluation of their angularity. A useful parameter for the shape factor is the roundness factor, κ:

$$\kappa = \frac{4\pi A}{p^2} \tag{5.15}$$

where A is the area and p the perimeter of the two dimensional projection of the particles. κ increases as roundness increases, and it theoretically assumes the value of 1 for particles with a spherical shape. Using a DSRW test (Sect. 4.5.5), it has been obtained that the abrasive wear rate of a steel (H: 140 $kg/mm^2$) increases from $5.5 \times 10^{-13}$ $m^2/N$ when using the Gambia sand to $1.4 \times 10^{-12}$ $m^2/N$ when using the more angular Ottawa sand [31].

A phenomenon that can further accelerate abrasion wear is *corrosion*. This may occur when the abrasive interaction takes place in wet or aggressive environments. Two models have been proposed when ferrous parts are involved [32]. The first is the *differential abrading cell model*, where the fresh abraded surface acts as an anode where iron is oxidized, and the oxide is then removed leading to additional wear. The second one is the *galvanic-cell model*, where the mineral granules are cathodic and the mating steel parts are anodic. In oxygen-rich sulphide environments, steels with high chromium content were found to perform better than those with low chromium content, because of the formation of a protective Cr-rich oxide [33].

In the choice of the most suitable material for a specific application, it is useful to distinguish between *low-stress abrasion* and *high-stress abrasion* [34]. When the contact stress between the hard particles and the abraded surface is high, wear by brittle contact may be induced (see Sect. 4.3.2). On the other hand, if contact stress is low enough, this contribution can be neglected and the materials that show the best behaviour are those with the highest hardness, including ceramic materials.

### 5.4.1 High-Stress Abrasion

High-stress abrasive wear is typical of different tribological systems, including:

- Crushing and grinding components in mills (the contact stresses between the mineral particles to be processed and the surface of liners and grinding media are very high, and they induce also the fracture of the granular material);
- blades and buckets of earth-moving machines;
- tillage tools used in agriculture (in very hard soils);
- machine parts subjected to grinding operations (in this case, advantage is taken from wear in obtaining very smooth and precise surfaces).

Figure 5.18 schematically shows the high-stress abrasive wear resistance of different materials as a function of their hardness [35]. From the observation of the figure, the following general information can be achieved:

(1) Polymers. Because of their low hardness, they show a very low abrasion resistance. Thermoplastics and thermoset polymers with high hardness also

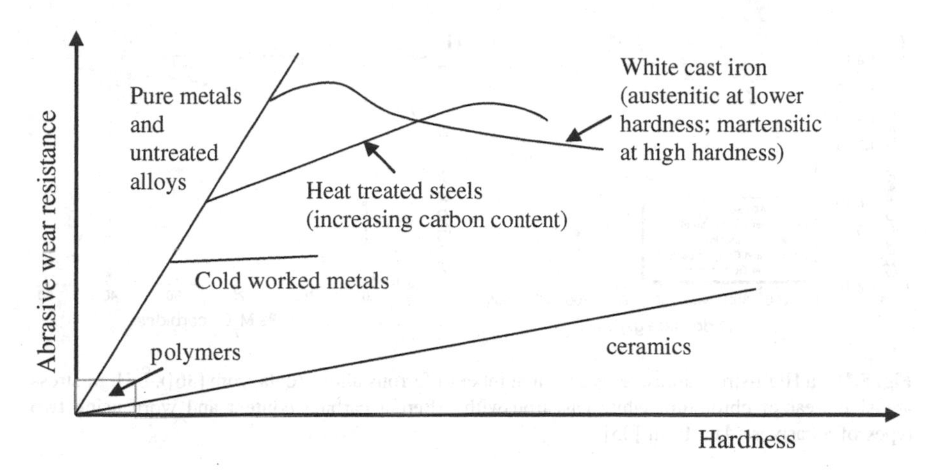


**Fig. 5.18** High-stress abrasion resistance of different classes of materials as a function of their hardness (modified from [35])

display low abrasion resistance because of their brittle behaviour. In some applications a minimum of abrasion resistance is required; the polymers that provide the best performance are high hardness polyurethanes (they are elastomers), and UHMWPE (a thermoplastic polymer).

(2) Pure metals and unhardened alloys. Such materials do not display high hardness (especially pure metals) but their abrasion resistance may be high because of their remarkable strain hardening ability that increases surface hardness during the abrasive interaction.

(3) Work-hardened metals. They possess high hardness but their abrasion resistance is less than expected. In fact, they have a limited in-service hardening capability, and also a rather low ductility.

(4) Heat-treated steels. They have high hardness and good ductility, and therefore display an appreciable abrasion resistance.

(5) Ceramic materials. They are characterized by very high hardness but their abrasion resistance is lower than expected, since it is limited by their low fracture toughness, which may favour abrasive wear by brittle fracture.

(6) White cast irons. They contain a large amount of hard carbides that are cemented together into a metal matrix that provides them with increased fracture toughness. Correspondingly, their performances may be very good.

It should be noted, moreover, that in metals abrasive wear resistance increases with hardness up to 200–300 kg/mm$^2$ and then reaches a plateau. This can be explained by considering Fig. 4.12, which shows that parameter $\Phi$ strongly increases with hardness in the range 250–400 kg/mm$^2$. As a consequence, W (compares Eqs. 4.8 and 4.9) becomes almost independent from H.

Figure 5.19a shows the results of different experiments carried out with the *Pin Abrasion Test* (PAT), which simulates the high-stress abrasion (Sect. 4.5.6), using

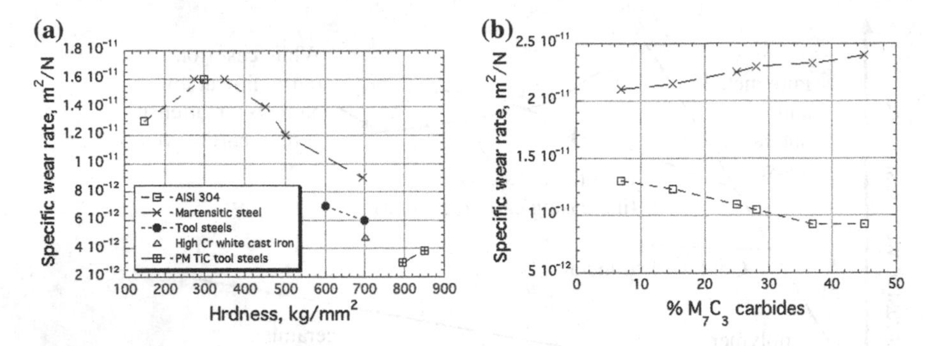


**Fig. 5.19** **a** High-stress abrasive wear of a number of ferrous alloys (data from [36]). **b** High-stress abrasive wear of chromium white cast ironswith different carbide content and worn using two types of abrasives (data from [35])

alumina particles as abrasives [36]. The tests were conducted on different ferrous materials:

- two austenitic stainless steels (AISI 304);
- a series of martensitic steels with medium carbon content, heat-treated to hardness values in the range between 300 and 700 kg/mm$^2$;
- two tool steels (AISI D2);
- a chromium white cast iron containing a high fraction of carbides;
- two tool steel obtained by powder metallurgy (PM) and reinforced with TiC particles.

In Fig. 5.19a the specific wear rates are reported, as given by the ratio between the recorded wear rates and the applied load. First of all, it can be noted that the recorded wear rates are all well above $10^{-12}$ m$^2$/N, showing that wear was very severe in every case. It is further noted that the austenitic stainless steel (AISI 304) with lower hardness displays a lower wear rate than the other one. This can be explained by considering the capability of the former steel to undergo noticeable strain hardening in the contact regions. As expected, martensitic steels show a decrease in the wear rate as their hardness increases. The tool steels have a hardness that is similar to that of the hardest martensitic steels, but their wear rates are lower. This can be explained by considering that such steels contain several hard and relatively large carbides in their microstructure, which are able to counteract the damaging action of the abrading particles. This behaviour is even more pronounced in the chromium white cast iron and in the two PM steels reinforced with very hard titanium carbides (H: 2000–3200 kg/mm$^2$). These latter results are quite general: the materials with the best resistance to high-stress abrasive wear will contain hard particles firmly embedded into a hard but sufficiently tough matrix.

It is also clear that a role is played by the dimension and amount of such hard reinforcing particles, as well as by their hardness with respect to that of the abrasive particles. This is shown by the experimental results reported in Fig. 5.19b, related to different chromium white irons with a carbide content ranging from 7 to 45 % (and a hardness between 720 and 840 kg/mm$^2$). The tests were carried out with a PAT tribometer, using two types of abrasive particles: alumina particles (hardness around 1600 kg/mm$^2$) and SiC particles (hardness around 2500 kg/mm$^2$) [35]. When using alumina particles, which have a hardness value that is comparable to that of the carbides, the increase in the carbide fraction in the white iron also increases the abrasion resistance. The abrasive action involves the martensitic matrix: carbides, as mentioned before, protect the material from excessive wear. However, when using SiC abrasive particles, a greater wear rate is obtained. In addition, wear rate was found to increase with increasing carbide content in the white iron. This behaviour can be explained by considering that cast iron carbides undergo brittle fracture under the action of the much harder SiC particles. Wear is thus greater than expected, and parameter $\Phi$ (Eq. 4.9) becomes greater than 1 since the fragmented carbides are expelled from the material during the wear process.

As already noticed, despite their high hardness ceramic materials display a lower high-stress abrasion resistance than expected. For example, it has been found that the specific wear rate under high-stress abrasion of alumina ceramics with a hardness of 1300 and 1970 kg/mm$^2$ was $4.4 \times 10^{-11}$ m$^2$/N and $8.54 \times 10^{-12}$ m$^2$/N, respectively [37]. The fracture toughness of ceramics can be improved by producing the so-called *cemented carbides* (or *hard metals*), which are made by hard ceramic particles (for example, WC particles) embedded in a ductile metal matrix (typically made of cobalt) that imparts the material a significant level of toughness. The high-stress abrasion resistance of a WC-10 vol. % Co with a hardness of 1780 kg/mm$^2$, was found to be $1.1 \times 10^{-12}$ m$^2$/N [38] (in substantial agreement with the trend displayed by the materials in Fig. 5.19a).

### 5.4.2 Low-Stress Abrasion

The low-stress abrasion wear typically occurs in:

- conveyors of mineral particles;
- sliding systems with entrapped abrasive particles;
- tillage tools used in agriculture (in normal working conditions).

Specific wear tests have shown that white irons with high carbide content, hard metals and ceramic materials can provide high resistance to low-stress abrasive wear, and are therefore excellent candidates in many applications in which this mechanism is active. In all cases, the interacting forces exerted by the hard particles are sufficiently low to avoid fracture and removal of the hard and brittle microstructural constituents. Figure 5.20 shows the results of DSRW tests for a number of materials and coatings (this type of test simulates low-stress abrasion, see Sect. 4.5. 5). The following points can be highlighted:

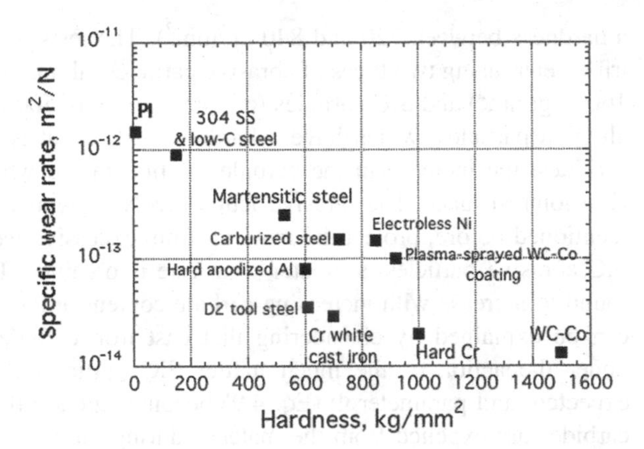


**Fig. 5.20** Low-stress abrasion behaviour of a number of materials and coatings (from different sources in the literature)

(1) The values of the specific wear rates are all much lower than those displayed in high-stress abrasion (compare data in Fig. 5.20 with those reported in Figs. 5.19).
(2) A general trend can be recognized: wear decreases as hardness is increased.
(3) As expected, materials containing hard phases in their microstructure, such as AISI D2 tool steel, chromium white cast iron, and WC-Co, display a very high abrasion resistance.
(4) Electroless nickel deposits, alumina grown on aluminium after hard anodizing, WC-Co coatings obtained by plasma spraying (all these surface engineering techniques will be described in Chap. 7), display lower wear resistance than expected, since they are usually quite defective, as they may contain pores and cracks. This shows that ceramics may provide high low-stress abrasion resistance if they are sufficiently free from embrittling defects.

### *5.4.3 Control Methods for Abrasive Wear by Hard, Granular Material*

To reduce or to keep abrasive wear under control, the following guidelines are suggested:

(a) Low-stress abrasion.

  (1) If possible, reduce or eliminate the content and size of the abrasive particles (using filters or protecting systems such as seals).
  (2) Choose materials with hardness 30 % greater than the hardness of the abrasive particles (or having a hardness that is greater than 150 kg/mm$^2$). The materials that provide the best wear resistance are, in decreasing order:

- hard metals, alumina and other ceramics;
- hard chromium plating;
- Ni-Cr or high chromium white cast irons;
- weld hard facings, made by hard metals or based on Cr-Mo-W-Nb alloys;
- high carbon, Cr-Mo martensitic steels;
- austenitic manganese steels (Hadfield steel, see Sect. 6.1).

(3) If hard particles are trapped between two mating surfaces, it may be suggested to use a hard and a soft surface; the hard particles get embedded into the soft surface preserving it from excessive wear, and they only abrade the hard counterface.

(4) Reconsider the design of the tribological system and reduce the contact loads.

(b) High-stress abrasion.

(1) Depending on the application, choose materials that combine high hardness and adequate fracture toughness. In a first instance, refer to the list of materials given above with the exception of ceramic materials. In the presence of impact loads, an increased fracture toughness is required, and the most suitable materials are Hadfield steels and martensitic steels.

(2) Since this type of wear may be particularly intense (linear wear can amount to some millimetres per day in rock crushers or mills), it is important that the components are easily replaceable and that the materials also have relatively low cost. In such cases, fine-grained HSLA steels and Hadfield steels can be a good choice.

## 5.5 Erosive Wear

Erosive wear takes place when solid particles or liquid droplets impact a surface, as schematically shown in Fig. 5.21a. In the case of *solid particle erosion* (SPE), wear damage is essentially due to the mechanism of abrasive wear. In case of *liquid droplet erosion* (LDE) and the *liquid jets erosion* (also called, *cavitation erosion*), wear is mainly due to contact fatigue, although a number of other damaging parameters may play an important and interrelated role.

### 5.5.1 Solid Particle Erosion

This wear process is characterized by the action of hard particles that are transported by a fluid and abrade a softer surface. The main tribological parameters

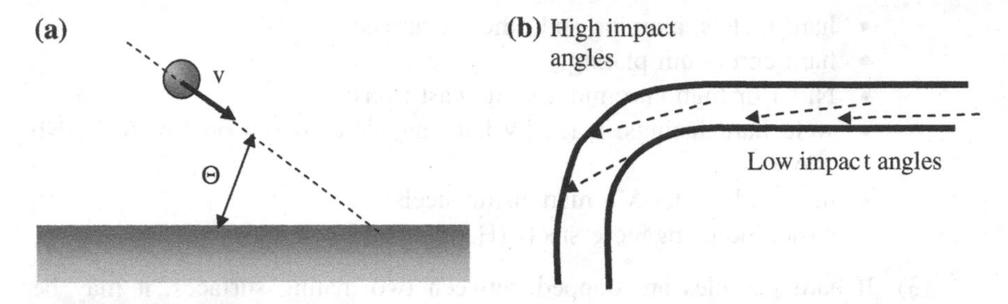


**Fig. 5.21** **a** Scheme of the erosive process: a *solid particle* is impacting a target surface with velocity, v, and an impact angle, Θ; **b** erosive interactions in a straight and bend pipe

affecting this wear process are the impact velocity and the impact (or attack) angle. Figure 5.21b schematically shows the abrasive interactions occurring in a pipe for hydraulic or pneumatic transportation. Other typical tribological systems that can face problems of solid particle erosion are cyclone separators, pump impeller blades, water turbines, transport systems for granulated material. The solid particle erosion process is also used to advantage in some technological processes, such as sand blasting, shot peening, cutting by abrasive jets. In some cases, the fluid may also exert a corrosive action, and corrosion can therefore become the predominant mechanism of surface degradation. A typical example is the erosion of a stainless steel or a steel painted surface: the surface damage, with the removal of the protective layers, may expose the metal to the aggressive action of the fluid carrying the abrasive particles.

The abrasive interaction with **ductile materials** can occur by microcutting or microploughing, as shown in Sect. 4.3.1 The wear volume, V, can be evaluated using Eq. 4.8, which can be rewritten in the following way (considering that $2\mathrm{tg}\Theta/\pi = \mu_{abr}$):

$$V = \Phi \cdot \mu_{abr} \cdot \frac{F_N \cdot s}{H} = \Phi \cdot \frac{F_T \cdot s}{H} = \Phi \cdot \frac{L}{H} \tag{5.16}$$

where L is the work done by the hard particles on the abraded surface. In the case of solid particle erosion, such a work is roughly given by ½ $mv^2$, where m is the total mass of erosive particles and v their average impact velocity. After substitution, the following relation is obtained [39]:

$$V = \Phi_{SPE} \cdot \frac{mv^2}{2H} \tag{5.17}$$

where $\Phi_{SPE}$ is a constant that is representative of the efficiency of the erosive process. It is also called *wear coefficient for erosive wear,* and varies between 0 and 1. It is theoretically 0 when the interaction is by microploughing and involves just a displacement of material, and it is 1 when all the material deformed by the particles is removed by microcutting. Therefore, the main factors that influence $\Phi_{SPE}$ are:

(1) The impact angle Θ. The erosive wear in ductile materials is expected to be maximum at small impact angles, in correspondence of which the shear stress exerted on the surface is maximum. Such a stress typically induces a considerable plastic deformation, which is incremented by repeated collisions up to the fulfilment of the critical conditions for fracture, either by low cycle fatigue or by ratcheting.
(2) Morphology of the particles. Angular particles are more effective than rounded particles to produce a microcutting interaction.
(3) The hardness of the target material. A mentioned in Sect. 4.3.1, microcutting is favoured in materials with low ductility and high hardness.

The two most used testing equipment's to directly evaluate solid particle erosion are the *gas-blast rig* and the *centrifugal erosion accelerator* (described in Sect. 5.6). The results are in general referred to the *erosion rate*, E, given by the ratio between the mass of eroded material and the mass of erosive particles striking the surface [39]. From Eq. 5.9, the following relation is therefore obtained:

$$E = \Phi_{SPE} \cdot \frac{\rho \cdot v^2}{2H} \tag{5.18}$$

where ρ is the density of the eroded material.

From a large number of experimental investigations, it has been obtained that $\Phi_{SPE}$ is quite low, and typically varies between $5 \times 10^{-3}$ and $10^{-1}$. In addition, $\Phi_{SPE}$, and thus E, were found to depend on several interrelated factors, showing that the model expressed by Eq. 5.10 is very simplified indeed. Let us consider first the role of impact velocity. If it is too low, the deformation of the target material is elastic, and wear may occur just by contact fatigue. If v is high (generally speaking, larger than 10 m/s), a very extensive and localized plastic deformation occurs, under a quite large strain rate (in the range $10^4$–$10^7$ $s^{-1}$ [40]). Such a high strain rate noticeably increases yield stress, and thus hardness, of the material, even if this effect may be partially counterbalanced by the softening induced by local thermal heating. Such a change in hardness modifies $\Phi_{SPE}$ and also renders rather unreliable the use of room temperature (static) hardness, H, in Eq. 5.10. When considering the effect of v, E is then expressed by: $E = E_0 v^n$, where $E_0$ is a constant. It has been experimentally obtained that the n-values would range between 2 and 3 in metals and polymers with a ductile behaviour, and between 3 and 5 in brittle materials [40]. As an example, Fig. 5.22a shows the erosion rate for two steels, showing the role of the impact angle and the impact velocity [41]. The $\Phi_{SPE}$–values were found in the range $6 \times 10^{-4}$–$2 \times 10^{-1}$, and the n-values were in the range 1.6–3.62. The relevant role of velocity is also obtained by comparing the data for 304L with those reported in Fig. 5.22b at room temperature [42].

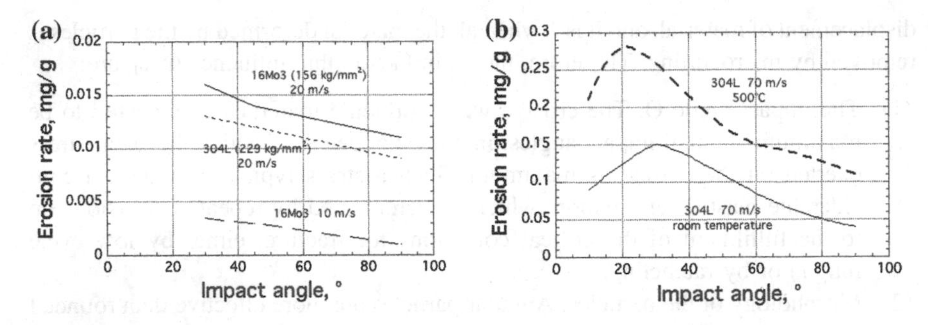


**Fig. 5.22** Erosion rate as a function of impact angle for a 304L stainless steel and a 16Mo3 steel, tested at room temperature (**a**) (modified from [41]), and a 304L stainless steel tested at room and high temperature (**b**) (modified from [42])

Another parameter that may play a role and is not considered in Eq. 5.10, is *particle feed rate*, which is the mass of eroding particles per unit area and time. As the feed rate increases, the erosive wear also increases since the probability of having local repeated contacts (and therefore plastic strain accumulations) increases too. But over a specific limit, the wear rate decreases since the arriving particles may interfere with the rebounding ones. Such a threshold was found to be around 10,000 $kg/m^2s$ in metals, and just 1000 $kg/m^2s$ for elastomers [40].

Solid particle erosion is also influenced by ambient temperature. Different systems operate at high temperature and may wear by erosion, such as heat exchangers, gas and steam turbines. In general, the erosion rate increases with increasing temperature, because of the corresponding decrease in hardness. Figure 5.22b shows the erosion rate in a stainless steel tested at room temperature and 500 °C [42]. In hardened steels, however, erosion rate may decrease if temperature rise is limited to 200–300 °C, since the hardness decrease induces an overwhelming decrease in $\Phi_{SPE}$. In stainless steels and high temperature alloys, such as Ni- or Co-alloys, the erosion rate increases continuously as temperature is increased up to 800 °C or more. In oxygen-rich environments, the metal surface may undergo direct oxidation, and erosion may thus induce a tribo-oxidative form of wear that synergically increases the overall wear rate.

Figure 5.23 shows the variation of the erosion rate with the impact angle for an Inconel 718 Ni-base superalloy tested at 538 °C using chromite particles at an impact velocity of 305 m/s [43]. The maximum wear rate is at about 20°, indicating the ductile behaviour of the material. In order to reduce the high erosion rates (compare the data with those of Figs. 5.22), a ceramic CVD titanium carbide coating was deposited on the superalloy. The results in Fig. 5.23 show that the improvement was really significant, with a reduction in the erosion rate of nearly two orders of magnitude. In addition, it is shown that the erosion rate increases with the impact angle and reaches a maximum for $\Theta = 90$ °C. This behaviour is typical of **brittle materials**, which undergo brittle contact during the abrasive interaction. A shown by Fig. 1.7, the normal component of the contacting load is responsible

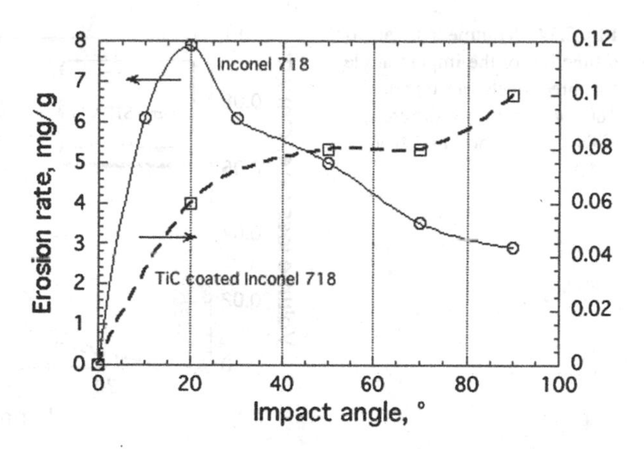


**Fig. 5.23** Erosion rate as a function of impact angle for uncoated and TiC CVD coated Inconel 718 (538 °C, 305 m/s) (modified from [43])

for the brittle contact by the Lawn and Swain mechanism that produces fragmentation by spalling.

To minimize the erosive wear it is first necessary to reduce the amount of erosive particles (by filtration, for example) and also reduce, if possible, their impact velocity (which strictly depends on the speed of the fluid that carries the particles). In the choice of the optimum material it is necessary to consider the impact angle and the impact velocity. If the impact angle is low, hard materials should be preferred, whereas when it is close to 90°, high-strength ductile materials would perform better. If possible, changing the equipment design in order to optimize the impact angle can provide a wear reduction without changing the wear resistant material. As far as the impact velocity is concerned, it has to be considered that the erosive wear at high impact velocity can be regarded as a high-stress abrasive wear, while the erosive wear at low impact velocity can be regarded as a low-stress abrasive wear. Table 5.8 lists some recommended materials for the different combinations of impact velocity and angle. Note the possibility of using elastomers in case of high impact angles and low impact velocity. Elastomers are capable of being subjected to high deformations and recover their initial shape on unloading, if the contact stress is not too high.

**Table 5.8** Recommended materials with high resistance to solid particles erosion

| | High impact angle | Low impact angle |
|---|---|---|
| High impact velocity | Martensitic steels with low and medium C content Ni-base alloys | Martensitic steels with high C content Ni-base alloys Cemented carbides |
| Low impact velocity | Elastomers | Ceramics Cemented carbides High Cr white cast irons Martensitic steels |

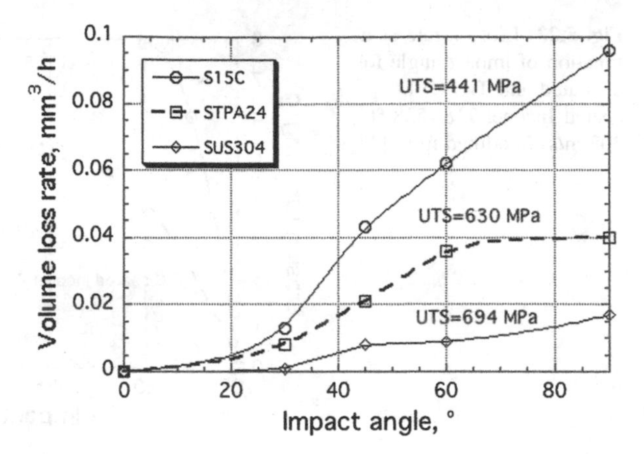


**Fig. 5.24** Volume loss rate as a function of the impact angle for three steels (jet nozzle, 180 m/s) featuring different UTS values (modified from [44])

## *5.5.2 Erosion by Liquid Droplets*

Erosive wear can be also induced by the impact of liquid droplets against the surface of a solid. Tribological systems that may display this wear process are pipes or valves used in water pipelines, blades of low pressure steam turbines (the abrasive action is due to the moisture drops) and the windows of airplanes flying in the rain.

High impact velocity (usually well above 100 m/s) induces high impact pressures due to the water-hammer effect, which may even exceed the yield strength of the impinged material. After an incubation period, wear typically increases in an almost linear wear with time. Wear damage is characterized by the formation of pits, that increase in number and depth as the exposure to the liquid impact increases. The wear mechanism resembles low-cycle contact fatigue. Therefore, materials that provide the best performance against liquid droplet erosion are those that combine high hardness with adequate fracture toughness.

Different factors influence wear rate, including flow velocity and impact angle. Figure 5.24 shows the wear rate as a function of impact angle for three steels [44]. Wear rate decreases as the ultimate tensile strength (UTS) of the target material is increased (note that UTS is generally proportional to fatigue strength). In addition, it is also observed that wear rate increases rapidly with the impact angle, and is maximum for $\Theta = 90°$. This is a general result, since only the normal component of the velocity of the impinging drops determines the impact pressure and is therefore responsible for the surface damage. Wear rate, $\dot{V}$, is often expressed by the following experimental relationship:

$$\dot{V} = K \cdot v^n \cdot \sin^m \Theta \tag{5.19}$$

where K, n and m are materials' constants. Note that a threshold velocity also exists, below which erosion does not take place.

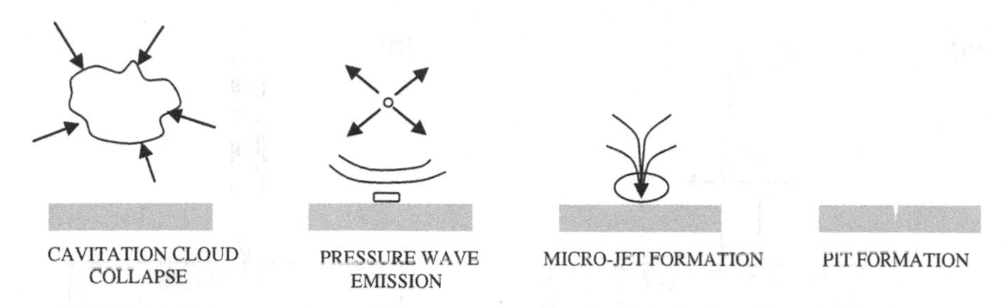


**Fig. 5.25** Events that lead to cavitation wear (modified from [45])

### *5.5.3 Cavitation Erosion*

Cavitation erosion is a particular case of erosion by liquid droplets. The damaging process is schematically shown in Fig. 5.25 [45]. Cavitation may occur in a liquid when the static pressure becomes lower than the vapour pressure of the liquid itself, and vapour clouds are therefore nucleated. Such clouds may then enter regions with higher pressure, and thus collapse emitting a shock wave in the liquid. One or more bubbles present near the solid surface are then reached by the wave, and start to oscillate, thus forming liquid micro-jets that impact at high velocity the solid surface. The impact pressure may be very high because of the water-hammer effect. After an incubation period, surface damage begins to occur with the formation of pits that increase in number and depth with time.

Regions with low and high static pressure may arise following an intense variation in the liquid velocity, such as when a liquid flows in a Venturi section. They may also form between surfaces under high-frequency oscillatory motion. During recession, a decrease in pressure is observed, whereas during the approach phase the local pressure increases. This wear process is typically encountered in turbine blades, pumps and high-speed lubricated bearings. Besides wear, cavitation produces vibration, loss of efficiency and noise.

## 5.6 Process-Oriented Wear Tests

In this paragraph some special laboratory testing devices are described, which are commonly used to simulate particular wear processes. In most cases, such testing rigs provide results that are in agreement with those produced by the tribometers of Table 4.1, but in other cases the results may be different since the basic tribometers may not be able to account for all particular testing conditions that may play an important role in wear dynamic.

In addition to the common pin-on-disc and block-on-ring testing rigs, the sliding wear process may be also investigated using the *reciprocating pin on a plate test*, schematized in Fig. 5.26a (ASTM G133). It is typically used to investigate sliding

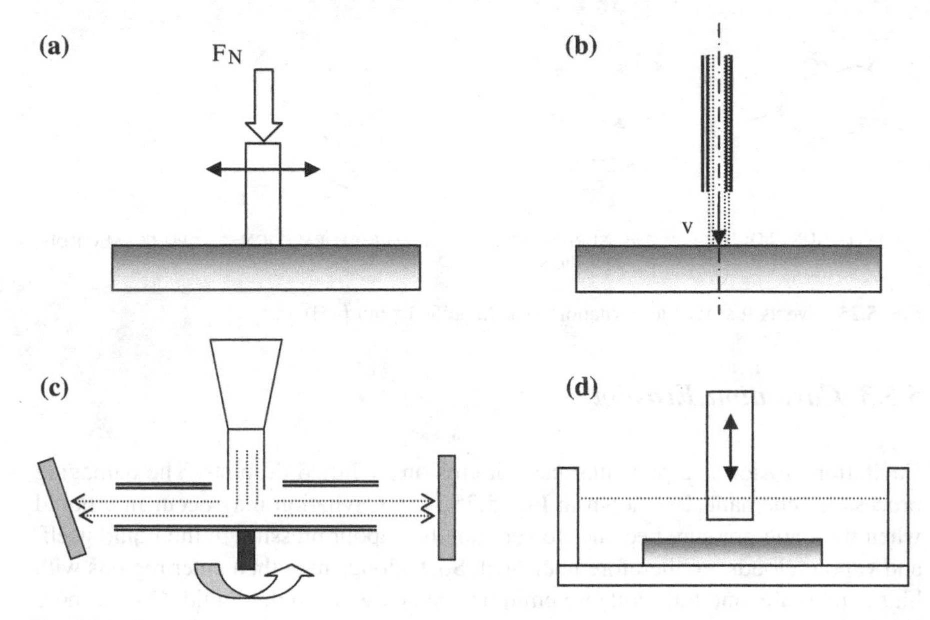


**Fig. 5.26** Schematic configuration of different process-oriented wear tests. **a** Reciprocating pin on plate; **b** gas blast erosion test; **c** centrifugal erosion accelerator; **d** vibratory apparatus for cavitation erosion

wear when the wear debris has to remain trapped in the contact region. It is also used to simulate fretting wear, by maintaining the amplitude of motion lower than 250 μm.

Wear by solid particle erosion can be investigated using the *gas-blast erosion test*, schematized in Fig. 5.26b. The abrading particles are accelerated by a high-pressure air stream, directed onto the specimen target (in a similar test device the abrasive particles are carried by a liquid jet). Adjusting geometrical and working parameters can vary the impact velocity and the impact angle. Figure 5.26c shows a schematized side view of a *centrifugal erosion accelerator*. It uses rotation to accelerate the particles against more than one specimen located in the rim perimeter and fixed at the desired impact angle.

The liquid impingement erosion can be investigated using a test apparatus that is in principle very similar to that schematized in Fig. 5.26b. A jet nozzle is used and liquid water, rather than a gas, can be directed against the specimen with a given velocity. The velocity is set by adjusting the water pressure in the inlet of the nozzle.

Figure 5.26d finally shows the configuration of a test apparatus for cavitation erosion. A cyclic variation of pressure in the liquid between two solid surfaces is induced by using a high-frequency oscillatory drive. Pressure is increased during the approach phase and it is decreased during the recession stage.

## References


1. K.H. Czichos, K.H. Habig, *Tribologie Handbook* (Reibung und Verlschleiss, Vieweg, 1992)
2. P.J. Blau, *Friction and Wear Transitions of Materials* (Noyes Publications, Park Ridge, 1989)
3. G. Straffelini, A. Molinari, Mild sliding wear of Fe-0.2 % C, Ti-6%Al-4%V and Al-7072: A comparative study. Tribol. Lett. **41**, 227–238 (2011)
4. S. Wilson, A.T. Alpas, Thermal effects on mild wear transitions in dry sliding of an aluminium alloy. Wear **225–229**, 440–449 (1999)
5. G. Straffelini, D. Trabucco, A. Molinari, Sliding wear of austenitic and austenitic-ferritic stainless steels. Metall. Mater. Trans. **33A**, 613–624 (2002)
6. A.F. Smith, The influence of surface oxidation and sliding speed on the unlubricated wear of 316 stainless-steel at low load. Wear **105**, 91–107 (1985)
7. S.C. Lim, M.F. Ashby, Wear-mechanism maps. Acta Metall. Mater. **35**, 1–24 (1987)
8. R.N. Rao, S. Das, D.P. Mondal, G. Dixit, S.L. Tulasi DEVI, Dry sliding wear maps for AA7071 (Al-Zn-Mg-Cu) aluminium matrix composite. Tribol. Int. **60**, 77–82 (2013)
9. J. Zhang, A.T. Alpas, Transitions between mild and severe wear in aluminum alloys. Acta Mater. **45**, 513–528 (1997)
10. M.M. Khonsari, E.R. Booser, *Applied Tribology—Bearing Design and Lubrication*, 2nd edn. (Wiley, New York, 2008)
11. A. Sethuramiah, *Lubricated Wear, Science and Technology* (Elsevier, New York, 2003)
12. S.E. Franklin, J.A. Dijkman, The implementation of tribological principles in an expert system (precept) for the selection of metallic materials, surface treatments and coatings in engineering design. Wear **181–183**, 1–10 (1995)
13. J.E, D.T. Gawne, Surface failure mechanisms of Ni-Cr-Mo steel under lubricated sliding. Wear **213**, 123–130 (1997)
14. O. Soderberg, D. Vingsbo, On fretting maps. Wear **126**, 131–147 (1988)
15. S. Fouvry, Ph. Kapsa, L. VINCENT, Quantification of fretting damage. Wear **200**, 186–205 (1996)
16. B. Bushan (ed.), *Modern Tribology Handbook,* Vol. 1. (CRC Press, Boca Raton, 2001)
17. S.E. Kinyon, D.W. Hoepper, Y. Mutoh (eds.), *Fretting Fatigue Advances in Basic Understanding and Applications, STP1425*. (ASTM International, West Conshohocken, 2003)
18. K. Endo, Practical observations of initiation and propagation of fretting fatigue cracks, in ed. by R.B. Waterhouse. *Fretting Fatigue*, (Applied Science Publisher, 1981), pp. 127–141
19. R.B. Waterhouse, Fretting fatigue. Int. Mater. Rev. **37**, 77–97 (1992)
20. M.B. Peterson, W.O. Winer (eds.), *Wear Control Handbook* (ASME, Fairfield, 1980)
21. H.K.D.H. Bhadeshia, Steels for bearings. Prog. Mater. Sci. **57**, 268–435 (2012)
22. N.A. Fleck, K.J. Kang, M.F. Ashby, The cyclic properties of engineering materials. Acta Metall. Mater. **42**, 365–381 (1994)
23. R.G. Bayer, *Wear Analysis for Engineers*, (HNB Publishing, New York, 2002)
24. Y. Kadin, A.V. Rychahivskyy, Modeling of surface cracks in rolling contact. Mater. Sci. Eng. A **541**, 143–151 (2012)
25. G. Donzella, C. Petrogalli, A failure assessment diagram for components subjected to rolling contact loading. Int. J. Fatigue **32**, 256–268 (2010)
26. G. Niemann, H. Winter, *Maschinen-Elemente, Band II*, (Springer, Berlin, 1986)
27. P. Clayton, Tribological aspects of wheel-rail contact: a review of recent experimental research. Wear **191**, 170–183 (1996)
28. J. Braendlein, P. Eschmann, L. Hasbargen, K. Weigand, *Ball and Roller Bearings, Theory, Design and Application*, 3rd Edn. (Wiley, New York, 1999)
29. G. Hoffman, K. Lipp, K. Michaelis, C.M. Sonsino, J.A. Rice, Testing P/M materials for high loading gear applications. Int. J. Powder Metall. **35**, 35–44 (1999)
30. Y. Wang, J.E. Fernandez, D.G. Cuervo, Rolling-contact fatigue lives of steel AISI 52100 balls with eight mineral and synthetic lubricants. Wear **196**, 110–119 (1996)

31. M. Woldman, E. Van Der Heide, D.J. Schipper, T. Tinga, M.A. Masen, Investigating the influence of sand particle properties on abrasive wear behaviour. Wear **294–295**, 419–426 (2012)
32. I. Iwasaki, R.L. Pozzo, K.A. Natarajan, K. Adam, J.N. Orlich, Nature of corrosive and abrasive wear in ball mill grinding. Int. J. Miner. Process. **22**, 345–360 (1988)
33. C. Aldrich, Consumption of steel grinding media in mills—a review. Miner. Eng. **46**, 77–91 (2013)
34. M.J. Neale, M. Gee, *Guide to Wear Problems and Testing for Industry*, (Professional Engineering Publishing, London, 2000)
35. K.H. Zum Gahr, *Microstructure and Wear of Materials*, (Elsevier, New York, 1987)
36. J.A. Hawk, J.H. Tylczak, R.D. Wilson, An assessment of the abrasive wear behavior of ferrous alloys and composites using small scale laboratory wear tests, in *Proceedings from Materials Solutions '97 on Wear of Engineering Materials*, ASM (1997)
37. C.P. Dogan, J.A. Hawk, Role of composition and microstructure in the abrasive wear of high-alumina ceramics. Wear **225–229**, 1050–1058 (1999)
38. A.Y. Mosbah, D. Wexler, A. Calka, Abrasive wear of WC-FeAl composites. Wear **258**, 1337–1341 (2005)
39. I.M. Hutchings, *Tribology*, (Edwald Arnold, London, 1992)
40. R. Kaundal, Role of process variables on the solid particle Erosion of polymeric composites: a critical review. Silicon **5**, 5–20 (2014)
41. E. Huttunen-Saarivita, H. Kinnune, J. Tuiremo, M. Uusitalo, M. Antonov, Erosive wear of boiler steels by sand and ash. Wear **317**, 213–224 (2014)
42. G. Sundararajan, M. Roy, Solid particle erosion behaviour of metallic materials at room and elevated temperatures. Tribol. Int. **30**, 339–359 (1997)
43. W. Tabakoff, V. Shanov, Protection of coated superalloys from erosion exposed to particulate flows, in *Proceedings from Materials Solutions '97 on Wear of Engineering Materials*, ASM (1997)
44. S. Hattori, M. Kakuichi, Effect of impact angle on liquid droplet impingement erosion. Wear **298–299**, 1–7 (2013)
45. M. Dular, B. Stoffel, B. Sirok, Development of a erosion model. Wear **261**, 642–655 (2006)

# Chapter 6
# Materials for Tribology

Different engineering parts are subjected to tribological loading, such as sliding at different loads and speeds, abrasive interactions with hard particles or repeated contact stresses. Therefore, it is necessary to select materials according to specific engineering requirements, and then to verify their ability to perform adequately under the acting wear conditions. In this respect, for applications with demanding tribological loadings, materials with specially designed tribological properties have been developed.

In the previous chapter, indications for the use of materials that better resist different wear processes were given. This chapter provides a more detailed overview of the engineering materials that are more frequently employed in tribological applications. For each class of materials, an outline of the relevant behaviour with respect to sliding wear, characterized by the mechanisms of adhesive and tribo-oxidative wear, wear by contact fatigue, and abrasive wear by hard granular material, will be provided. Specific surface treatments may be also adopted to improve the tribological properties of materials; they will be considered in the next chapter.

## 6.1 Steels

Steels are widely used in tribological applications, in like-on-like couplings or in combination with other materials. Particular steel grades have been developed for specific applications. Some examples are:

(1) tool steels;
(2) bearing steels;
(3) Hadfield steels (austenitic steels with high abrasion resistance);
(4) martensitic stainless steels for applications in aggressive environments.

By varying the chemical composition and/or by appropriate heat treatments (including surface treatments), it is possible to obtain steels with a wide a range of



G. Straffelini, *Friction and Wear*, Springer Tracts in Mechanical Engineering,
DOI 10.1007/978-3-319-05894-8_6

**Table 6.1** Surface phenomena that can occur at a steel surface during tribological interaction (modified from [2])

| Phenomenon | Type of steel | Strain induced | Temperature induced |
|---|---|---|---|
| (1) Stress relieving | All | – | X |
| (2) Recristallization | All | X | X |
| (3) Tempering | Martensitic | – | X |
| (4) Strain hardening | All | X | – |
| (5) Strain-induced martensitic transformation | • Hadfield steel<br>• Some austenitic stainless steels<br>• Martensitic steels containing retained austenite | X | – |
| (6) Thermal martensitic transformation | Quench hardenable steels | – | X |
| (7) Precipitation hardening | High alloy steels | – | X |
| (8) Oxidation | All | – | X |
| (9) Melting | All | – | X |

properties, as concerns mechanical strength and fracture toughness [1, 2]. In particular, hardness values up to 800–1000 kg/mm$^2$ can be attained.

During the tribological interactions, plastic deformations and energy dissipation with related temperature rise usually take place at the contacting asperities. In steels, both phenomena may produce local microstructural changes that are very important from the viewpoint of tribology. They are summarized in Table 6.1. The phenomena numbered from 1 to 3 all cause a reduction in hardness at the contacting asperities, while a material hardening is induced by the phenomena from 4 to 7. Oxidation (no. 8 in the table) induces the formation of a surface oxide layer, which, under particular conditions, may decrease friction and wear. Melting (no. 9 in the table) may only occur under particularly severe sliding conditions.

### 6.1.1 Sliding Wear

In case of mild wear steels display specific wear coefficients ($K_a$) typically ranging from $10^{-15}$ to $10^{-14}$ m$^2$/N. For adhesive wear, $K_a$ takes much higher values, typically around $10^{-12}$ m$^2$/N. The iron oxides have a hardness in the range 300–600 kg/mm$^2$ that strongly decreases as temperature is increased [3]. Therefore, the use of a steel with high hardness is very important. In fact, a suitably hard steel would be particularly capable to support the oxide layer and, in addition, not to be abraded by the oxide fragments during sliding. This is well shown in Fig. 6.1a, for a C60 steel (a non-alloy steel containing 0.6 % C) after different thermal treatments, like: normalizing, to obtain a hardness of 200 kg/mm$^2$; quenching and tempering to

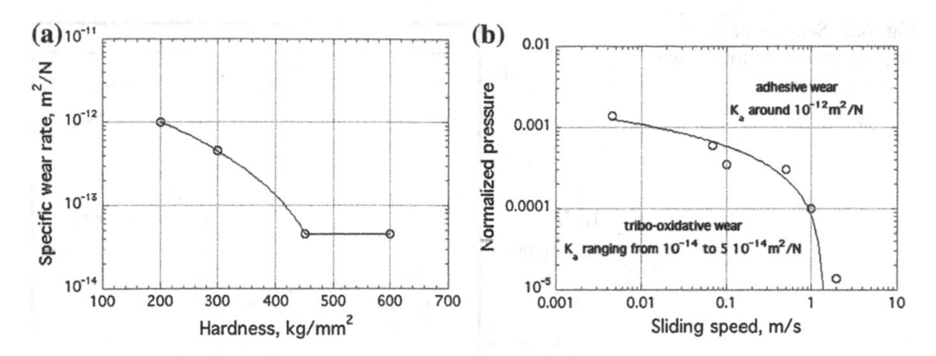


**Fig. 6.1** **a** Specific wear rate as a function of hardness for a C60 steel, subject to dry sliding with $p_0 = 1.3$ MPa and v = 0.1 m/s. The steel hardness was changed by adopting suitable heat treatments (modified from [4]). **b** Wear map for steels, in case of relatively low sliding speeds and applied pressures (from different literature references)

obtain a hardness of 300 and 450 kg/mm$^2$; and quenching and stress relief to obtain a hardness of 600 kg/mm$^2$. When hardness exceeds 450 kg/mm$^2$ wear is mainly by tribo-oxidation and the corresponding $K_a$-values are typical of mild wear. It is worth recalling that this process is fully effective when fragments are allowed to remain in the contact region.

In order to be able to compare the behaviour of materials with different hardness values, the concept of *normalized pressure* has been introduced. It is defined as the ratio between the nominal pressure and the materials' hardness (that of the softer one, if the mating materials are different):

$$\bar{p} = \frac{F_N}{A_n H} = \frac{p_0}{H} \tag{6.1}$$

For example, the data in Fig. 6.1a show that at v = 0.1 m/s the transition between adhesive and tribo-oxidative wear occurs for hardness values between 400 and 450 kg/mm$^2$, i.e., for a normalized pressure about $3 \times 10^{-4}$. Using this information and additional data taken from the literature (referring to pin-on-disc tests with the fragments able to remain in the contact region), it is possible to construct the wear map shown in Fig. 6.1b. It can be observed that at low sliding speeds, typically lower than 0.1 m/s, the boundary between tribo-oxidative (mild) wear and adhesive (severe) wear is mainly dependent on the normalized pressure. In fact, it is determined by the ability of the steel (in relation to the applied pressure) to support the layer of protective oxides forming on its surface. For sliding speeds in excess of 0.1 m/s, the frictional heating gains in importance and the boundary between tribo-oxidative and adhesive wear is controlled both by normalized pressure and sliding speed, i.e., by the attainment of a critical surface temperature, at which softening is so intense to prevent the steel substrate to properly sustain the surface oxide layer (item 1 in Table 6.1).

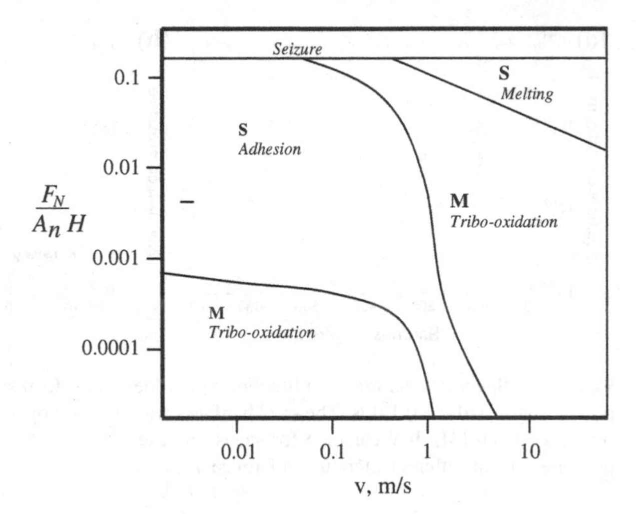


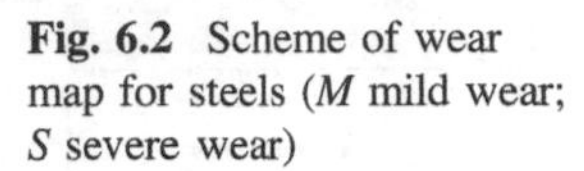

**Fig. 6.2** Scheme of wear map for steels (*M* mild wear; *S* severe wear)

Using the data collected by Lim and Ashby [5], it is possible to construct a complete wear map for steels under dry sliding. It is schematically shown in Fig. 6.2. It may be noted that for high normalized pressures (greater than about 0.2), *seizure* occurs regardless of the sliding speed. Under these conditions, the actual area of contact approaches the nominal one, and wear (by adhesion) becomes extremely severe, with intense material transfer. A new region is introduced at high sliding speeds and high normalized pressures, i.e., *melting wear*. In this region, frictional heating is so high to induce melting of the asperities, and wear is consequently severe. At high sliding speeds (greater than 1 m/s) and lower normalized pressure, melting cannot occur and (mild) tribo-oxidative wear by direct oxidation occurs (Sect. 4.2.1). If the sliding speed is between 1 and 10 m/s, martensite can form in the contact regions during sliding (item 5 in Table 6.1). Such a formation is a consequence of the attainment of high flash temperatures that induce local austenitization, and by the following rapid cooling caused by the intense heat removal exerted by the bulk material. Such a martensite layer (which is sometimes called *white layer* because of its appearance after metallographic etching), helps supporting the surface oxide layer, counteracting the thermal softening. For particularly high speeds (v > 10 m/s), the oxide layer becomes very thick and it plastically spreads onto the contact surface, thereby contributing to an efficient dissipation of heat and thus preventing the attainment of high average surface temperatures.

To ensure the best performance in dry sliding, steels should then possess high hardness and they should also be able to maintain a relatively high hardness even with increasing contact temperature. The best candidates for dry sliding applications are therefore:

- nitrided steels;
- tool steels;
- carburized steels;

**Table 6.2** Chemical compositions and typical mechanical properties of some tool steels

| Type of steel | Hardness (kg/mm$^2$) | Charpy-V impact energy (J) | Material (AISI code) | Typical chemical composition | | | | | |
|---|---|---|---|---|---|---|---|---|---|
| | | | | % C | % Mn | % W | % Cr | % Mo | Other |
| Cold work tool steel | 650–850 | 3 | O1 | 0.9 | 1 | 0.5 | 0.5 | – | – |
| | | | D2 | 1.5 | 0.3 | – | 12 | 1 | 1 % V |
| Hot work tool steel | 400–700 | 15 | H11 | 0.35 | – | – | 5 | 1.5 | 1 % Si, 0.4 % V |
| | | | H13 | 0.35 | – | – | 5 | 1.5 | 1 % V |
| High-speed steel | 800–1000 | 8 | T15 | 1.5 | – | 12 | 4 | – | 5 % V<br>5 % Co |
| | | | M2 | 0.8 | – | 6 | 4 | 5 | 2 % V |

- heat treated (martensitic) steels;
- bearing steels.

Nitrided and carburized steels will be described in the next chapter. Here the attention will be firstly focussed on *tool steels*, which have been specially developed for the production of processing tools, such as dies and moulds, that are intensively subjected to sliding wear (mainly under dry or boundary lubricated conditions). Tool steels are classified into three groups: *cold work* steels, *hot work* steels, and *high-speed* steels. Table 6.2 shows the chemical compositions, typical hardness and impact fracture energy values of some common tool steels (after quenching and tempering). The steels are indicated according to the AISI (American Iron and Steel Institute) code.

After machining and heat treatment, tool steels will generally have a microstructure comprising a martensitic matrix and a carbide dispersion of the added elements. As an example, Fig. 6.3 shows the microstructure of two widely used steels, AISI D2 and AISI M2. Cold work tool steels (series W, A, O, and D) have a

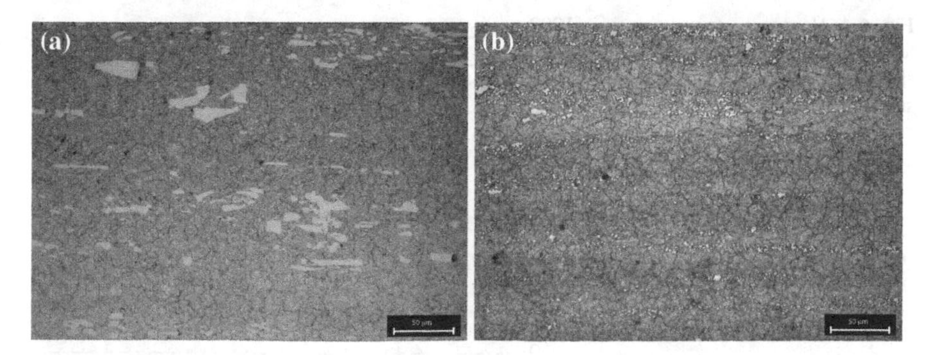


**Fig. 6.3** Microstructures of the AISI D2 (**a**) and AISI M2 (**b**) tool steels. Note the presence of quite large primary carbides in AISI D2

high carbon content (>1 %), and contain other alloying elements such as chromium, tungsten and manganese. Chromium, which is particularly present in steels of D series, and tungsten form relatively large primary carbides (typically around 10 μm in size), while manganese is added to increase matrix hardenability. These steels are fairly cheap and have a maximum operating temperature of 200–300 °C. Hot work tool steels have a low carbon content and correspondingly display a high fracture toughness. If tempered at about 600 °C after quenching, they acquire a high hardness, due to the precipitation of fine carbides (*secondary hardening*). Because of this, they are able to maintain a high hardness up to temperatures of about 540 °C. Finally, high-speed steels feature a high carbon content, and other alloying elements such as tungsten, chromium and vanadium. They are characterized by high hardness, retained even at high temperatures, as they also display a secondary hardening. Fracture toughness of high-speed steels is clearly lower than that of hot work tool steels. Some high-speed steels also contain cobalt, which provides an increased hardness at high temperatures.

In order to obtain mild tribo-oxidative wear in dry sliding, it is suggested to use steels containing small and uniformly dispersed carbides, to avoid any possible abrasive action from the largest carbides. This is also important in case of sliding under boundary lubrication conditions.

Figure 6.4 schematically shows the dependence of hardness on the environmental temperature, in the case of hot work tool steels and high-speed steels. Note that even the AISI D2 cold work tool steel is able to maintain a hardness of about 400 kg/mm$^2$ up to 600 °C, if it has been previously quenched and tempered to a hardness of about 700 kg/mm$^2$.

Heat-treated martensitic steels may achieve quite high hardness values (typically up to 800 kg/mm$^2$), depending on carbon content, alloying elements and heat-treatment cycle. Steels with a high amount of alloy elements may contain some *retained austenite* (RA) in their microstructure after the heat treatment. The role of RA on the sliding resistance of steels (as well as on their contact fatigue behaviour) has not yet completely clarified. It is believed that if RA is unstable (i.e., it contains

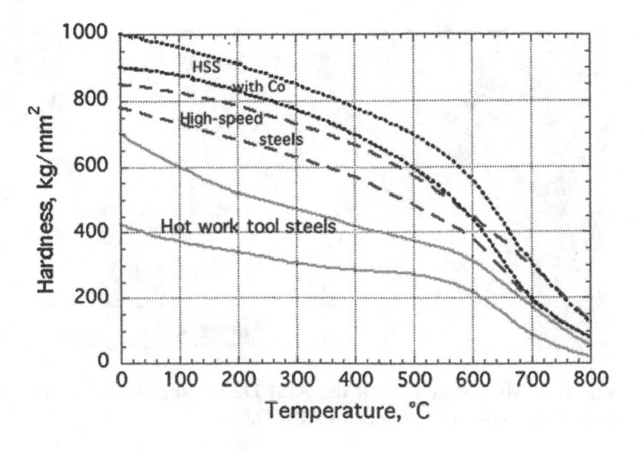


**Fig. 6.4** Hardness versus temperature for hot work tool steels and high-speed tool steels

a small amount of alloying elements) and can then easily transform into martensite by surface shear deformation, its role is positive in that it induces a local strength increase (the amount of RA, however, has to be controlled to avoid excessive distortions associated with the austenite to martensite transformation). If RA is stable, however, it does not harden during sliding, and it may become a weak microstructural region, where an intense plastic deformation can concentrate.

In applications where high fracture toughness and ductility are required, or when thermal treatment is difficult to be performed, martensitic steels are not suitable, and *pearlitic steels* may be preferable. Due to the intense surface plastic deformation during sliding, the cementite lamellae of these steels are oriented in the direction of sliding, significantly increasing the mechanical strength and thus the wear resistance. Typical pearlitic steels contain 0.7 % carbon and 1.5 % manganese that reduces the pearlite interlamellar spacing. Pearlitic steels with a hardness between 300 and 350 kg/mm$^2$, display a sliding wear resistance that is comparable to that of bainitic and martensitic steels [6].

### 6.1.2 Wear by Contact Fatigue

The steels that guarantee the best resistance against contact fatigue are the so-called *bearing steels*, which are specially developed to produce the rolling elements and raceways of bearings. These steels are characterized by high hardness, excellent dimensional stability in service, and high degree of homogeneity and microstructural quality. These steels are produced with a very low content of inclusions, such as oxides, sulphides and nitrides, and a high surface finish. As said, the reduction in concentration and dimension of the inclusions decreases the probability that an inclusion of a critical size is found in the most stressed area, i.e., close to the distance $z_m$ from the surface (Table 1.1). Steels with extreme cleanliness are now produced using special secondary metallurgical technologies, such as vacuum induction melting (VIM) and vacuum arc remelting (VAR).

The most common bearing steel is the *AISI 52100* (also indicated with 100Cr6), which contains 1 % carbon and 1.45 % chromium [7]. This steel is quenched and tempered at about 160 °C to ensure the obtainment of high hardness, typically between 650 and 1000 kg/mm$^2$. To increase its dimensional stability, the steel is often subjected to a soaking at −80 °C before tempering, aiming at completely eliminate residual austenite present after quenching (usually about 6 %). It is clear that the in-service temperature should not exceed 200 °C to avoid microstructural softening. In order to obtain higher surface hardness values, and thus higher contact fatigue resistance, specific steels may be exposed to a *carburizing* (or carbonitriding) treatment, obtaining a graded microstructure with a carbon enrichment at the surface and consequent introduction of compressive residual stresses. The M50 NiL steel (typically containing 0.13 % C, 3.5 % Ni, 4 % Cr, 4.25 % Mo, 1.2 % V) is commonly used in the carburized state, with a surface hardness of about 750 kg/mm$^2$ and a hardness greater than 550 kg/mm$^2$ for a depth of at least 1.5 mm [8].

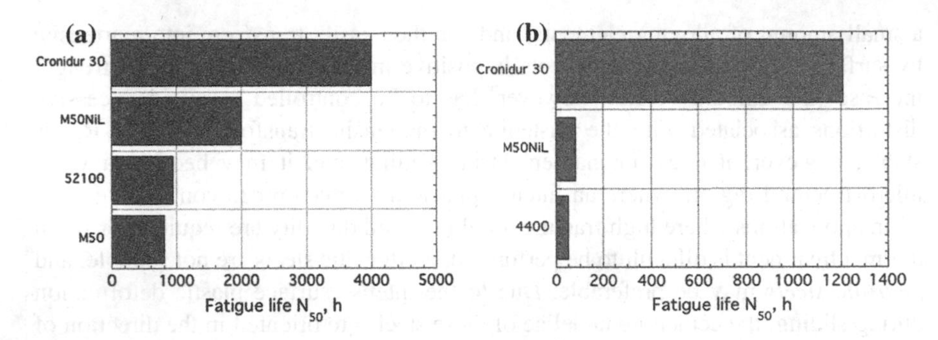


**Fig. 6.5** Contact fatigue life of different high strength steels tested in (**a**) fluid film lubrication, and in (**b**) boundary lubrication. The Hertzian pressure was 2800 MPa and the oil temperature was maintained at 90 °C (**a**) and 95 °C (**b**) (modified from [9])

Surface treated steels retain a good fracture toughness in the bulk and are thus very suitable for engineering applications that require both high contact fatigue resistance and high fracture toughness. In some applications, such as in advanced aerospace engines, wind turbines and high-speed railways, a further increase in hardness and fatigue strength may be required. It can be achieved by using *high nitrogen martensitic steels*, where the sum of the carbon and nitrogen content is tuned between 0.6 and 0.8 %. Figure 6.5 shows the results of bearing life tests carried out on several high-strength steels with a rather high contact fatigue resistance (Cronidur 30, X30CrMoN15-1, containing 0.38 % N, 15 % Cr and 1 % Mo; it is used in the martensitic state) [9].

In case of ambient temperatures exceeding 200 °C, other types of steels are required, such as the M series of the high-speed steels (e.g., AISI M2) that display the secondary hardening phenomenon. In the presence of corrosive environments (commonly encountered, for example, in aircraft engines and in the paper industry), martensitic stainless steel containing at least 17 % chromium are employed. An example is the AISI 440C, with nominal chemical composition: 1 % C, 0.4 % Mn, 0.3 % Si, 17 % Cr, 0.5 % Mo. With a suitable heat treatment, it can reach a hardness level similar to that of tool steels. Because of the presence of carbon, however, this steel also contains large eutectic carbides in its microstructure, which may reduce corrosion resistance and also fatigue strength. For these reasons, steels with lower chromium content may be used instead. When using the abovementioned high nitrogen martensitic steels (containing 13 % Cr approx.), the formation of large eutectic carbides is avoided and the corrosion resistance, in specific environments, in maintained [7].

### 6.1.3 Abrasive Wear by Hard, Granular Material

According to what reported in Sect. 5.4, the steels with the best abrasive wear resistance are those with a microstructure constituted by high carbon martensite and hard and large carbides. It should be noted that martensitic steels (without large carbides) can reach maximum hardness values of about 1000 kg/mm$^2$, inadequate for many abrasive particles, including silica particles that are the most common natural abrasives. The presence of large and hard carbides in the microstructure is then paramount for steels to gain an excellent resistance to abrasive wear (see, for example, the behaviour of the AISI D2 tool steel in Fig. 5.19a).

Steels are commonly used in applications where abrasive wear is present, because they can retain some ductility even when they possess high hardness, and also because of their relatively low cost. The most commonly used steels are, in descending order [10]:

(1) Cold work tool steels and high-speed steels, containing 1–1.6 % C (hardness about 800–900 kg/mm$^2$);
(2) Hot work tool steels, containing about 1 % C (hardness about 600–900 kg/mm$^2$);
(3) Martensitic steels, containing about 0.7 % C and possibly other alloying elements to increase their hardenability (hardness about 600 kg/mm$^2$);
(4) Pearlitic Cr-Mo steels containing 1 % C (hardness about 350 kg/mm$^2$ only, but greater ductility than tool steels);
(5) Austenitic manganese steels, like the *Hadfield steels* containing 12–14 % Mn and 1 % C. These steels have initial hardness values around 200 kg/mm$^2$, but due to the strain-induced martensitic transformation (item 5 in Table 6.1) they achieve hardness values in the range 400–500 kg/mm$^2$ in the contact region.

The *austenitic manganese steels*, such as the Hadfield steels, are used if high fracture toughness is required, for example when the mechanical parts have to stand impact loads. These steels are therefore mainly used in applications characterized by high-stress abrasion wear. Since they are quite difficult to machine by chip removal (due to the hardening effect during cutting), they are mainly used in the as-cast state, or as deposited surface layers. Manganese steels harden during plastic deformation because of the shear-induced transformation of austenite into martensite. If such a transformation can take place during the tribological interaction, the abrasion resistance of these materials increases dramatically. The martensitic transformation is favoured if the austenite is unstable, that is, if relatively low concentrations of carbon and manganese are present [11]. It should be noted that manganese steels contain a certain amount of carbides (iron and chromium carbides) in their microstructure after solidification. Therefore, they should be submitted to a solution treatment at about 1050 °C before being placed in service in order to obtain a fully austenitic microstructure.

Figure 6.6 summarizes, in a simplified way, the abrasive wear resistance of various ferrous alloys, including white irons that will be discussed in the next section.

In many applications, abrasion occurs in the presence of an aggressive environment. When the environment is very aggressive and the abrasive action is limited, austenitic stainless steels can be used: next to a very high corrosion resistance in specific environments, they can ensure a sufficient abrasion resistance

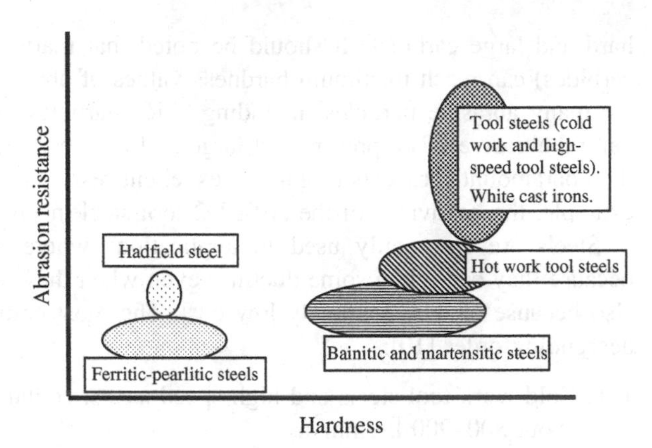


**Fig. 6.6** Schematization showing the abrasion resistance of different ferrous alloys (modified from [12])

exploiting their ability to work harden during deformation (see Fig. 5.19a). In other applications, however, higher hardness may be required, and the martensitic stainless steels, such as AISI 440, should be employed.

## 6.2 Cast Iron

Cast irons are Fe–C alloys with a carbon content between 2 and 4 % and a silicon content between 1 and 3 %. They are very suitable for the production of castings and are then commonly used for manufacturing particular mechanical components (including large parts or components with very complex shapes). High levels of silicon allow carbon to solidify in the form of graphite. On the other hand, if silicon content is relatively low and the cooling rate, during solidification, is high, carbon forms carbides, whose composition depend on the alloying elements. Table 6.3

**Table 6.3** Main characteristics of cast irons

| Cast iron | Hardness | % C | Microstructure | Observations |
|---|---|---|---|---|
| Grey | Moderate ($\approx$200 kg/mm$^2$) | <3 | Lamellar graphite; matrix: F, P, B, M | Most common iron |
| Ductile (or nodular) | Moderate ($\approx$350 kg/mm$^2$) | 2.5–4 | Spheroidal graphite; matrix: F, P, B, M, AF | High fatigue resistance; complex shapes |
| White | High (420–600 kg/mm$^2$) | 4.2–5.6 | Large eutectic and secondary carbides<br>Matrix: M, B | Alloying elements can be profitably added (Ni, Cr) |
| Malleable | Moderate ($\approx$250 kg/mm$^2$) | $\approx$2.5 | "Flowerlike" nodular graphite<br>Matrix: F, P | Properties similar to steels |

*F* ferrite, *P* pearlite, *B* bainite, *M* martensite, *AF* ausferrite; modified from [1]

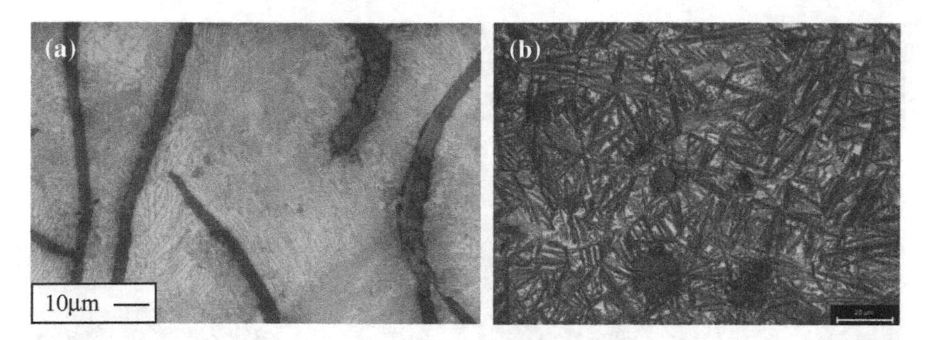


**Fig. 6.7** Microstructure of **a** a pearlitic grey iron, and **b** an ADI characterized by a graphite fraction of 7.2 %, an average nodule count of 195, a mean nodule diameter of 15 μm, and an amount of retained austenite of 25 % [14]

summarizes the main types of cast irons and relevant features. The matrix microstructure of graphitic cast irons is typically similar to that of steels. Within the family of ductile (or nodular) irons, *austempered ductile irons* (ADIs) have recently reached a widespread diffusion. They are obtained following an austempering treatment that usually comprises austenitisation between 850 and 950 °C, followed by quenching in a salt bath furnace at 250–400 °C. The material is then kept at this temperature for a time suitable to stabilize austenite, and then it is cooled down to room temperature. An *ausferritic* (AF) microstructure is obtained, characterized by the presence of bainitic ferrite plus carbon stabilized austenite, that increase strength and ductility [13]. As an example, Fig. 6.7 shows the microstructure of a pearlitic grey iron and an austempered ductile iron.

Cast irons are characterized by an attractive combination of mechanical properties and advantages, concerning cost and manufacturing. The high hardness that is obtainable with thermal treatments and the addition of alloying elements, allow cast irons to be used in several tribological applications. For example, if sliding wear problems are of concern, grey or ductile cast irons are widely used, in like-on-like applications or against steel or copper alloys (such as bronzes). In the case of abrasive wear processes, white cast irons are profitably used, indeed.

### 6.2.1 Sliding Wear

Grey and ductile cast irons are commonly employed in tribological systems characterized mainly by sliding wear, both in dry and in boundary lubricated conditions. Typical applications are disc brakes, piston rings, cylinder liners, cam-follower systems and gears. The applications relate mostly to situations where lubrication, particularly fluid film, is difficult to obtain, or, on the contrary, undesirable.

The high sliding wear resistance of grey or ductile iron is due to the presence of graphite, which acts as a solid lubricant during sliding. In fact, during the run-in

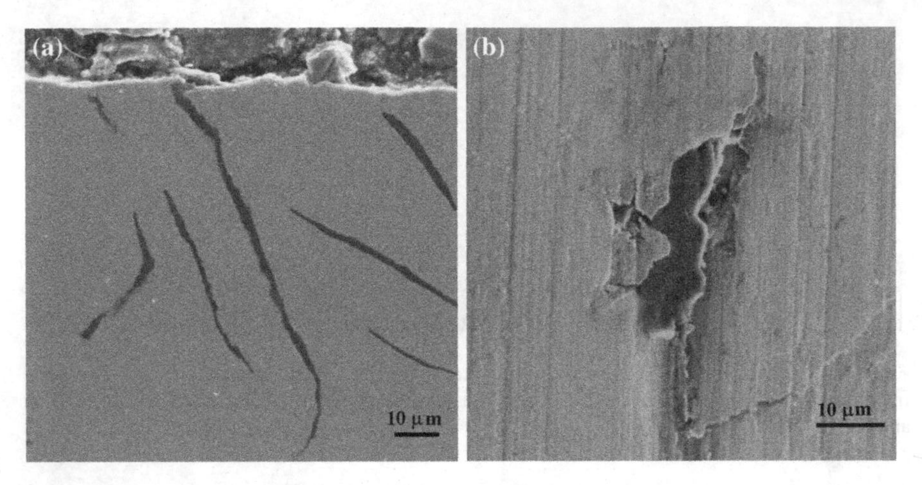


**Fig. 6.8** **a** Cross section showing a graphitic lamella emerging at the surface during dry sliding; **b** graphite is rubbed on the wear surface thus exerting a solid lubricant effect [15]

phase graphite protrudes from the surface and is interposed between the two mating bodies. This is shown, for example, in Fig. 6.8, referring to a pearlitic grey cast iron after dry sliding against a friction material [15]. This phenomenon may thus work together with the formation of a surface oxide, which ensures the establishment of a mild tribo-oxidative wear. As a result, a reduction in friction and sliding wear are attained. As shown in Table 5.2, the pearlitic or martensitic grey cast irons are characterized by specific wear coefficients similar to those of tool steels, although their hardness values are significantly lower.

As observed in steels, a transition to severe adhesive wear may occur if the contact pressure increases above a critical value. Figure 6.9 shows the experimental dependence of the nominal pressure, at the transition from mild to severe wear, on the carbon content, i.e., on the fraction of graphite flakes, in the case of a pearlitic

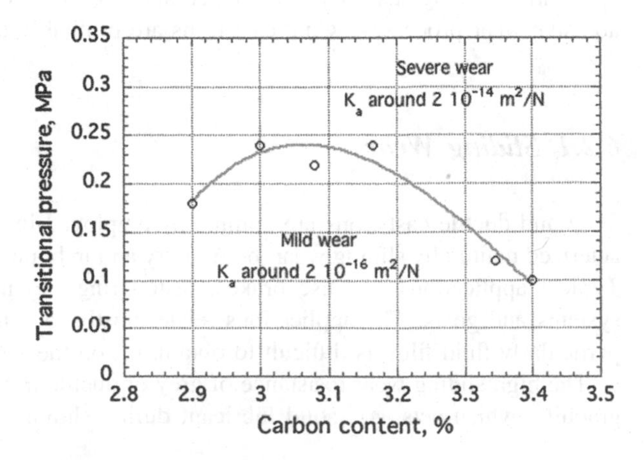


**Fig. 6.9** Nominal contact pressure at the transition between mild and severe wear as a function of carbon content, in a pearlitic gray cast iron dry sliding against in high strength steel (modified from [16])

grey cast iron (with microhardness of the pearlite matrix around 440 kg/mm$^2$) [16]. It can be noted that the transition pressure displays a maximum at about 3.05 % C. This maximum is due to the fact that an increase in graphite content increases the solid lubricant effect and, at the same time, induces a decrease in the load-bearing area, i.e., in the fraction of metal that supports the contact stress. For a given grey cast iron, the transition pressure also decreases with increasing the sliding speed, as observed in steels and in other alloys, because of the thermal softening effect at the contact regions.

If the graphite is present in the form of nodules (or *spheroids*, ductile or nodular cast iron) instead of lamellae the sliding wear resistance is higher in the presence of relatively high nominal pressures. Indeed, while the lubricating effect of the graphite remains almost unchanged, the spheroids give rise to a lower stress concentration thus reducing the tendency to subsurface damage influencing adhesive wear. However, at relatively low loads the geometry of the lamellar graphite is preferable, since the feeding of the contact surface by graphite flakes is easier. In addition, lamellar graphite realizes a greater thermal conductivity than nodular graphite, inducing a better capacity of remove the frictional heat from the contact surface.

Regarding the influence of the matrix microstructure, it has to be observed that ferritic or austenitic matrices are generally to be avoided because they are too soft. Cast irons are then preferably to be used with pearlitic, martensitic or austempered matrix microstructures. In general, if ambient temperature is lower than 250 °C, ADI is superior to quenched and tempered ductile irons with comparable hardness values, because the strain-induced transformation of retained austenite and the strain hardening of bainitic ferrite contribute to the oxidation wear resistance by better supporting the surface oxide layer [17].

As mentioned before, grey and ductile irons are also satisfactory employed in the case of lubricated sliding under boundary conditions. In this respect, it has to be further observed that ductile iron with an austempered microstructure may display an higher scuffing resistance than heat-treated steels, since free graphite reduces friction and hence the temperature rises at the contacting asperities.

### 6.2.2 Wear by Contact Fatigue

Gray and ductile cast iron with a pearlitic or austempered microstructure are employed in many applications where contact fatigue damage may occur, such as in gears or cams. The main advantages are the possibility of obtaining near-net shapes with casting techniques, and also the high machinability by chip removal.

Grey cast irons display a resistance to contact fatigue that is only about 60 % that of the fatigue resistance of steels with the same hardness level. This reduction is due to the negative effect of the graphite lamellae, which exert a stress concentration at their edges, facilitating the nucleation of fatigue cracks. In addition, the lamellae also provide an energetically favourable path for crack propagation. Because of this, grey cast irons are typically characterized by a value of the coefficient n in Eq. 5.8 that is definitely higher than that of steels. Ductile iron displays a better contact fatigue resistance, especially after austempering. The graphite nodules, in fact, facilitate the nucleation phase, although to a lower extent than the lamellae, but they do not provide any favourable path for the subsequent crack propagation. Indeed, when the propagating crack intersects graphite nodules it may be stopped or at least slowed down in its propagation rate. The contact fatigue endurance limit of ductile irons is therefore only 10–15 % lower than that of steels of the same hardness [18]. Thanks to their ausferritic microstructure, ADIs are characterized by a relatively high fracture toughness that makes them particularly attractive for the production of gears.

A way to improve further the contact fatigue resistance of nodular cast irons is to increase the density of graphite nodules, and, consequently, to reduce their average size. Typically the nodular cast irons contain a density of nodules between 200 and 250 nodules/mm$^2$. By increasing the cooling rate during solidification (and, simultaneously, using high levels of silicon which promotes graphitization), cast irons containing more than 1000 nodules/mm$^2$ can be obtained [19].

### 6.2.3 Abrasive Wear by Hard Particles

As mentioned, white cast irons possess high resistance to abrasive wear. In particular, *high-chromium white cast irons* with high content of primary carbides of the type $(Fe, Cr)_7C_3$ are widely used. These materials typically contain: 1.2–3.5 % C, 0.3–1.5 % Si, 0.4–1.5 % Mn, 15–32 % Cr, 0.5–3 % Mo, 1–3 % Cu and have hardness ranging from 850 to 900 kg/mm$^2$. The *nickel-chromium martensitic white cast irons* are also commonly used in practice. The addition of nickel, which does not form carbides, increases hardenability of the alloy, i.e., its attitude to form martensite even at relatively low cooling rates. These irons typically contain: 2.6–3.6 % C, 0.4–1 % Si, 0.4–0.8 % Mn, 1–2.5 % Cr, 3.5-5 % Ni and have hardness values between 560 and 700 kg/mm$^2$.

After solidification, white cast irons achieve a microstructure consisting of carbides in an austenitic matrix. Because of the high content of alloying elements austenite is stabilized at room temperature. These irons may be then used in the as-cast condition in applications with low-stress abrasive wear. In fact, during the abrasive interactions the austenite transforms into martensite, as already observed in the manganese steels. However, to get the best abrasive wear resistance, including the high-stress abrasion resistance, it is necessary to heat treat the alloys to obtain a fully martensitic matrix microstructure. Since austenite is stabilized by chromium, a soaking treatment is initially carried out at 900–1000 °C to promote the precipitation of chromium carbides thus destabilizing the austenite. Subsequently, the quenching and relieving treatment is performed to obtain the martensitic matrix.

Figure 6.10 shows a typical microstructure of a martensitic white cast iron. Note the presence of primary carbides that form a three-dimensional network giving rise

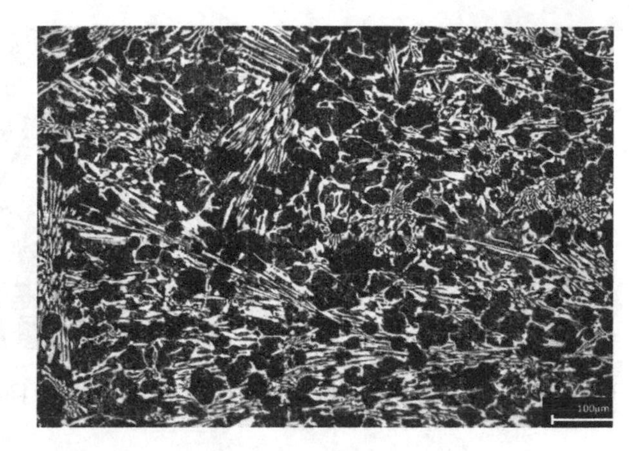


**Fig. 6.10** Typical microstructure of a martensitic white cast iron

to high hardness but also noticeable brittleness. In many applications, as in the case of the cylinders for rolling mills, the solidification aims at obtaining a surface rim of white cast iron and a core of nodular (ductile) cast iron, able to provide the component an adequate fracture toughness (the process used is *centrifugal casting*).

The abrasive resistance of white cast irons depends on several factors, such as the type and volume fraction of the carbides, the matrix hardness, the hardness, size, and the angularity of the abrasive particles. As already remarked, chromium carbides have a hardness of about 1500–1800 kg/mm$^2$ and therefore may well resist the abrasive action of particles with lower hardness (such as silica or alumina). However, they do not offer adequate resistance in the case particles, such as those of SiC, with higher hardness (see also Fig. 5.19b). To increase the abrasive wear resistance of the white irons in the latter case, it is therefore necessary to introduce alloying elements, such as vanadium, capable of forming harder carbides (Table 2.1).

## 6.3 Copper Alloys

Copper alloys are used in several tribological applications characterized by sliding wear or wear by contact fatigue in mixed or boundary lubrication (such as in gears, bearings, seals, just to mention some important examples), especially when the counterface is made of steel and the environment is aggressive. Beside quite excellent tribological properties (except the abrasion resistance, which is very low), copper alloys are characterized by other attractive engineering properties, such as high thermal conductivity and high corrosion resistance in many environments. The main copper alloys used in tribology are:

(1) Cu-Zn alloys (at the base of the brass family);
(2) Cu-Sn alloys (conventional bronzes);
(3) Cu-Al alloys (aluminium bronzes);

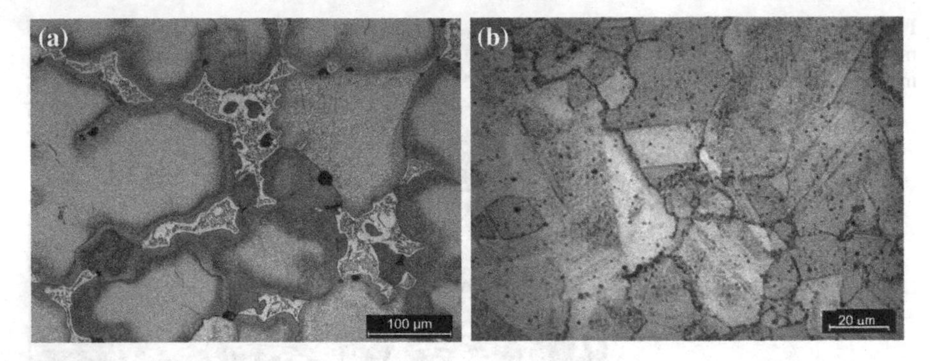


**Fig. 6.11** Typical microstructure of an as cast CuSn12 alloy (**a**) [20], and of an extruded Cu-2 % Be-0.5 % (Co + Ni) alloy (**b**) [21]

(4) Cu-Be alloys (beryllium bronzes);
(5) Cu-Ni alloys (nickel bronzes);

Other elements used in copper alloys are lead, manganese and phosphorus, which remains from the refining operation. Copper alloys are used both in the as-cast and wrought condition. They are also easily machined by chip removal. As an example, Fig. 6.11, shows the microstructure of the widely used as cast CuSn12 alloy, and of an extruded Cu-Be alloy. The as-cast alloy is segregated with an interdendritic solidification phase that is made by the α phase, a copper rich solid solution of tin in copper, and the δ phase, an intermetallic compound with composition $Cu_{31}Sn_8$. This latter compound, characterized by a jagged shape in Fig. 6.11a, is hard and brittle. The microstructure of a Cu-2 % Be alloy (containing also 0.5 % Co + Ni), Fig. 6.11b, displays α grains and Co-Be-Ni precipitates (*beryllides*). The matrix hardness is given by a fine dispersion of metastable Cu-Be precipitates.

Copper alloys can be hardened by work hardening and, in some cases, also by heat treatment. The maximum attainable hardness is about 300 kg/mm$^2$, except for Cu-Be alloys that can reach values around 400 kg/mm$^2$ after solution treatment and age hardening.

Bronzes are widely used in tribological applications in the case of sliding wear. The coupling between bronzes and steel is characterized by $K_a$-values as low as $10^{-15}$ (Table 5.2). This excellent sliding resistance is due to several reasons: the easy formation, during sliding, of a compact oxide layer that is adequately supported by the underlying matrix; the low compatibility between copper and iron that reduces the tendency to adhesive wear (and its contribution when present); the high thermal conductivity that avoids reaching high contact temperatures that could soften the material. For example, by means of dry pin-on-disc tests, it has been obtained that pure copper (H: 102 kg/mm$^2$) displays a $K_a$-value of $3.2 \times 10^{-14}$ m$^2$/N [22]. When using a age hardenable CuBe alloy (H: 342 kg/mm$^2$), $K_a$ decreases to $6.16 \times 10^{-15}$ m$^2$/N. But when using an age hardenable CuBeNi alloy, characterized by lower hardness (H: 242 kg/mm$^2$) but higher thermal conductivity, the recorded

$K_a$ was $4 \times 10^{-15}$ $m^2/N$. The resistance to adhesive wear is important during the run-in stage and also when wear fragments do not stay in the contact region and thus mild tribo-oxidative wear cannot be fully established.

Bronzes too have good performances under boundary lubrication. In particular, they afford a high resistance to scuffing thanks to their low compatibility versus steel. A typical application is constituted by the coupling between a bronze crown and a hardened helical screw in worm gearing. Here the contact is characterized by prevailing sliding conditions, and a boundary lubrication is attained. Figure 6.12a shows the wear evolution for a cast CuSn12 alloy, during rolling–sliding test carried out at three Hertzian pressures, with a sliding velocity of 3.74 m/s (a synthetic polyglcycolic oil containing anti-scuffing additives was used) [20]. In all cases, a run-in stage was observed. After run-in, wear rate was almost negligible at 260 MPa. At 325 MPa, $K_a$ resulted typical of boundary lubrication. At 350 MPa, after run-in, a steady state stage was firstly observed, but after about 1300 km of sliding, a transition with a noticeable increase in the wear rate was recorded. Such

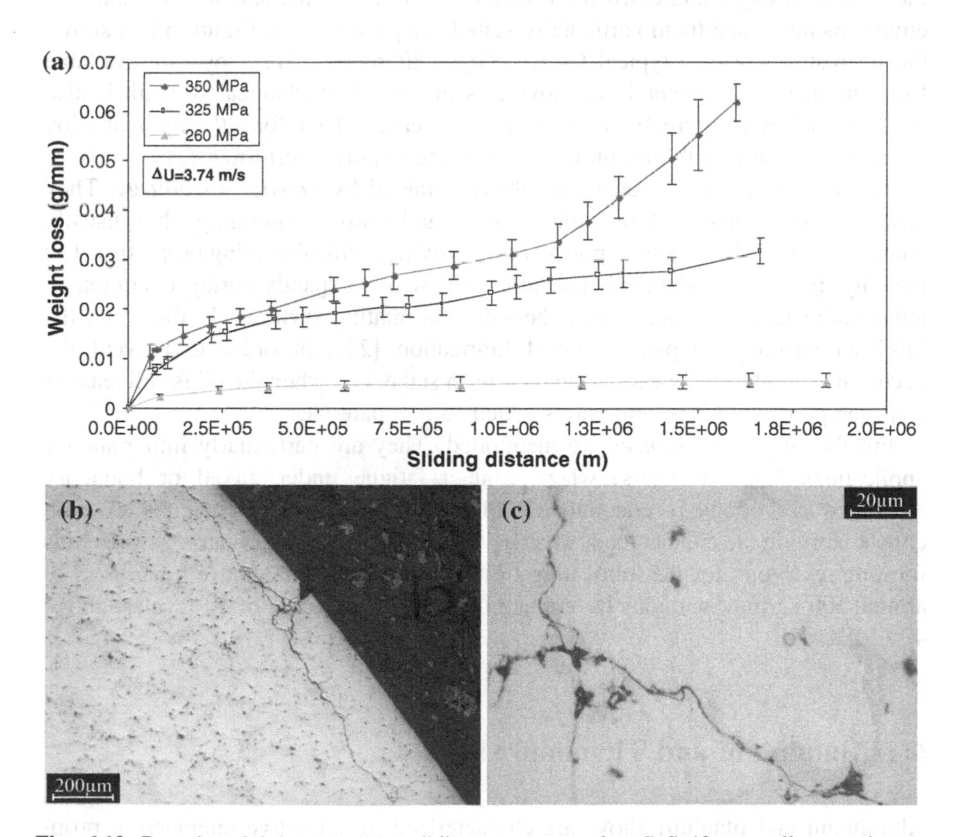


**Fig. 6.12** Boundary lubricated rolling-sliding wear test of a CuSn12 cast alloys against a hardened steel (sliding speed: 3.74 m/s; synthetic polyglycolic oil). **a** Wear curves obtained at three load levels; **b** subsurface crack paths; **c** preferential crack path through brittle dendritic phases [20]

an increase was found to be due to the onset of damage by contact fatigue. Surface and sub-surface fatigue cracks were found to nucleate and propagate (Fig. 6.12b, c), aided in this by the brittle inter-dendritic phases rich in tin and lead.

Copper alloys containing lead between 25 and 40 % are widely used in *self-lubricating bearings*. Lead is virtually insoluble in copper (Fig. 1.13), and during sliding it is spread onto the surface by lowering the coefficient of friction, with a mechanism similar to that shown by graphite in grey irons. Better mechanical properties are displayed by lead bronzes (here the amount of lead is usually between 1 and 10 %) or by aluminium bronzes (5–11 % Al). These alloys are suitable for the realization of bearings loaded by high nominal pressures (above 1 MPa) or for the manufacturing of gears. Similar to the Cu-Sn-Al alloys are the Cu-Sn-Mn alloys, containing up to 1 % Mn.

A special mention deserve the so-called *beryllium bronzes* (Cu-Be alloys), which can be precipitation hardened if the concentration of beryllium is greater than 1.3 %, thus achieving a tensile resistance comparable to that of steels. The high mechanical strength and corrosion resistance, both in marine and several industrial environments, make them particularly suited for power gears. Figure 6.11b shows the microstructure of a typical Cu-Be alloy. Different Cu-Be alloys are available from the market. In general, as alloying is increased, mechanical strength is also enhanced whereas thermal conductivity is decreased. Therefore, the optimal alloy composition can be selected on the basis of the required performances.

A particular class of bronzes are those obtained by *powder metallurgy*. They have a typical content of tin around 10 % and, most importantly, they usually contain up to 10 % of residual porosity that provides self-lubricating properties. The porosity, in fact, is infiltrated with lubricant, which expands during operation as temperature increases and spreads between the mating surfaces. In this way, the lubricant ensures a typically mixed lubrication [23]. In order to prevent the occurrence of adhesive wear during the run-in stage, i.e., when the oil is still leaking from the pores, 1–3.5 % graphite is added to the material.

Finally, *phosphor bronzes* are mentioned. They are particularly important for applications (such as gears) where contact fatigue under mixed or boundary lubricated conditions is encountered. Phosphorus forms hard precipitates with copper carrying the contact load, whereas the surrounding softer matrix is worn thus forming reservoirs for the lubricating oil [24]. These materials have a hardness of around 100 kg/mm$^2$ and may have a fatigue strength (for a life of $10^7$ cycles) above 400 MPa.

## 6.4 Aluminium and Titanium Alloys

Aluminium and titanium alloys are characterized by attractive engineering properties. Both have a relatively low density (around 2.7 g/cm$^3$ for aluminium alloys and 4.5 g/cm$^3$ for titanium alloys) and high corrosion resistance in many aggressive environments. Titanium alloys are also biocompatible, and can be used as

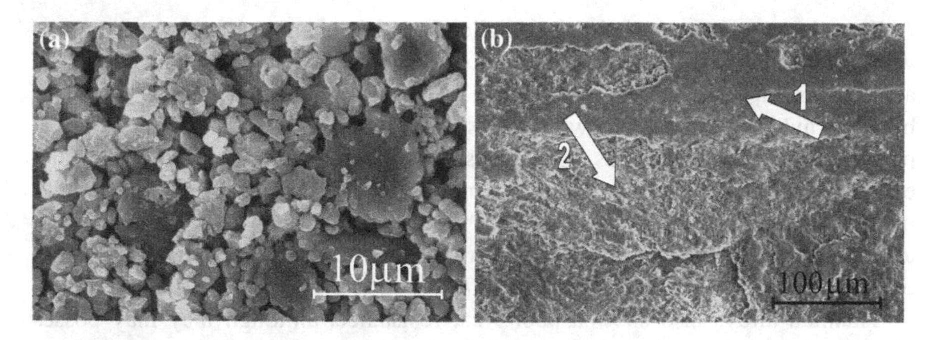


**Fig. 6.13** Morphology of the wear debris (**a**) and of the worn track (**b**) in the case of an Al-7072 alloy after dry sliding at 1 MPa and 0.2 m/s against a 52100 steels [27]. The tribolayer is very compact and dense in some regions (see *arrow 1*), whereas it is rather fragmented in others (see *arrow 2*)

orthopaedic and osteosynthesis materials. However, albeit for different reasons, both have a limited resistance to nearly all wear mechanisms.

Aluminium alloys are characterized by a relatively low hardness (they reach hardness values of about 250 kg/mm$^2$) that considerably decreases as temperature increases. Aluminium alloys, in fact, begin intense thermal softening at a temperature around 100 °C. Consequently, at this temperatures both their friction coefficient and specific wear coefficient start to noticeably increase. The transition from mild tribo-oxidative wear to severe adhesive wear occurs at relatively low values of applied pressure and sliding speed. It has to be further noted that the protective action of alumina, which is at the basis of tribo-oxidative wear, is not as efficient as, for example, the Fe-oxides in steels, since it does not display any ductility, i.e., any ability to spread onto the wear surface. Moreover, its adhesion to the substrate is low [25]. In Al-alloys the protective layer that is formed under tribo-oxidative wear is typically a mixture of oxides originating also from the counterface. As an example, Fig. 6.13 shows the morphology of the wear debris (a) and of the worn track (b) in the case of an Al-7072 alloy after dry sliding at 1 MPa and 0.2 m/s against an AISI 52100 steel [26]. The fragments appear to be small and equiaxed; they originate from the brittle fragmentation of the tribolayer, i.e., from regions 2 in Fig. 6.13b, that is formed by aluminium and iron oxides.

In aluminium alloys adhesive wear is particularly severe, since it is characterized by intense transfer phenomena. These alloys are then unsuitable for sliding applications that may give rise to relatively high contact temperatures. For example, they are poorly suitable for non-conformal contacts, which are typically characterized by high contact pressures [27]. Aluminium alloys are therefore mainly used in conditions of conformal contact, such as in *sliding bearings*, that work under relatively mild loading conditions. A common alloy used in plain bearings is the Al-20 % Sn alloy, consisting of a ductile phase (tin, practically insoluble in aluminium) in a relatively hard matrix. Lead can be also added in place of tin.

The sliding resistance of aluminium alloys can be enhanced by reinforcing them with ceramic particles. *Aluminium matrix composites* typically contain 20 % of alumina or SiC particles. The presence of these particles shifts the boundary between mild tribo-oxidative wear and severe adhesive wear to higher nominal pressures and sliding speeds [28]. These composites, however, are still quite expensive, especially for the complex production process they require. Another method that can be employed to improve the sliding wear resistance of aluminium alloys, consists in growing a surface layer of aluminium oxides by hard anodizing or in the deposition of thin films (Chap. 7).

Titanium alloys also show a poor wear resistance, although they may posses high mechanical strength and they do not undergo thermal softening as aluminium alloys (note that titanium has a melting temperature of 1670 °C compared to 660 °C of aluminium). In case of dry sliding wear, titanium alloys show typical $K_a$-values that are around $10^{-13}$ $m^2/N$, even if the wear mechanism is tribo-oxidation [29]. This is due to several reasons: titanium oxides possess low protection properties since the ratio between their specific volume and that of the metal is less than 1. In addition, the oxides are not adequately supported by the underlying material that, during sliding, it is not able to strain-harden enough. Indeed, it undergoes considerable plastic deformation that localizes easily in shear bands that weaken the microstructure [30]. This last tendency my also promote intense material transfer during adhesive wear, making seizure quite easy to occur.

Also for titanium alloys various surface treatments have been proposed to improve wear resistance, including thermal oxidation (like aluminium, titanium displays a high affinity for oxygen), anodizing, nitriding, ion implantation, or the deposition of ceramic coatings, which will be presented in the next chapter. It is worth noting that thermal oxidation is a simple and cost effective treatment, able to reduce the friction coefficient and the wear rate in different tribological systems. However, the thickness and compactness of the oxide layer has to be optimized since it is quite brittle and shows a tendency to damaging and removal even at quite low applied loads [31].

Another method for reducing friction and wear in sliding or sliding-rolling is the use lubrication. However, in titanium alloys the use of conventional lubricating oils is not quite effective, and specific research is still ongoing for the development of lubricants that could be suitable for these alloys. Table 6.4 shows the experimental results of tribological tests carried out in a disc-on-disc configuration using discs made of the Ti-6 % Al-4 % V alloy (with a hardness of 350 kg/mm$^2$ and a yield strength of 900 MPa) and a common SAE 15 W/40 lubricating oil [32]. At a

**Table 6.4** Results of rolling-sliding tests (10 % sliding) performed using Ti-6 % Al-4 % V discs, lubricated with common SAE 15 W/40 oil (data from [32])

| Hertzian pressure, MPa | Λ factor | Coefficient of friction | Wear |
|---|---|---|---|
| 640 | 0.4 | 0.065 (0.3) | Scuffing |
| 320 | 0.5 | 0.034 | $K_a = 2.64 \times 10^{-15}$ $m^2/N$ |
| 226 | 0.6 | 0.034 | $K_a = 2.12 \times 10^{-15}$ $m^2/N$ |

Hertzian pressure of 640 MPa, the coefficient of friction was initially around 0.065, but, soon after, a transition to a much greater value, around 0.3, was observed. This is a typical value for dry sliding. As a matter of fact, after the transition wear became particularly intense with values of $K_a$ around $10^{-14}$ $m^2/N$. This result is quite unexpected since at this load level the ratio $p_{max}/\tau_Y$ is very low (less than 2). In this condition, steels would show a contact fatigue life in excess of $10^8$ cycles. The observed transition was caused by scuffing. In fact, during the initial stages of the test, the thin natural surface oxide layer (1.5–10 nm thick) is removed since lubrication is mixed ($\Lambda = 0.4$). As a consequence, the lubricant is no longer able to properly wet the surface of the metal, since the ionic bonds of the oxide layer favour the anchoring of the polar molecules of the lubricant, and scuffing may start. The low thermal conductivity of titanium and its alloys (around 6.7 W/mK) plays clearly an important role in favouring the onset and propagation of scuffing.

In the tests at lower Hertzian pressures, scuffing was not observed. However, wear was still present and the specific wear coefficient went up to around $2 \times 10^{-15}$ $m^2/N$, which is about two orders of magnitude lower than the typical value for dry conditions. This means that a boundary lubrication regime was attained during the tests, even if the lambda factor was around 0.5–0.6, a typical value for mixed lubrication. Again, this behaviour is due to the poor wettability of titanium alloys by lubricating oils.

## 6.5 Advanced Ceramics

Several *natural ceramics*, such as rocks (silica or aluminosilicates), marble (mainly made of calcium carbonate), or *traditional ceramics*, such as glass, clay products and concrete, have been used for centuries, and are used also today, for tribological applications, such as in floors or as grinding tools. In the past decades, the so-called *advanced ceramics*, such as *alumina*, *zirconia*, *silicon nitride*, *silicon carbide* and *sialon* (containing alumina and silicon nitride in different compositions), have been developed. They are ideal for wear resistant applications, thanks to their high hardness, high chemical inertness in many environments, and low density. In particular, their ability to retain high hardness at high temperatures, as shown in Fig. 6.14, is particularly important in tribology.

However, all ceramics possess quite low fracture toughness, typically around 5 MPa $m^{1/2}$, and can display brittle contact and wear by brittle fragmentation. To minimize this problem, advanced ceramics are produced in a very controlled manner, in order to reduce the content of defects, especially on the surface. In addition, special ceramic composites have been also developed, containing toughening phases in their microstructure or having laminated structures designed to induce compressive stresses at the surface. It is clear that the use of highly controlled production routes and/or special materials has a significant impact on the manufacturing costs. Therefore, it is often preferred to deposit a ceramic layer on a different substrate, such as a metal, which is less expensive and is also tougher.

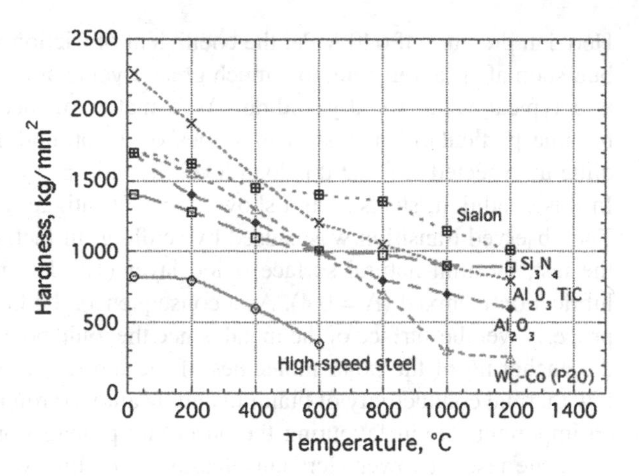


**Fig. 6.14** Hardness as a function of temperature for different advanced ceramics and, for comparison, for WC-Co hardmetal, and a high-speed steel (from different literature sources)

Monolithic advanced ceramics are typically produced by powder routes; the most used techniques are liquid phase sintering, hot isostatic and dry pressing, injection moulding and slip casting. With these techniques, finished components with complex shapes can be manufactured. Standard parts with simple geometries, e.g. plates or bars, are also produced. They can be then machined and assembled together and/or with metallic parts to obtain the final part [27].

Table 6.5 lists some advanced ceramics that are used in tribological applications, along with relevant physical-mechanical properties. In the table, the constant β is also included. It expresses the dependence of the material hardness on temperature according to the following relationship:

**Table 6.5** Physical and mechanical properties of some advanced ceramics (data obtained from different literature sources)

| Material | Density ($g/cm^3$) | Elastic modulus (GPa) | Hardness ($kg/mm^2$) | β ($10^{-4}$, °C) | $K_{IC}$ ($MPa\ m^{1/2}$) | Thermal shock resistance ($\Delta T_c$, °C) |
|---|---|---|---|---|---|---|
| Alumina ($Al_2O_3$) | 3.9 | 380 | 1700 | 8.5 | 2–4 | 200 |
| Partially stabilized zirconia (PS $ZrO_2$) | 5.8 | 200 | 1300 | | 10 | 500 |
| Sialon ($Si_3Al_3O_3N_5$) | 3.2 | 300 | 1430–1850 | 4.4 | 6–7.5 | 510 |
| Silicon nitride ($Si_3N_4$) | 3.2 | 310 | 1400 | 3.8 | 4 | 700 |
| Silicon carbide (SiC) | 3.2 | 410 | 3100 | 10 | 4 | 300–400 |

$$H = H_0 \cdot e^{-\beta T} \quad (6.2)$$

where T is the temperature in °C, and $H_0$ is the hardness at room temperature. It should be noted that all the listed values are only indicative, because the materials properties depend very much on the manufacturing process and on the compositional characteristics (such as the residual porosity, grain size, purity, and so on).

Advanced ceramics are particularly suitable in applications characterized by sliding wear and low-stress abrasive wear. Examples include cutting tools, sealing rings, manufacturing dies, which are mainly made by alumina, or zirconia when some fracture toughness is required, as in presence of impacts. Other examples are ball bearings (especially operating at high temperature) and engine valves, which are mainly made by silicon nitride or silicon carbide [33]. Ceramic particles are also used as abrasives for grinding wheels.

### 6.5.1 Sliding Wear

In case of dry sliding against a steel counterface, the specific wear coefficient, $K_a$, of ceramics is quite low, typically between $10^{-16}$ and $10^{-15}$ $m^2/N$ (see Table 5.2). In the case of couplings between ceramic materials, $K_a$ can vary between $10^{-12}$ and even $10^{-18}$ $m^2/N$. Therefore, by a proper selection of a tribological pair it is possible to achieve very low wear rates. The dry sliding behaviour of ceramics is characterized by three main damage mechanisms [34, 35]:

(1) For small loads and low sliding speeds, wear is mild and $K_a$ typically assumes values that are less than $10^{-15}$ $m^2/N$. During running in, the brittle fragmentation of the highest asperities produces wear debris that are compacted and possibly oxidized during sliding, forming protective scales such as those shown in Fig. 2.14. As a consequence, wear is mild and possibly by tribo-oxidation.
(2) As load is increased, a transition to severe wear may occur. Wear is dominated by macroscopic brittle contacts, with the formation of wear fragments due to the propagation of cracks under the action of the surface tensile stress originated by friction (see Sect. 4.1.2). The fracture is often intergranular, as shown, again, in Fig. 2.14.
(3) At high sliding speeds, a transition to severe wear may occur even if the contact stresses are relatively small. In fact, the low thermal conductivity of ceramics favours the attainment of high flash temperatures. They can induce the formation of surface tensile thermal stresses that may in turn form surface cracks, which then produce wear fragments.

The possibility of having severe wear of type (2) is lower in materials with high fracture toughness (see Eq. 4.3), while the possibility of having severe wear of type 3 is lower in materials with high *thermal shock resistance*. The thermal shock resistance is expressed by the critical temperature interval, $\Delta T_c$, the material can withstand without breaking. In a simplified view, an abrupt temperature change, $\Delta T$, can induce a tensile stress given by

$$\sigma_t = E \cdot \alpha \cdot \Delta T \quad (6.3)$$

where $\alpha$ is the coefficient of thermal expansion. To estimate $\Delta T_c$, the relation 1.10 can be used:

$$\Delta T_c \approx \frac{K_{Ic}}{1.12 \cdot E \cdot \alpha \cdot \sqrt{\pi c}} \quad (6.4)$$

where c is the size of a pre-existing surface microcrack that may trigger the brittle fracture. As mentioned in Sect. 4.1.2, c is most often equal to the grain size. Typical values for $\Delta T_c$ are included in Table 6.5.

It is therefore clear that the materials with the largest values of both $K_{Ic}$ and $\Delta T_c$ are those with the best sliding wear resistance, since the transition to severe wear is displaced to high Hertzian pressures and high sliding speeds. From Eq. 4.3, it is obtained that the transition to type (2) severe wear is avoided if:

$$\frac{(1 + 10\mu) \cdot p_{\max} \sqrt{d}}{K_{Ic}} \leq 3 \quad (6.5)$$

where d is the average grain size. It has been obtained that the transition to type (2) severe wear is avoided if [36]:

$$\frac{\gamma\mu}{\Delta T_c} \sqrt{\frac{vF_N H}{k\rho c}} \leq 0.04 \quad (6.6)$$

where $\gamma$ is here the fraction of heat that enters the sliding body (it is defined by Eq. 2.23). Table 6.6 shows the experimental transition values for the Hertzian pressure and sliding velocity in the case of $Al_2O_3/Al_2O_3$ and SiC/SiC pairs, obtained using a pin-on-disc test in a sliding point contact [36].

To complete the picture, it has to be considered that sliding wear of ceramics is also affected by ambient *humidity*. If it is increased, it may induce a decrease in $K_{Ic}$ thus favouring wear. Most importantly, humidity may interact with the ceramic surface and produce oxide/hydroxide molecules by tribochemical reactions. Such

**Table 6.6** Transition values for the Hertzian pressure and sliding speed from mild to severe wear for two ceramic pairs

| Coupling | Hardness (kg/mm$^2$) | Hertzian pressure (MPa) | v (m/s) |
|---|---|---|---|
| $Al_2O_3/Al_2O_3$ | 1660 | ≈1820 | ≈0.3 |
| SiC/SiC | 3140 | ≈1600 | ≈1.5 |

Data obtained with a pin-on-disc test, operating in dry sliding and in a point contact (data from [36])

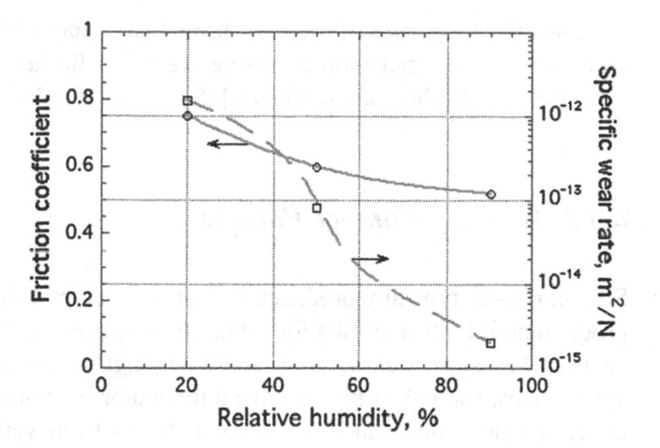


**Fig. 6.15** Experimental dependence of friction and wear on the ambient humidity, in case of the SiC/SiC coupling (modified from [38])

molecules may act as soft lubricants, reducing both friction and wear (see also Sect. 2.6) [37]. These effects have been throughout recorded in alumina, silicon nitride and silicon carbide. Figure 6.15 shows, as an example, the experimental dependence of friction and wear on the ambient humidity, in case of the SiC/SiC coupling [38]. The same beneficial effects are obtained in the case of sliding in water [39]. It is also reported that the formation of a tribochemical layer on ceramics may greatly help with water lubrication, improving in this respect wear resistance of mechanical parts that operate in the presence of water, such as seals or bearings in water pumps.

A reduction of the transition pressure from mild to severe wear is typically observed as temperature is increased up to 300–500 °C, as shown in Fig. 6.16 for an alumina/alumina coupling. This is mainly due to the increase in the friction coefficient (see Eq. 6.5), due to the desorption of the tribochemical layer. However, at particularly high temperatures, i.e., above about 800 °C, wear becomes mild again. As shown in Fig. 2.21, at these temperatures the incipient melting of the sintering additives and impurity inclusions, mainly those at the grain boundaries, takes place.

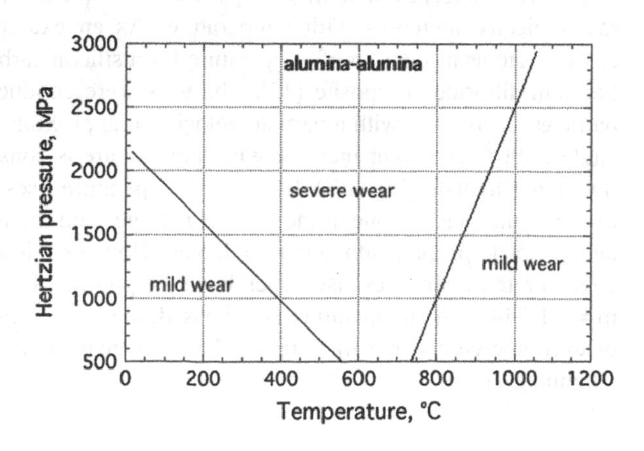


**Fig. 6.16** Influence of ambient temperature and Hertzian pressure on the dry sliding behaviour of the alumina/alumina pair (obtained from different literature data)

It causes the formation of a glassy layer that reduces friction and, in turn, makes more difficult the transition to severe wear. To further reduce wear at high temperatures, solid lubricants such as $MoS_2$ can be employed.

### 6.5.2 Wear by Contact Fatigue

Silicon nitride, typically produced by hot isostatic pressing, is currently used in the production of balls and rolls for rolling bearings, especially of the hybrid type, i.e., bearings having a steel raceway. These bearings are particularly suitable in extreme applications, such as at high temperature and/or high speed (the low density of the material reduce the centrifugal forces). It has been verified that rolling bearings made of silicon nitride can be effectively lubricated using mineral oils, obtaining a lubrication degree that is similar to that found in steel components. The contact fatigue life of these materials is comparable or superior to that of the best bearing steels [40]. In order to reduce the possibility of surface nucleation of fatigue cracks, all ceramic parts must be carefully produced with a very low roughness and low defectiveness, such as the residual pores or grinding cracks, which should be smaller than 2 μm (see Sect. 8.10).

### 6.5.3 Abrasive Wear by Hard, Granular Material

Ceramics are particularly suited for applications with low-stress abrasion such as in conveyors of mineral particles where alumina is widely used. Ceramics are also used in the case of erosive wear, when both the impact velocity and the impact angles are relatively low (Table 5.8). For example, alumina and silicon carbide (with higher hardness than alumina) are used for gas turbine parts, sinter plants, sealing bearings, powdered coal lines, just to mention some applications where high erosion resistance, even at high temperature, is required [41]. In general, the erosion rate typically increases with temperature. As an example, Fig. 6.17 shows the erosion rate as a function of temperature for a silicon carbide and a silicon carbide-titanium diboride composite [42]. The tests were conducted using silicon carbide particles as erodent, with a particle velocity of approximately 70 m/s and an impact angle of 90°. The wear increase with temperature is consistent with the Lawn and Swain mechanism (Fig. 1.7). In fact, as temperature rises, the materials hardness is lowered, the plastic zone at the contact is larger and the residual stresses that drive lateral crack propagation are also larger. However, it should be noted that the erosion rate of ceramics also depends on the product: $H^n K_{Ic}^m$, where $n = -0.15$ and $m = -1.3$ [43]. As temperature is increased, the erosion rate may remain unaffected or even decrease if the fracture toughness increase with temperature has an overwhelming effect.

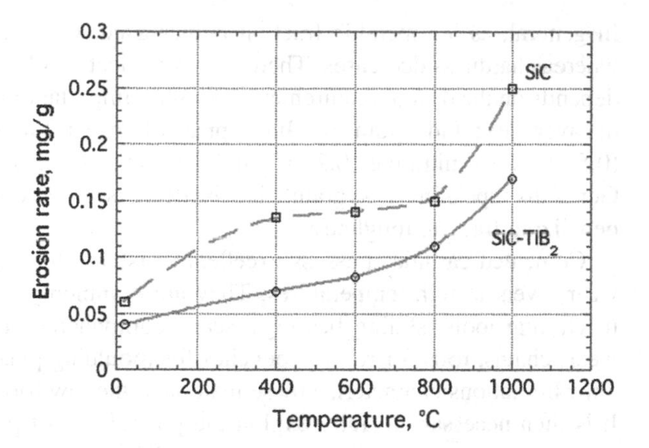


**Fig. 6.17** Erosion rate as a function of temperature for a silicon carbide and a silicon carbide-titanium diboride composite (modified from [42])

## 6.6 Cemented Carbides

Cemented carbides, also called *hardmetals* or *cermets*, have been developed with the objective of obtaining ceramic materials with improved fracture toughness. These materials are produced by sintering powder mixes of metal carbides, such as WC, TiC and TaC, and about 3–30 % of metal powder, such as cobalt or nickel. During sintering, the metallic powder melts forming a liquid film that wets the carbide particles thus activating the sintering process. The result is a material characterized by high hardness and stiffness, coupled with a fracture toughness of about 15 MPa $m^{1/2}$, i.e., definitely larger than that of advanced ceramics. This is because the metallic film (also called: *binder*) renders more difficult both crack nucleation and propagation.

Most cemented carbides contain tungsten carbides in a cobalt matrix. Materials containing different fractions of other carbides, such as TiC and TaC (the so-called *ternary carbides*) are also produced. Table 6.7 summarizes some types of cemented carbides, indicated with the ISO designation, with their compositions and main mechanical properties. It can be noted that these materials display high hardness, high values of the elastic modulus (it is about three times that of steel), and a high density.

**Table 6.7** Composition and mechanical properties of some common cemented carbides

| Designation ISO | WC (%) | TiC + TaC (%) | Co (%) | Density ($g/cm^3$) | Hardness ($kg/mm^2$) | Elastic modulus (GPa) |
|---|---|---|---|---|---|---|
| P20 | 79 | 13 | 8 | 12.1 | 1580 | 530 |
| P40 | 85 | 5 | 10 | 13.4 | 1420 | 540 |
| K20 | 94 | | 6 | 14.8 | 1650 | 610 |
| K40 | 89 | | 11 | 14.1 | 1320 | 560 |
| M20 | 82 | 10 | 8 | 13.3 | 1540 | 560 |
| M40 | 86 | 6 | 10 | 14 | 1380 | 530 |

In general, as the metallic fraction is increased, fracture toughness also increases whereas hardness decreases. Therefore, the selection of a suitable cemented carbide depends on the design requirements. Another important microstructural parameter is the average carbide grain size. It is typically between 1 and 6 μm, but also submicron (0.5–0.8 μm), ultrafine (0.2–0.5 μm) and nano (<0.2 μm) powders can be used. Generally speaking, as grain size is decreased, hardness is increased without penalizing fracture toughness.

Cemented carbides possess excellent resistance to sliding wear and to abrasion wear, even at high temperatures. They are commonly used in the construction of machining tools, sliding bearings, seals, components or parts subject to abrasive wear, chains, rolls for rolling, punches for moulding plates, matrices for extrusion. Two limitations characterize these materials: the low formability and the high cost. It is then necessary to fully exploit the potentiality of powder metallurgy for producing near net shape components. The allowed machining processes are electro-discharged machining (EDM) and grinding. Because of this, cemented carbides are often deposited as surface layers by the HVOF technique (Fig. 7.18).

### *6.6.1 Sliding Wear*

Cemented carbides display a very high sliding wear resistance. Table 5.2 shows that they are characterized by $K_a$ values around $5 \times 10^{-16}$ m$^2$/N when dry sliding against steel. In general, wear rate decreases as hardness is increased, i.e., as the amount of metallic binder and the average carbide grain size are decreased. Figure 6.18 shows the experimental dependence of hardness and $K_a$ with the binder content, for a WC-Co hardmetal with a carbide grain size of around 1 μm, dry sliding against steel (H: 200 kg/mm$^2$) [44]. It has been observed that wear was characterized by the progressive removal of the cobalt binder, accompanied by the brittle fracture and fragmentation of the carbide grains.

As shown in Fig. 6.14, cemented carbides are able to retain a relatively high hardness also at high temperature. Therefore, they are also used in dry sliding,

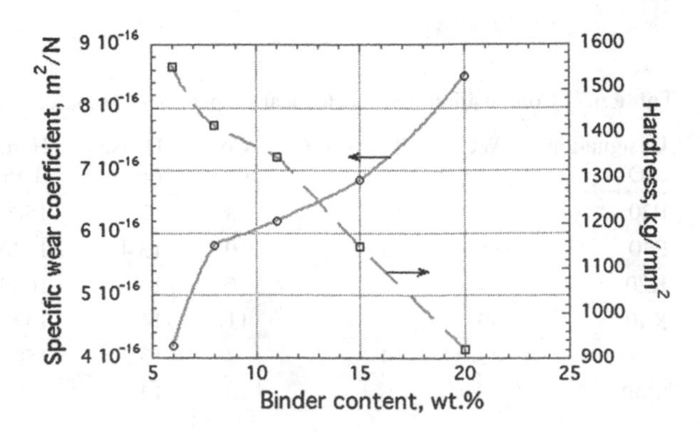


**Fig. 6.18** Specific wear coefficient and hardness versus binder content in case of WC-Co dry sliding against a steel counterface (modified from [44])

where high temperatures are reached in the contact regions, such as in cutting tools. Under these conditions these materials exhibit excellent wear resistance when mating cast iron or non-ferrous alloys, but they do not afford a good performance in sliding against steel. In fact, at high contact temperatures (above 800 °C) tungsten carbide grains decompose and then carbon diffuses easily into the austenitic steel phase of the counterface (the chip in case of cutting tools). This leads to a considerable reduction in mechanical strength with a corresponding increase in adhesive wear. However, it has been verified that the addition of titanium and tantalum carbides significantly increases the wear resistance, because these carbides are more stable at high temperature.

## *6.6.2 Abrasive Wear by Hard, Granular Material*

Cemented carbides display a high abrasion resistance under low-stress as well as high-stress abrasion (Sect. 5.4). Figure 6.19 shows the results of abrasion tests carried out using silica particles in the size range 125–180 μm and a testing apparatus similar to the DSRW (but using a metal wheel without a rubber rim) [45]. The results are in the typical range of low-stress abrasion, and highlight the role of the fraction of binder and, most importantly, the role of the grain size. In particular, it is noted that the specific wear rate is very low when the grain size is below 1 μm. Authors observed that the ultrafine grades behave as a one-phase material, without the fragmentation and displacement of the individual WC grains during the abrasive interaction. Ultrafine materials were obtained by sintering nanocrystalline powder produced by the spray conversion process and contained 0.8 % vanadium carbide as grain growth inhibitor.

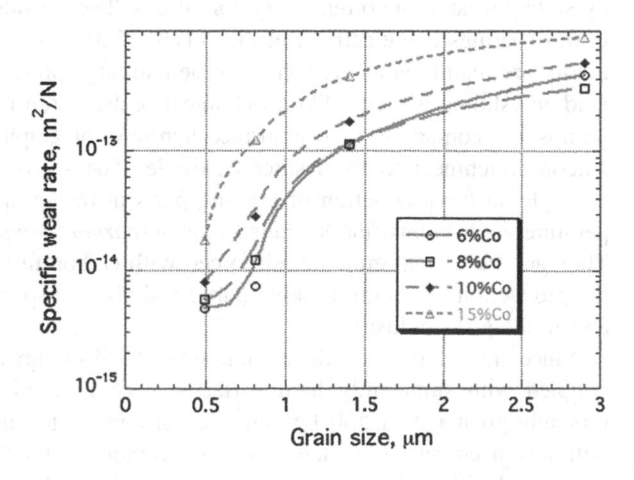


**Fig. 6.19** Effect of grain size and binder content on the abrasive wear resistance of WC-Co cemented carbides (modified from [45])

## 6.7 Graphite and Diamond

Graphite and diamond are two allotropic forms of carbon. They are characterized by very different properties but both are quite important in tribology.

*Graphite* has the crystal structure shown in Fig. 3.2a. The adsorption on the carbon planes of molecules from the external environment, such as water vapour, provides this material the capability of a solid lubricant. As a solid lubricant, graphite is typically employed as a powder (Sect. 3.1.1). In many applications it is also used as monolithic component. In this latter case, the process route consists in mixing fine natural coal or coke with natural graphite (in the typical ratio 80/20) together with pitch. The mix is then extruded or compression moulded to obtain finished parts with simple geometry, or blanks such as rods, plates and tubes to be further machined and assembled. The products are sintered at about 1000 °C, and the remaining porosity is removed by hot isostatic pressing or it is impregnated with a phenolic resin or a metal.

The main properties of graphite are:

- Low density (about 1.8 g/cm$^3$);
- Low hardness (less than 100 kg/mm$^2$);
- Low elastic modulus (between 15 and 20 GPa, depending on the porosity content;
- High thermal conductivity (which varies from 20 to 180 W/mK, and can even reach values of 400 W/mK in products with a high graphite content).

The properties of industrial graphite depend much on the characteristics of the starting materials and the production route. For example, by sintering at high temperature (above 2500 °C) and for long times, the degree of graphitization is increased, and materials with increased thermal conductivity are obtained [1].

Industrial graphite is employed in several tribological applications characterized by sliding wear, quite often in dry conditions. The extruded grades are mainly used in dry bearings and electrical bushes (Table 5.3), while the compression moulded grades are mainly employed in more demanding applications, in terms of applied load and sliding velocity, like mechanical seals, piston rings and vanes in vacuum pumps and compressors. The surface hardness of graphite can be increased by a silicon enrichment to form silicon carbide. The siliconized grades are used, for example, in the production of braking parts in racing applications. For high temperature applications, the so-called *carbon-carbon composites* are also produced. They are made from graphite reinforced with carbon fibres, and are widely used in the production of aircraft braking parts, and also for special aerospace applications and in the glass industry.

Since the materials usually contain up to 15 % of abrasive ash, graphite has to be coupled with sufficiently hard surfaces, such as hardened steel (with hardness typically greater than 500 kg/mm$^2$), chromium plated materials, materials coated with advanced ceramics such as silicon carbide. At relatively small loads and low sliding velocities, sliding wear (by adhesion) is mild and the friction coefficient is low. Typical values of friction coefficient are in the range 0.06 and 0.2, and the specific wear rate is around $10^{-15}$ $m^2$/N. If the contact temperature reaches a critical value of about 350 °C, the surface desorption of the water molecules takes place and the so-called *dusting wear regime* is entered. The friction coefficient increases to 0.4–0.5, and also wear rate increases [40]. For the estimation of contact temperature, it has to be considered that the specific heat of graphite ranges between 0.71 and 0.83 kJ/kg °C, and is around 1.4 kJ/kg °C for carbon-carbon composites [46].

*Diamond* is characterized by the highest hardness (up to 10,000 kg/$mm^2$) of all known materials, the highest modulus of elasticity (around 1000 GPa), the highest thermal conductivity (around 2000 W/m °C) and the lowest coefficient of thermal expansion ($0.8 \times 10^{-16}$ °$C^{-1}$) [47]. In industrial applications, the synthetic *polycrystalline diamond* (also referred to as PCD) is typically used. PCD powders are obtained from graphite, by heating it at high temperature and high pressure in the presence of a catalyst. Diamond powder is used as a super-abrasive in grinding and polishing. It may be also sintered with metallic binders (such as cobalt and nickel) to obtain particular parts, such as machining tools where PCD is typically sintered on a hard metal. The applications of PCD machining tools are substantially similar to those of cemented carbides, with the difference that the high hardness allows working extremely abrasive materials, such as metal matrix composites and ceramics. PCD tools are also not suitable for cutting ferrous alloys, since the high contact temperatures allow carbon to diffuse into the ferrous alloy that is typically austenitic, and the tool may be severely damaged.

The polycrystalline diamond is also produced as a coating by CVD (Chemical Vapour Deposition). During deposition, the temperature of the substrate is variable between 700 and 1000 °C. The coatings cannot be realized on steels since the deposition temperature is too high. They are typically realized on cemented carbides. A limitation for the use of these coatings is given by the poor adhesion with substrate that can be improved by suitable surface treatments before diamond deposition, such as mechanical scratching or chemical etching to remove the cobalt from the surface in the case of deposition on Co-based cemented carbides [48]. Another possible limitation is given by the high surface roughness, due to the faceted morphology of the coatings (Fig. 6.20 shows an example).

The polycrystalline diamond is characterized by a very low coefficient of friction (between 0.02 and 0.05) in the case of dry sliding against itself or a ceramic material. Figure 6.21 shows the evolution of friction coefficient in the case of dry sliding of a PCD against granite. The tests were conducted in a block on ring configuration, at three loads: 200, 400 and 600 N. It is noted that during a short running in stage the friction coefficient is higher than 0.1 but then it decreases reaching very low values. Such low values are determined by a combination of high hardness and low surface energy. It is believed that the low surface energy of PCD, which gives rise to a low work of adhesion, is due to adsorption of water molecules from the external environment [40]. The specific wear rates are also very low, between $10^{-17}$ and $10^{-16}$ $m^2$/N.

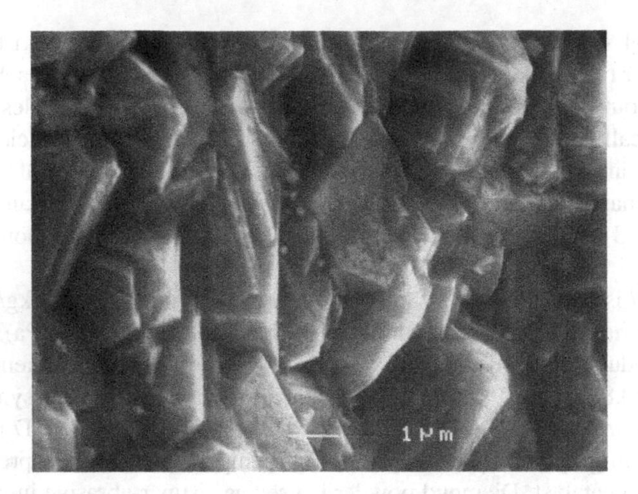


**Fig. 6.20** SEM micrograph showing the faceted morphology of a CVD PCD deposited on WC-Co [48]

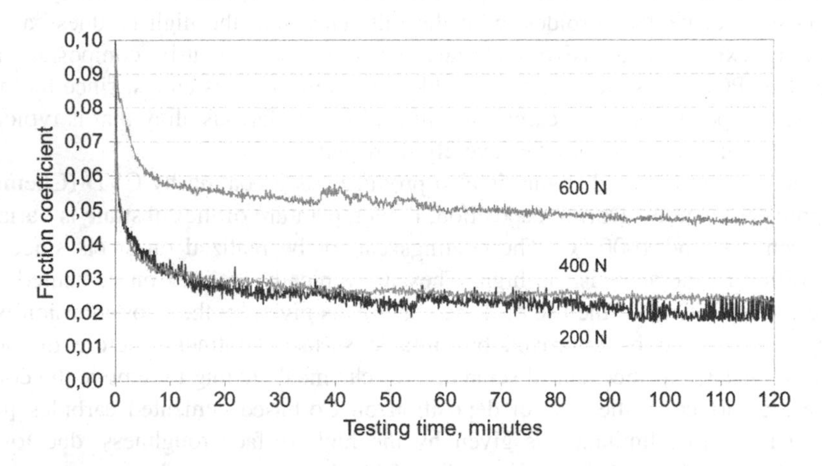


**Fig. 6.21** Evolution of the friction coefficient in the case of dry sliding of a PCD against granite

## 6.8 Polymers

Polymeric materials are widely used in tribology, especially in applications where dry sliding wear occurs (in particular against a metal counterface). They are often produced with the addition of short fibres or particles, to increase their strength, or to provide them with self-lubricating properties. Among the advantages of using polymers are: low compatibility versus metals (due to the low surface energy), high chemical inertness in several aggressive environments, self-lubricating properties, low density (around 1 g/cm$^3$), ability to absorb vibrations and shocks. In addition, polymers can be quite easily shaped, mainly by extrusion and injection moulding. Semi-finished rods, tubes, plates and sheets are also produced, and they are easily machined and joined together. Examples of applications include plain bearings (typically operating in the absence of lubrication), various types of gears (with the additional characteristic of being little noisy due to their damping capability), piston rings (running dry), seals, biomedical applications (several polymers show excellent biocompatibility). In comparison to metals, polymers have the disadvantage of a lower strength (partially limited by the use of reinforcements) and a lower temperature resistance.

Polymers are divided into two groups: *thermoplastics* and *thermosets*. Thermosetting polymers are stronger and stiffer than thermoplastics but they do not show a melting phenomenon and are more difficult to process and shape. Thermosets include phenolic materials, epoxy resins and polyimides. Thermoplastic polymers are most widely used in tribological applications. In the solid state they can be *amorphous* or *partially crystalline*. Figure 6.22 provides a schematic representation of the main mechanical properties of thermoplastics with respect to temperature ($T_g$: glass transition temperature; $T_f$: melting temperature), in the case of amorphous polymers and 100 % crystalline polymers (partially crystalline polymers display intermediate behaviour). Depending on the position of $T_g$ (and $T_f$) with respect to room temperature, four classes of materials can be recognized:

- A: Amorphous materials with high $T_g$-values (including polycarbonate, PC, with $T_g$ = 150 °C, and polymetyl methacrilate, PMMA, with $T_g$ = 105 °C). At room temperature they are mainly hard and brittle since room temperature is far below their $T_g$. These polymers display brittle behaviour since their chains are interlocked, making plastic flow difficult. In the presence of surface cracks, these materials exhibit brittle contact.

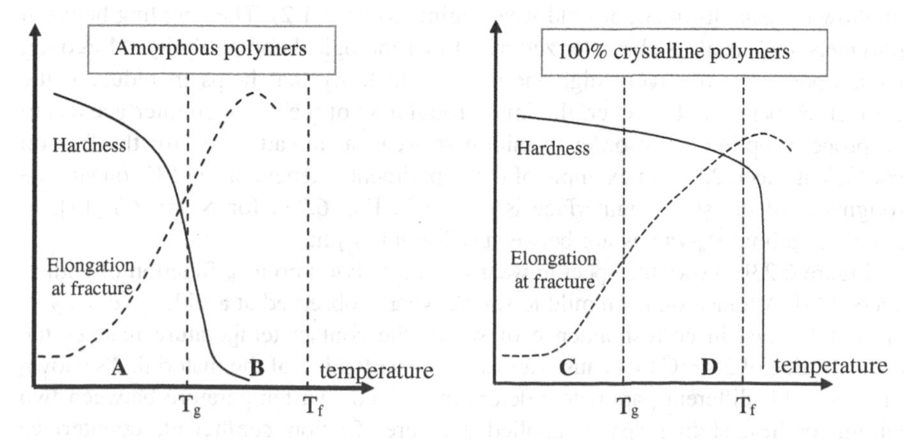


**Fig. 6.22** Effect of temperature on hardness and ductility of polymers. Regions *A*, *B*, *C* and *D* identify four classes of polymers that at room temperature show a distinctive mechanical behaviour (see text for more details)

- B: Amorphous materials with low $T_g$-values (including rubber with $T_g = -73$ °C). At room temperature they are largely above their $T_g$ and they are therefore very soft and display high fracture strain.
- C: Partially crystalline materials with high $T_g$-values (including polyamides, PA, like Nylon 66 with $T_g = 55$ °C, polyether ether ketone, PEEK, with $T_g = 148$ °C). At room temperature they are quite hard and strong, and they fracture at a limited elongation (of the order of 5 %). These materials my give brittle contact at low temperature.
- D: Partially crystalline materials with low $T_g$-values (including high-density poliethylene, HDPE, with $T_g = -100$ °C, polyoximethylene, POM, with $T_g = -80$ °C and typically a low content of amorphous phase, polytetrafluoroethilene, PTFE, with $T_g = -90$ °C). At room temperature they are quite hard and ductile. They typically give a viscoplastic contact.

The most common polymers employed in tribology are: polyamide (PA, like Nylon 66); polyacetals or polyoximethylene (POM); polytetrafluoroethilene (PTFE); high-density poliethylene (HDPE) and ultrahigh molecular weight polyetylene (UHMWPE); polyimide (PI). In most engineering applications they are employed with the addition of different types of particles or short fibers, i.e., in the form of *composites*. Table 6.8 lists some characteristics of polymers used in tribology. The data are only indicative because they may vary greatly depending on the type of polymer and production process.

### 6.8.1 Sliding Wear

Polymeric materials display a high sliding wear resistance against metals, and in particular against steel, as long as the contact temperature remains sufficiently low to allow the establishment of mild wear regime (Sect. 5.1.2). The coupling between polymers and metals is characterized by a low tribological compatibility. Moreover, metals possess a relatively high thermal conductivity that helps in reducing the contact temperature. However, the initial roughness of the metal counterface has to be properly optimized in order to minimize wear, as already seen for the friction coefficient (Sect. 2.9). An example of the experimental dependence of $K_a$ on surface roughness of the steel counterface is shown in Fig. 6.23a for Nylon 66 [17]. In general, optimal $R_a$-values are between 0.2 and 0.4 μm.

Figure 6.23b shows the specific wear rate of a Nylon 66 as a function of sliding speed [49]. A transition from mild to severe wear is observed at a sliding velocity of around 10 m/s, in correspondence of which the contact temperature reaches the critical value of 250 °C that causes an excessive softening of the material. As shown in Sect. 2.11, different parameters determine the contact temperature between two sliding bodies: sliding speed, applied pressure, friction coefficient, counterface material, ambient temperature and system geometry. Figure 6.24 shows the dry sliding wear map of a polyimide containing 15 % of graphite, obtained using a

**Table 6.8** Main characteristics of some polymers widely used in tribological applications

| Material | Density ($g/cm^3$) | Elastic modulus (GPa) | Maximum operating temperature (°C) | $K_a$ ($m^2/N$) | Observations |
|---|---|---|---|---|---|
| HDPE | 0.96 | 0.4–1.2 | 120 | $10^{-14}$ to $10^{-15}$ | Used for piping, toys, houshold ware |
| | | | | | UHMWPE is used in orthopedic implants |
| Polyamide (nylon 66) | 1.14 | 3.3 | 180 | $10^{-14}$ to $10^{-15}$ | Widely used in gears, bearings, bushes |
| Polyamide + graphite | 1.2 | 4 | 180 | $\approx 10^{-15}$ | Graphite reduces friction and wear |
| Polyacetal (POM) | 1.4 | 2.9–3.3 | 140 | $\approx 10^{-15}$ | Excellent dimensional stability; high mechanical properties; recommended for precision parts |
| PTFE (teflon®) | 2.1–2.2 | 0.48–0.76 | 260 | $\approx 10^{-13}$ | Produced by powder sintering; excellent self-lubricating properties; high chemical inertness |
| PTFE + glass or carbon fibers | | | 260 | $\approx 10^{-14}$ | Reinforced to increase strength |
| Polyimide | 1.42 | 2–3 | 320 | $\approx 3 \times 10^{-15}$ | Excellent mechanical properties (up to 250 °C). Quite expensive |
| Polyimide + graphite | 1.5 | 5–14 | 320 | $\approx 3 \times 10^{-16}$ | Graphite reduces friction and wear |
| Phenolics | 1.4 | 50 | 200 | $\approx 10^{-14}$ | |
| Phenolics + PTFE | 1.4 | 50 | 200 | $\approx 2 \times 10^{-16}$ | |

The specific wear rate, $K_a$, refers to dry sliding tests against steel; data obtained from various sources in the literature

thrust-bearing tester with a steel counterface [50]. For this material, the transition from mild to severe wear occurs when a surface critical temperature of 395 °C is reached. In the investigated tribological conditions, the limit value of the product between the applied pressure and sliding speed at which the contact temperature reaches 395 °C is 1 MPa m/s at 14 m/s, and 12 MPa m/s at 1.71 m/s. Table 6.9 lists some *maximum PV limits* (whose meaning is different to the $PV_{limit}$ concept introduced in Sect. 5.1.4) for several polymeric materials dry sliding against a steel counterface and obtained using a similar testing geometry [50].

As said, polymeric composites containing particles or short fibres are commonly used in different applications. To reduce friction and wear (both determined by adhesion), lubricants such as PTFE and graphite flakes are typically employed,

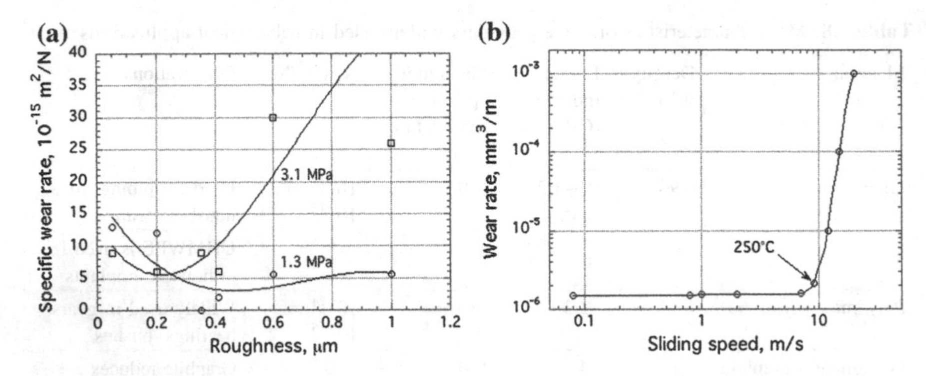


Fig. 6.23 Specific wear rate of Nylon 66: **a** role of the counterface roughness (counterface: 100Cr6, modified from [17]); **b** role of sliding velocity (steel counterface with $R_a$ = 0.15 µm, modified from [49])

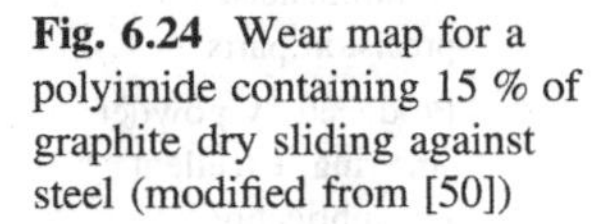
Fig. 6.24 Wear map for a polyimide containing 15 % of graphite dry sliding against steel (modified from [50])

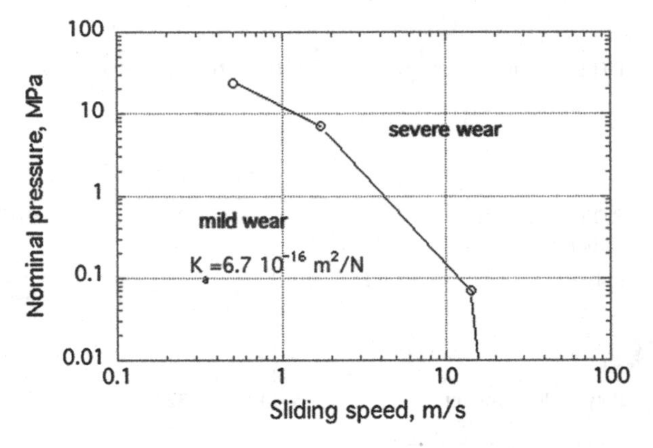


whereas glass, carbon and aramid fibres are used to increase the mechanical strength and the Young's modulus. In some cases, both types of fillers are added together, to optimize the properties. For example, polyamide 66 (Nylon 66) composites containing both reinforcing and self-lubricating additions are often used. Typical compositions are: 20 % glass + 20 % PTFE; 30 % carbon + 15 % PTFE/ silicon oil; 15 % aramid fibres + 10 % PTFE. As a further example, consider the polyimide composites listed in Table 6.9. The graphite containing composites display the best wear resistance (high PV limit), but they also display quite a high friction coefficient (around 0.31) during the run-in stage. With the addition of PTFE, the friction coefficient is 0.1 from the beginning of sliding. It should also be noted that polymers reinforced with hard fibres, such as glass fibres, could exert an abrasive action against the counterface material. Glass has a hardness of about 500 kg/mm$^2$ (Table 2.1) and the antagonists should therefore possess an hardness in excess of 700 kg/mm$^2$ in order to avoid abrasive wear. When using steels, they should be heat-treated or nitrided.

**Table 6.9** Maximum PV limits for several polymeric materials (modified from [50])

| Material | Maximum PV limit (MPa m/S) | Maximum contact temperature (°C) | Material | Maximum PV limit (MPa m/s) | Maximum contact temperature (°C) |
|---|---|---|---|---|---|
| PI-15 % graphite | 12 | 395 | PTFE | 0.064 | 260 |
| PI-40 % graphite | 12 | 395 | PTFE 15-25 % glass | 0.45 | 260 |
| PI-15 % graphite-10 %PTFE | 3.6 | 260 | PTFE-25 % carbon | 0.71 | 260 |
| Acetal | 0.27 | 120 | PTFE-60 % bronze | 0.66 | 260 |
| Acetal-PTFE | 0.12 | 120 | Nylon 66 | 0.14 | 150 |

A special mention has to be devoted to PTFE-based composites. Because of the high viscosity of PTFE even at high temperature, these materials are produced by mixing PTFE and fillers, pressing and sintering. Unreinforced PTFE is characterized by very high wear rates (Tables 6.8 and 6.9), but with proper additions wear resistance is greatly increased while maintaining the self-lubricating properties. It is supposed that fillers increase the load bearing capacity of the material and also favour the formation of a stable and compact transfer layer [51]. Recently it has been shown that the addition of low amounts of nano-fillers can decrease the PTFE wear by 4 orders of magnitude [52]. As an example, PTFE nanocomposites reinforced with α-alumina display an ultra-low steady state specific wear rate of around $10^{-16}$ $m^2/N$, after a run-in period necessary to form a high performing transfer layer [53].

A particular class of polymeric composites are the *organic friction materials*. They are used in automotive brake pads typically sliding against a grey pearlitic cast iron disc. These materials contain a thermosetting phenolic polymer (10–40 % in volume) that holds together a large amount of different additives [54]:

- Reinforcements, such as metal (steel, copper and bronze), carbon, glass and aramid fibres;
- Friction modifiers, such as graphite and metal sulphides particles that act as solid lubricants, and $Al_2O_3$, MgO and silica particles that act as abrasives;
- Fillers, such as vermiculite (a Mg silicate), barium sulphate and mica particles that improve the manufacturability and reduce cost.

At low loads and sliding speeds, dry sliding wear is mild. The specific wear rate of the organic friction materials is typically between $5 \times 10^{-15}$ and $5 \times 10^{-14}$ $m^2/N$. The friction coefficient is between 0.3 and 0.6, depending on the material composition. The so-called *low-metallic pads* contain a low amount of metal reinforcements and a high amount of abrasives and they are thus characterized by a higher friction coefficient and also a higher wear than the so-called *NAO pads*, that constitutes an other important class of organic friction materials.

At high loads and sliding speeds, severe wear occurs. The transition takes place when a critical contact temperature in the range of 200–300 °C is attained [54]. At this temperature, the thermal degradation of the phenolic polymer occurs. This causes an increase in the wear rate, which then exponentially rises with contact temperature, and a decrease in the friction coefficient. Such a decrease has to be reduced to a minimum by carefully selecting the materials ingredients, in order to avoid jeopardizing the braking effectiveness. In applications where the attainment of quite high temperatures is usual (such as in train or aeroplane brakes), it is necessary to employ more resistant materials, such as sintered metals and carbon-carbon composites.

### 6.8.2 Abrasive Wear by Hard, Granular Material

Because of their low hardness, a very low abrasion resistance characterizes polymeric materials. In applications where a minimum abrasion resistance is required, it is appropriate to employ ultra-high molecular weight polyethylene (UHMW PE, a thermoplastic polymer) or elastomeric polyurethanes with high hardness.

*Elastomers* are a class of polymers with an adequate resistance to low-stress abrasion (at least in some applications). In particular, they display a good resistance to solid particle erosion wear in case of low impact velocity. This behaviour is mainly due to their low elastic moduli and high elastic limit, resulting in a high elastic deformation and rebound capability. Beside natural rubber, the elastomers that are mainly used are styrene butadiene rubber (SBR), for example in the production of automobile tires; chloroprene (CR), for example in the production of gaskets or conveyors; polyurethanes, for example in the production of idlers, cyclones and screens [27].

In case of relatively low impact velocities, elastomeric materials display $\Phi_{SPE}$-values (representative of the efficiency of the process), in the range $10^{-3}$–$10^{-2}$ [55], and therefore similar, or lower, than those of steels. Table 6.10 shows the results of

**Table 6.10** Normalized erosion rates for different engineering materials (data from [56])

| Material | Hardness (kg/mm$^2$) | Relative erosion resistance |
|---|---|---|
| AISI 1018 | 254 | 1 |
| Natural rubber | 40 Shore A | 14 |
| Natural rubber | 60 Shore A | 3.6 |
| Polyurethane | 90 Shore A | 1.1 |
| 35 % Cr 5 % C white iron | 780 | 5.8 |
| 27 % Cr 3 % C white iron | 621 | 3.1 |
| Martensitic stainless steel | 308 | 1.1 |
| WC-3.3 % Co | 2300 | 59 |

erosive tests carried out at low impact angles (lower than 10°), an impact velocity ranging from 14 to 24 m/s and using SiC erosive particles with high angularity [56]. The tests were conducted on two natural rubbers, a polyurethane and several materials for use in comparable applications. The results are expressed in terms of the relative erosion resistance (RER) by dividing the erosion rate of the reference AISI 1018 steel by that of the tested material. The results show that the natural rubbers, in particular the softer one, exhibit better erosion resistance than the polyurethane and also the martensitic stainless steel. The softest rubber also shows a better performance than the two cast irons under study. This behaviour was attributed by the Authors to the higher propensity of natural rubber to elastic recovery. As expected, the hard metal displays the best behaviour.

The applicability of elastomers in engineering applications is limited to a defined temperature range, between their $T_g$ and their melting or decomposition temperature, which is typically in the range 80–130 °C.

## References

1. W.A. Glaeser, *Materials for Tribology* (Elsevier, Amsterdam, 1992)
2. D.A. Rigney (ed.), *Fundamentals of Friction and Wear of Materials* (ASM, Metals Park, 1981)
3. M. Takeda, T. Onishi, S. Nakakubo, S. Fujimoto, Physical properties of iron-oxide scales on Si-containing steels at high temperature. Mater. Trans. **50**, 2242–2246 (2009)
4. H. Czichos, K.H. Habig, Wear of medium carbon steel—a systematic study of the influences of materials and operating parameters. Wear **110**, 389–400 (1986)
5. S.C. Lim, M.F. Ashby, Wear-mechanism maps. Acta Metall. Mater. **35**, 1–24 (1987)
6. P. Clayton, Tribological aspects of wheel-rail contact: a review of recent experimental research. Wear **191**, 170–183 (1996)
7. H.K.D.H. Bhadeshia, Steels for bearings. Prog. Mater. Sci. **57**, 268–435 (2012)
8. A. Bhattacharyya, G. Subhash, N. Arakere, Evolution of subsurface plastic zone due to rolling contact fatigue of M-50 NiL case hardened bearing steel. Int. J. Fatigue **59**, 102–113 (2014)
9. I. Hucklenbroich, G. Stein, H. Chin, W. Trojhan, E. Streit, High nitrogen martensitic steel for critical components in aviation. Mater. Sci. Forum **318–320**, 161–166 (1999)
10. A.R. Lansdown, A.L. Price, *Materials to Resist Wear* (Pergamon Press, Oxford, 1986)
11. Z.Y. Yang, M.G.S. Naylor, D.A. Rigney, Sliding wear of 304 and 310 stainless steels. Wear **105**, 73–86 (1985)
12. K.H. Zum Gahr, *Microstructure and Wear of Materials* (Elsevier, Amsterdam, 1987)
13. N.S. Tiedje, Solidification, processing and properties of ductile cast iron. Mater. Sci. Technol. **26**, 505–514 (2010)
14. G. Straffelini, M. Pellizzari, L. Maines, Effect of sliding speed and contact pressure on the oxidative wear of austempered ductile iron. Wear **270**, 714–719 (2011)
15. G. Straffelini, L. Maines, The relationship between wear of semimetallic friction materials and pearlitic cast iron in dry sliding. Wear **307**, 75–80 (2014)
16. P.W. Leach, D.W. Borland, The unlubricated wear of flake graphite cast iron. Wear 85, 247–256 (1983)
17. K.H. Czichos, K.H. Habig, *Tribologie Handbuch, Reibung und Verlschleiss* (Vieweg, Wiesbaden, 1992)
18. N.A. Waterman, M.F. Ashby, *Elsevier Materials Selector* (Elsevier, Oxford, 1992)

19. R.C. Dommarco, J.D. Salvande, Contact fatigue resistance of austempered and partially chilled ductile irons. Wear **254**, 230–236 (2003)
20. V. Fontanari, M. Benedetti, G. Straffelini, Ch. Girardi, L. Giordanino, Tribological behavior of the bronze-steel pair for worm gearing. Wear **302**, 1520–1527 (2013)
21. G. Straffelini, L. Maines, M. Pellizzari, P. Scardi, Dry sliding wear of Cu-Be alloys. Wear **259**, 506–511 (2005)
22. C.T. Kwork, P.K. Wong, H.C. Man, F.T. Cheng, Sliding wear and corrosion resistance of copper-based overhead catenary for traction systems. IJR Int. J. Railway **3**, 19–27 (2010)
23. M.M. Khonsari, E.R. Booser, *Applied Tribology*, 2nd edn. (Wiley, West Sussex, 2008)
24. M.J. Neale, M. Gee, *Guide to Wear Problems and Testing for Industry* (Professional Engineering Publishing, London, 2000)
25. W.M. Rainforth, A.J. Leonard, C. Perrin, A. Bedolla-Jacuinde, Y. Wang, H. Jones, Q. Luo, High resolution observations of friction-induced oxide and its interaction with the worn surface. Tribol. Int. **35**, 731–748 (2002)
26. G. Straffelini, A. Molinari, Mild sliding wear of Fe-0.2 % C, Ti-6 % Al-4 % C and Al-7072: a comparative study. Tribol. Lett. **41**, 227–238 (2011)
27. K.G. Budinski, M.K. Budinski, *Engineering Materials, Properties and Selection* (Prentice Hall, Upper Saddle River, 2002)
28. S. Wilson, A.T. Alpas, Wear mechanism maps for metal matrix composites. Wear **212**, 41–49 (1997)
29. A. Molinari, G. Straffelini, B. Tesi, T. Bacci, Dry sliding wear mechanisms of the Ti6Al4 Va alloy. Wear **208**, 105–112 (1997)
30. A. Molinari, G. Straffelini, B. Tesi, T. Bacci, G. Pradelli, Effects of load and sliding speed on the tribological behaviour of Ti-6Al-4V plasma nitrided at different temperatures. Wear **203–204**, 447–454 (1997)
31. R. Bailey, Y. Sun, Unlubricated sliding friction and wear characteristics of thermally oxidized commercially pure titanium. Wear **308**, 61–70 (2013)
32. G. Straffelini, A. Andriani, B. Tesi, A. Molinari, E. Galvanetto, Lubricated rolling-sliding behaviour of ion nitrided an untreated Ti-6Al-4V. Wear **256**, 346–352 (2004)
33. B. Bushan, *Introduction to Tribology* (Wiley, New York, 2002)
34. H.S. Kong, M.F. Ashby, Wear mechanisms in brittle solids. Acta Metall. Mater. **40**, 2907–2920 (1992)
35. S.M. Hsu, M. Shen, Wear prediction of ceramics. Wear **256**, 867–878 (2004)
36. K. Adachi, K. Kato, N. Chen, Wear map of ceramics. Wear **203–204**, 291–301 (1997)
37. K. Kato, Wear in relation to friction—a review. Wear **241**, 151–157 (2000)
38. J. Takadoum, *Materials and Surface Engineering in Tribology* (Wiley, London, 2008)
39. J. Xu, K. Kato, Formation of tribochemical layer of ceramics sliding in water and its role for low friction. Wear **245**, 61–75 (2000)
40. B. Bushan (ed.), *Modern Tribology Handbook*, vol. 2 (CRC Press, Boca Raton, 2001)
41. J.G. Chacon-Nava, F.H. Stott, S.D. de la Torre, A. Martinez-Villafane, Erosion of alumina and silicon carbide at low-impact velocities. Mater. Lett. **55**, 269–273 (2002)
42. A.F. Colclough, J.A. Yeomans, Hard particle erosion carbide and silicon carbide-titanium diboride from room temperature to 1000 °C. Wear **209**, 229–236 (1997)
43. I.M. Hutchings, *Tribology* (Edward Arnold, London, 1992)
44. J. Pirso, S. Letunovits, M. Voljus, Friction and wear behaviour of cemented carbides. Wear **257**, 257–265 (1997)
45. J. Pirso, M. Voljus, K. Juhani, S. Letunovits, Two-body dry abrasive wear of cermets. Wear **266**, 21–29 (2009)
46. B. Vankataraman, G. Sundararajan, The influence of sample geometry on the friction behaviour of carbon-carbon composites. Acta Metall. Mater. **50**, 1153–1163 (2002)
47. F. Cardarelli, *Materials Handbook*, 2nd edn. (Springer, New.York, 2008)
48. G. Straffelini, P. Scardi, A. Molinari, R. Polini, Characterization and sliding behaviour of HF CVD diamond coatings on WC-Co. Wear **249**, 461–472 (2001)

49. D.C. Evans, J.K. Lancaster, *Wear, Treatise on Materials Science and Technology* (Academic Press, London, 1979)
50. Vespel line Design Handbook, Du Pont. http://dupont.com/vespel
51. J. Khedkar, I. Negulescu, E. Meletis, Sliding wear behavior of PTFE composites. Wear **252**, 361–369 (2002)
52. K. Friedrich, Z. Zhang, A.K. Schlarb, Effects of various fillers on the sliding wear of polymer composites. Compos. Sci. Technol. **65**, 2329–2343 (2005)
53. J. He, H.S. Khare, D.L. Burris, Quantitative characterization of solid lubricant transfer film quality. Wear **316**, 133–143 (2014)
54. A.E. Anderson, Friction and wear of automotive brakes. ASM Handb. **18**, 569–577 (1992)
55. S. Arjula, A.P. Harsha, Study of erosion efficiency of polymers and polymer composites. Polym. Test. **25**, 188–196 (2006)
56. L.C. Jones, Low angle scouring erosion behaviour of elastomeric materials. Wear **271**, 1411–1417 (2011)

# Chapter 7
# Surface Engineering for Tribology

The functional modification of the surfaces is in many cases the best way to control the tribological damage of a component. Following a *surface engineering* approach, it is possible to choose the base material, or *substrate*, with tailored properties (for example, special mechanical or workability properties) and delegate to the modified surface the role of counteracting the tribological loadings. In most cases, the choice of a proper surface treatment results in cost saving, since it allows using cheaper substrate materials. The use of a low-cost substrate may compensate the additional costs of the surface treatment, required to improve system performances [1–4].

The choice of the optimal surface treatment is a complex task, given the high number of parameters that are involved and also the high number of available technologies. The choice should be primarily driven by the need of counteracting a specific wear mechanism that is expected to operate in the tribological system under consideration. Note, in this regard, that a particular surface treatment might be also requested to improve the corrosion resistance of the treated part in specific environments.

In Sect. 3.1 some surface treatments specifically targeted at reducing the friction coefficient are described. This chapter is mainly focussed on the available surface treatments for improving the wear resistance of materials. Relevant technological aspects of each treatment and the performances of the treated surfaces under different wear conditions will be outlined, to help the designer in the correct selection of a proper surface treatment.

## 7.1 Surface Functional Modifications: General Aspects

Surface engineering treatments can be broadly divided into four categories:

- Treatments for microstructural modification;
- Thermochemical diffusion treatments;
- Chemical conversion coatings;
- Surface coatings (including: metallic plating, thin and thick coatings).



G. Straffelini, *Friction and Wear*, Springer Tracts in Mechanical Engineering,
DOI 10.1007/978-3-319-05894-8_7

In Table 7.1 the main surface treatments used in tribology are listed. Some special aspects, to be thoroughly considered in the selection process are:

(1) the types of materials that have to be treated with the selected technology;
(2) the process temperature, i.e., the temperature reached by the substrate during the treatment;
(3) the maximum hardness of the surface layer;
(4) the thickness of the hardened surface layer.

The knowledge of the maximum temperature reached by the substrate during the treatment cycle is important to verify if the substrate can actually sustain the treatment. For example, steels should not be subjected to a coating treatment with the CVD technique at a temperature of 1000 °C, because they would loose their mechanical strength (as a matter of fact, CVD coatings are generally realized on hard metals). Furthermore, it should also be considered that treatments carried out at temperatures above around 500 °C easily induce geometrical distortions in the treated parts. If necessary, appropriate preheating or post-heating cycles should be performed, to minimize the effects of too high heating and cooling rates.

The hardness and thickness of the treated layers are definitively the two most important parameters from the tribological point of view. In some treatments, such as the thermochemical diffusion treatments, a hardness gradient is attained rather than a uniform hardness in the surface layer. As an example, Fig. 7.1 shows the typical hardness profiles resulting from carburizing and nitriding treatments of steels. The profiles were obtained from microhardness tests on metallographically prepared cross sections. It is shown that hardness is maximum at the surface and then decreases moving towards the interior of the material, reaching finally the typical value of the substrate. The treatment thickness is defined by the *effective thickness*, which is the thickness of the material layer with hardness greater than or equal to a given value, which is typically fixed at 550 kg/mm$^2$.

In the case of coatings, the maximum surface hardness is that of the coated material, and the treatment thickness is given by the coating thickness.

In case of wear by adhesion, tribo-oxidation and abrasion, the maximum contact stress is reached at the asperities. To counteract the damage associated to these mechanisms, it is therefore necessary to select treatments able to induce a very high hardness right in the outer layer of the treated part. For example, ceramic coatings on high-speed steels obtained with the PVD technique are particularly suitable for improving the tribological resistance of cutting tools that, in operation, undergo adhesive (and, possibly, abrasive) wear. For abrasion, it is obviously necessary to consider also the hardness of the abrasive particles and choose a surface treatment able to produce a sufficiently high surface hardness (along with an adequate fracture toughness in case of high-stress abrasive wear). In case of damage by contact fatigue, the maximum contact stress is located at a certain depth (the depth $z_m$, as illustrated in Sect. 1.1), and it is therefore necessary to choose a treatment capable of giving an adequate effective thickness. As a rule of thumb, a high hardness should be assured for a depth of at least $2z_m$.

**Table 7.1** Main characteristics of some widely employed surface treatments

| Treatment | Metals treated | Processing temperature (°C) | Maximum surface hardness (kg/mm$^2$) | Typical thickness (mm) |
|---|---|---|---|---|
| *Treatments for microstructural modification* | | | | |
| Surface rolling | Steels, Ti, and Ni alloys | Room temperature | | |
| Shot-peening | Steels, Ti, and Ni alloys | Room temperature | | |
| Flame hardening | Hardenable steels and cast irons | 850–1100 | 500–600 | 1–6 |
| Induction hardening | Hardenable steels and cast irons | 850–1100 | 500–700 | 0.2–2 |
| Laser hardening | Hardenable steels and cast irons | 850–1100 | 500–700 | 0.1–0.6 |
| *Thermochemical diffusion treatments* | | | | |
| Carburizing and carbonitriding | Low carbon steels | 800–1100 | 700–900 | 0.05–1.5 |
| Nitriding | Nitriding steels Tool steels (hot working and HSS) | 500–600 | 800–1200 | 0.025–0.5 |
| Ion implantation | All | 200–600 | 600–1100 | $<10^{-3}$ |
| *Chemical conversion coatings* | | | | |
| Phosphate coatings | Steels, Al alloys | 25–95 (drying at 200 °C) | – | Up to 0.01 |
| Hard anodizing | Mainly Al alloys | 0 | >1100 | 0.04–0.05 |
| **Surface coatings** | | | | |
| Hard chromium | Most ferrous and non ferrous alloys | <70 | 700–1200 | 0.001–0.5 |
| Electroless nickel | Most ferrous and non ferrous alloys | Room temperature (ageing between 200 and 500 °C) | Up to 1000 | 0.001–0.025 |
| Physical vapour deposition (PVD) coatings | All metals | 150–500 | 2000–2500 (TiN) | 0.001–0.003 |
| Chemical vapour deposition (CVD) coatings | All metals (with the limitation of the process temperature) | 800–1000 (lower temperatures in new techniques) | 2800–3000 (TiC) | 0.001–0.01 |
| Thermal spray coatings | All metals | <200 | 700–2000 | 0.5–1 |
| Hardfacing by welding | Steels and non ferrous alloys with melting point greater than 1100 °C | 1200–1600 | 800–2000 | 3–10 |

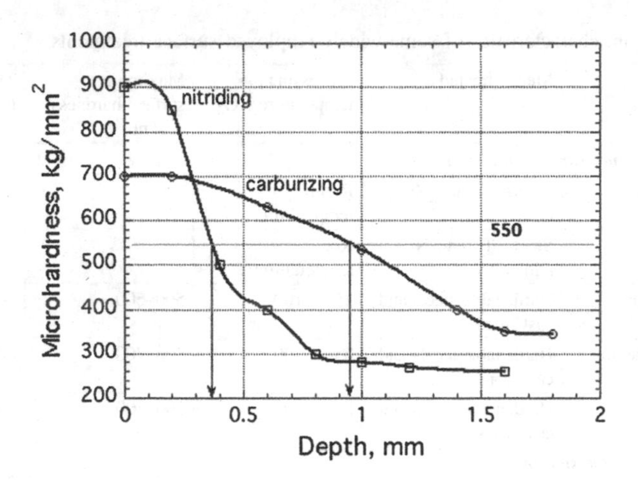


**Fig. 7.1** Typical microhardness profiles in carburized and nitrided steels

A special aspect to consider in surface engineering is the possible presence of residual stresses in the treated areas. Residual stresses are elastic in nature and they can be compressive or tensile. Generally speaking, compressive residual stresses are beneficial while tensile stresses may accelerate the tribological damage. Surface modification and thermochemical diffusion treatments mainly induce compressive residual stresses at the surface, while coatings may feature either compressive or tensile residual stresses [3, 4].

In the case of coatings, two additional parameters should be also carefully addressed:

(1) the ability of the substrate to adequately support the coating;
(2) the adhesion of the coating on to the substrate.

Wear-resistant coatings are harder and less deformable than the substrates. If they have to stand high and very localized stresses, as it occurs in non-conformal or angular contacts, the substrate may deform plastically, and the coatings may not be able to accommodate such a deformation. Coatings may then fracture in a brittle manner, as schematized in Fig. 7.2. This phenomenon can be limited by increasing

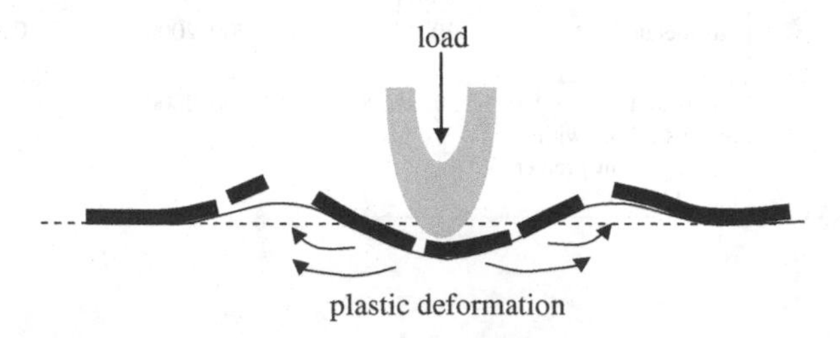


**Fig. 7.2** Fragmentation of a hard coating deposited onto a soft substrate after the application of a concentrated contact load (modified from [2])

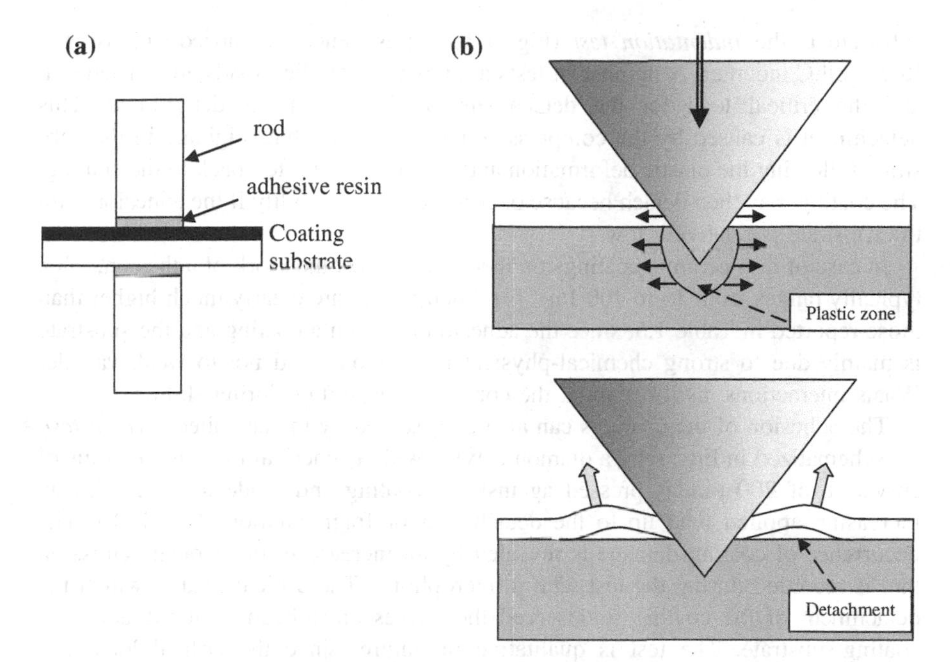


**Fig. 7.3** Schematizations illustrating the pull-off (**a**) and indentation tests (**b**) tests

the hardness of the substrate, that is, by reducing the difference in hardness and deformability between the coating and the substrate. As a rule of thumb, such a difference should not be in excess of 800 kg/mm$^2$ in the case of ceramic coatings. Another possibility is to increase the thickness of the coating. However, it must be considered that an excessive thickness might not be acceptable, especially if the coating is ceramic, since the probability of finding a critical defect for brittle contact would increase with the coating thickness.

Surface coatings can be fully effective if they are well anchored to the substrate, i.e., if the *adhesion* coating-substrate is sufficiently high. The adhesion depends on the physic-chemical interactions, which take place at the interface, and are greatly influenced by the presence of contaminations and defects. From a mechanical point of view, the adhesion can be quantified through a number of tests, including pull-off, indentation, scratch or fracture mechanics tests. In the *pull-off* test (see the schematic in Fig. 7.3a), the coated specimen is bonded on two rods, and subjected to a tensile test up to the detachment of the coating. The stress for the detachment is representative of the coating-substrate adhesion. If the adhesion is very high, the detachment occurs inside the adhesive resin (typically, an epoxy resin with a strength of around 300 MPa is used). This is clearly a limitation of this test. This test is widely used to determine the adhesion of coatings obtained by thermal spraying techniques, which are often characterized by a low adhesion.

A very simple test often used in the evaluation of the adhesion between thin ceramic coatings (such those obtained with the PVD and CVD techniques) and the substrate is the *indentation test* (Fig. 7.3b). It is generally carried out using a Rockwell C indenter. A number of tests at increasing applied loads are carried out, and the critical load for the detachment of the coating is determined. This detachment is caused by the compressive radial stresses that originate in the substrate following the plastic deformation and that are transmitted back to the coating. The coating may then detach because of compressive instability if the adhesion with the substrate is relatively low.

In case of thin ceramic coatings on metal substrates, the work of adhesion, $W_{12}$, typically ranges from 15 to 100 J/m$^2$ [4]. Such values are clearly much higher than those reported in Table 1.5, since the adhesion between a coating and the substrate is mainly due to strong chemical-physical interactions, and not to weak van der Waals interactions, as it occurs at the contacting asperities during sliding.

The adhesion of the coatings can also be assessed by the so-called *scratch test*. As schematized in Fig. 7.4a, a diamond stylus with a spherical tip, with a radius of curvature of 200 μm, is pressed against the coating and made to slide with an increasing applied load up to the detachment or fragmentation (Fig. 7.4b). The occurrence of coating damage is revealed by an increase in the acoustic emission that is recorded during the test with a microphone. The critical load at which the detachment of the coating is observed then gives an indication of the adhesion coating-substrate. The test is qualitative in nature, since the critical load also depends on the hardness of the substrate (see Fig. 7.2), the nature and intensity of the residual stresses, the friction coefficient between the stylus and the coating, and the fact that the coating can be also damaged by fragmentation, i.e., by brittle contact.

*Fracture mechanic tests* avoid all these problems. With these tests a crack is propagated at the interface between the coating and the substrate, and the energy spent in the propagation is evaluated. These tests are clearly quite complex and costly [5].

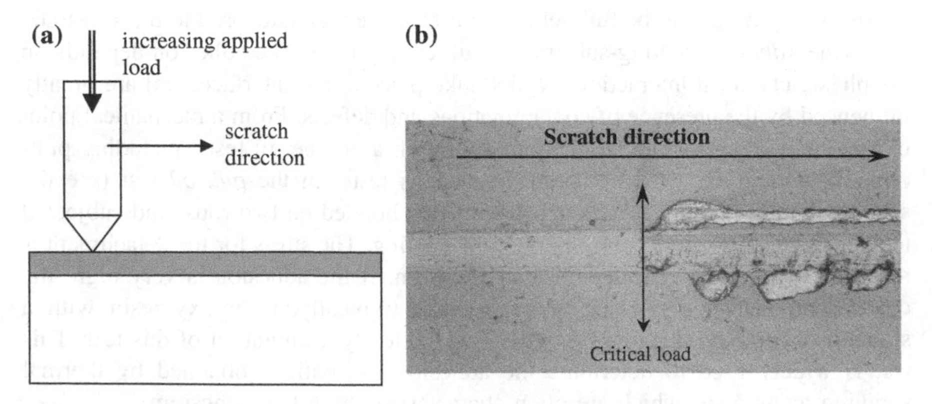


**Fig. 7.4** Schematization of the scratch test to determine the adhesion of coatings

## 7.2 Treatments for Microstructural Modification

In this type of treatments, the surface microstructure of a material is modified without altering its chemical composition. The objective is usually achieved by work hardening with mechanical treatments such as rolling or shot peening, or by inducing a martensitic transformation with surface heat treatment.

### 7.2.1 Mechanical Treatments

In *surface rolling* (a process that is however rarely used in tribology), plastic deformation is induced by a hard cylinder that rolls on the surface with an applied load able to induce a fully plastic contact (Sect. 1.1.3). In *shot peening*, a jet of small spheres (with a diameter in the range 0.2–2 mm) made by steel, ceramic or glass, is directed towards the surface to be treated with a velocity typically ranging from 40 to 120 m/s. Each ball induces a surface elastic-plastic or fully plastic contact. By adjusting the intensity of the treatment, a proper degree of surface hardening can be obtained. Shot peening is not much used to work harden soft metals, but rather to treat case hardened steel parts, with the aim of transforming the residual austenite into martensite, as will be seen in the next paragraph. Shot peening and surface rolling find some application in the closure of surface pores of sintered components obtained by powder metallurgy, to improve their sliding or contact fatigue resistance. However, in order to be really effective a re-sintering process, to weld the surface of pores that have been put in close contact by the plastic deformation, should follow the treatment. Otherwise, the closed pores may still behave as cracks and therefore easily trigger the wear by contact fatigue.

### 7.2.2 Surface Heat Treatments

Surface heat treatments are generally conducted on hardenable steels that contain a sufficient amount of carbon (0.3–0.5 %) to provide surface hardening by *martensitic transformation*. With *flame hardening*, relatively large parts can be manually treated, while with the automated *induction hardening*, parts of lower dimensions are usually treated. In induction hardening, the surface of a steel part is rapidly austenitized by the magnetic field of a water-cooled copper coil, and then quenched by spraying of water or other fluids. In case of *laser* or *electron beam* heating, the subsequent forced cooling is not necessary. The heating, in fact, is very intense and confined to a thin surface layer, so that quenching is directly induced by the surrounding material, which behaves as a low-temperature thermal reservoir.

Surface heat treatments are commonly used for critical machine parts such as gears, shafts and bearings that may be damaged in service by contact fatigue. In

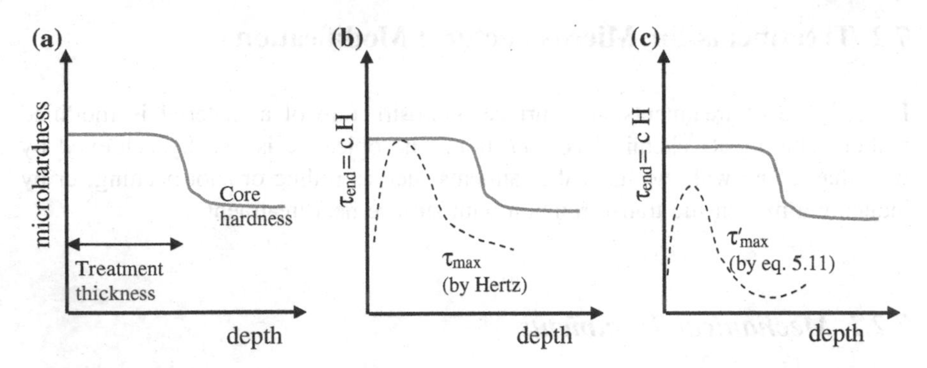


**Fig. 7.5** Surface hardening: **a** typical microhardness profile, **b** endurance limit profile with superimposed Hertzian $\tau_{max}$-profile, **c** role of compressive residual stress, $\sigma_{rs}$ (schematization)

fact, the quite high surface hardening accompanied by a relatively high case depth (see Table 7.1) produce beneficial effects on contact fatigue resistance, which can be appreciated by considering the schematization of Fig. 7.5. Figure 7.5a shows a typical microhardness profile obtained by surface hardening. It can be converted into an *endurance limit profile*. In fact, the endurance limit, usually given in terms of the Hertzian pressure, can be also expressed in terms of $\tau_{max}$, by setting $\tau_{end}$ = c H, where c is a constant and H is the material's hardness. In heat-treated steels, the endurance limit at $10^8$ cycles and $\Lambda > 3$ is given by $p_{end}$ = 3 H (see Table 5.6). In the case of a line contact, $\tau_{max} = 0.3\ p_{max}$ and thus $\tau_{end} \approx$ H, i.e., c ≈ 1 (in the case of mixed lubrication, $p_{end}$ = 2.25 H, and c = 0.75). It can be then stated that a life of $10^8$ cycles is exceeded as long as, at each depth z, $\tau_{max} < \tau_{end}$ [6].

But this is not the whole picture. To the high contact fatigue resistance of surface hardened steels and cast irons contributes also the presence of compressive residual stresses possibly present in the outer layer. They are determined by the surface martensitic transformation, which occurs with volume expansion with respect to the neighbouring phases. The thermal cycle may contribute a further mechanical term to building up the compressive stress field [7]. For example, Fig. 7.6 shows the

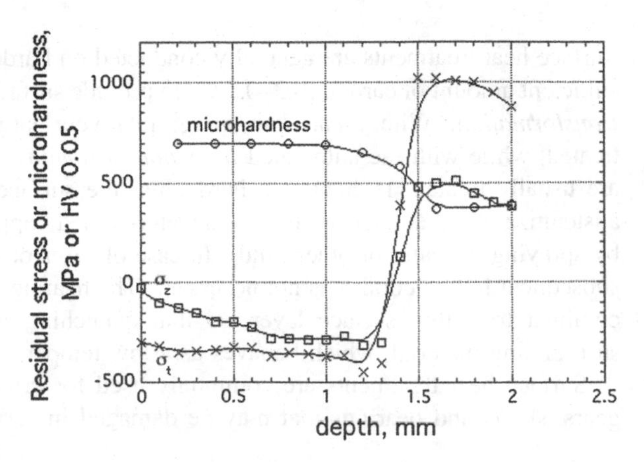


**Fig. 7.6** Microhardness profile and residual stress profile for an induction hardened AISI 4340 steel (modified from [8])

microhardness profile and the residual stress pattern in a disc made of an AISI 4340 steel firstly heat treated to achieve a quenched and tempered microstructure with hardness 367 HV and then fast induction heated and hardened [8]. The evolution of the circumferential, $\sigma_t$, and longitudinal, $\sigma_z$, stress components is reported (the radial component is 0 at the surface, and it is positive, although quite low, moving towards the interior of the disc). In the case region, both $\sigma_t$ and $\sigma_z$ are compressive and at a certain depth they assume comparable values. In the transitional region between the case and the core, i.e., between a fully martensitic and a tempered microstructure, both $\sigma_t$ and $\sigma_z$ increase in value and become tensile. In general, higher compressive stresses are obtained at increasing the difference between case and core hardness. On the basis of a number of experimental data, Lang have proposed the following relations [9]:

$$\begin{aligned} \sigma_{rs} &= -1.25 \cdot (H - H_c) \quad \text{for} \quad (\mathrm{H} - \mathrm{H_c}) < 300 \\ \sigma_{rs} &= 0.2857 \cdot (H - H_c) - 400 \quad \text{for} \quad (\mathrm{H} - \mathrm{H_c}) > 300 \end{aligned} \tag{7.1}$$

where $\sigma_{rs}$ is the residual stress in MPa (setting $\sigma_{rs} = \sigma_t = \sigma_z$), H is the hardness (in $kg/mm^2$), and $H_c$ is the hardness of the core (in $kg/mm^2$). These approximate relations are nonetheless useful for a preliminary evaluation, since the experimental determination of the residual stress profile is time consuming and expensive. The beneficial effects of residual stresses on the contact fatigue behaviour can be appreciated by considering Eq. 5.11 and the schematization of Fig. 7.5c. To a first approximation it can be supposed that no stress relaxation takes place during the fatigue loading. This assumption decreases in validity if the $\Lambda$-factor is decreased, since the plastic deformations at the contacting asperities can relieve the residual stresses.

Because of the heat cycle and the phase transformations, the surface hardened parts display geometrical distortions after the treatment, and they must submitted to a final grinding operation. This should be carefully conducted to avoid introduction of residual tensile stresses at the surface.

Consider now the wear damage by adhesion and tribo-oxidation. The performance of surface heat treated steels can be considered quite similar to that displayed by through hardened steels (Sect. 6.1.1). In fact, the presence of residual stresses at the surface does not play an important role. Such stresses are elastic and they are relieved during sliding by the high deformations at the asperities and the consequent local heating. The same reasoning is valid for the abrasive wear. Considering the maximum hardness that can be obtained at the surface, surface hardening is clearly not sufficient to solve the problem of the abrasion of steels by hard particles such as silica contaminants. However, it might be a very good choice in the case of low-stress abrasion.

## 7.3 Thermochemical Diffusion Treatments

The main thermochemical diffusion treatments of steels are carburizing and nitriding.

### 7.3.1 Carburizing

In this treatment, the steel part is heated in the austenitic field (at about 900 °C) in an appropriate medium and the surface is enriched in carbon by diffusion, up to values not exceeding 1 %. It is subsequently quenched in oil, to obtain high surface hardness by the martensitic transformation. After quenching the steel parts are tempered at about 170–180 °C. The treatment is indicated for steels containing less than about 0.4 % of carbon, to ensure high fracture toughness in the core. Figure 7.1 shows a typical microhardness profile obtained after carburizing.

Like surface heat treatments, carburizing is well suited to withstand the adhesive and tribo-oxidative wear, and wear by low-stress abrasion. However, carburizing finds its best application in mechanical parts that experience contact fatigue damage, because of the high surface hardness, the high effective thickness (Table 7.1), and the presence of compressive residual stresses in the case region. The onset of residual stresses is due to the martensitic transformation and depends on many interrelated factors, including steel composition, size and geometry of the treated part, and treatment parameters [7]. Figure 7.7a schematizes a typical residual stress profile and its role on the evolution of $\tau_{max}$. Figure 7.7b shows a comparison of $\tau_{max}$ with two $\tau_{end}$-profiles that can be obtained from the microhardness profiles as

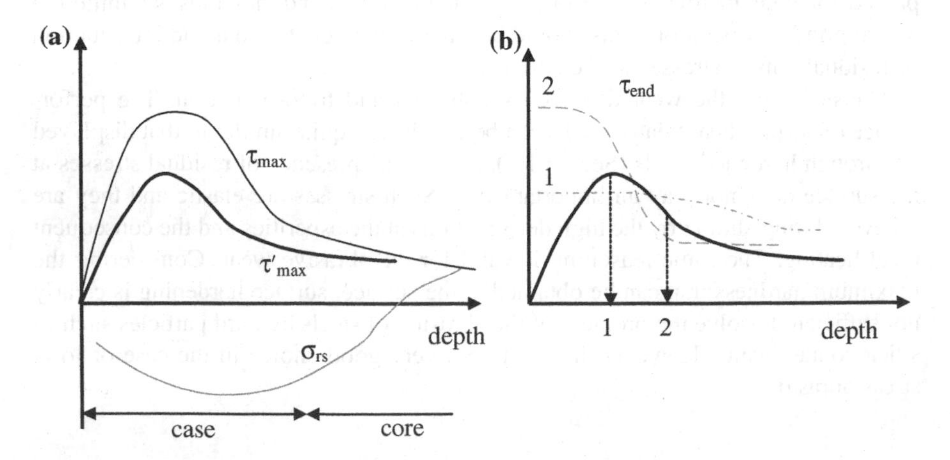


**Fig. 7.7** Schematizations showing: **a** the influence of residual stresses on the Hertzian stress profile; **b** two typical unsuitable endurance limit profiles that produce contact fatigue damage by spalling (profile 1; damage is by pitting if the Λ-factor is low) or case-crushing (profile 2)

outlined in the previous paragraph. The expected contact fatigue life is attained if $\tau'_{max} < \tau_{end}$ over the entire depth. Two critical situations can be encountered:

(1) When the surface hardness, i.e., the surface endurance limit, is too low in comparison to the $\tau'_{max}$-profile. In this case, a crack prematurely nucleates in the subsurface region, giving damage by spalling. Residual stresses move between the surface and $z_m$ the region of crack nucleation and propagation. In case of low Λ-factors, the crack nucleates at the surface giving damage by pitting.
(2) When the effective thickness is too low in comparison to the $\tau'_{max}$-profile. In this case, the crack prematurely nucleates close to the boundary between case and core. This type of damage is a form of spalling that is commonly called *case crushing*.

As already mentioned, the process parameters have to be carefully optimized. For example, by increasing the treatment time a larger hardness thickness may result, well above 1 mm. In case of gas carburizing, however, *internal oxidation* in the surface layers can be too severe, hence reducing the fatigue properties [10]. In addition, the increased carbon content at the surface would induce larger residual austenite content. The role of residual austenite on the contact fatigue resistance is not clear as yet and conflicting results are reported in the literature [11]. In case of *carbonitriding*, i.e., surface enrichment with carbon and nitrogen, the amount of retained austenite can be quite high. In this case, the treated parts are often submitted to a shot peening treatment, with the objective of inducing the transformation of residual austenite into martensite. In this way, an increase in hardness and, above all, in the compressive residual stresses is attained. As shown in Fig. 7.8, the compressive residual stress after shot peening is maximum at a certain depth into the material, owing to the elastic-plastic contact between the spheres and the surfaces [12].

The carburized parts may undergo geometrical distortions, a problem that can be solved by mechanical finishing to obtain the correct final geometry. Such operations involve the removal of surface layers and must be carried out with particular care.

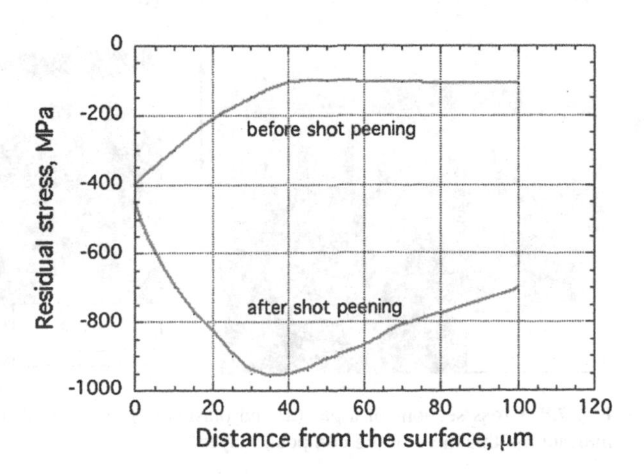


**Fig. 7.8** Residual stress evolution in carburized 16MnCr5 steel before and after shot peening (modified from [12])

### 7.3.2 Nitriding

Another important thermochemical diffusion treatment for steels is nitriding. In this case, the surface hardening is obtained by nitrogen diffusion, which hardens the steel by solid solution and for the formation of a fine dispersion of nitrides. Therefore, the nitriding steels contain nitride former elements, such as chromium, molybdenum or aluminium. The treatment is carried out over a temperature range in between 500 and 590 °C, with the steel in the ferritic state, and it takes longer time than carburizing (a typical nitriding treatment would last for 30 h). This sort of treatment is very attractive indeed, since it allows obtaining very high surface hardness values (greater than those obtainable with carburizing, see Table 7.1 and Fig. 7.1) and it might not require finishing operations.

Figure 7.9 shows the typical morphology of the surface layers of a steel after *gas* and *plasma* nitriding [13]. The surface is characterized by an outer *compound layer*, also called *white layer*, consisting of a mixture of ε and γ′ iron nitrides, and a sub-surface layer, called *diffusion layer*. The thickness and properties of the *white layer* depend on the treatment parameters. Generally, gas nitriding produces a thick layer, with a thickness of around 10–15 μm or more, which is also porous in its outer part. Plasma nitriding produces a narrow and compact layer, with a thickness of about 3 μm. Figure 7.1 shows a typical microhardness profile in the diffusion layer. The effective thickness is lower than in carburizing, but the diffusion layer displays higher thermal resistance. In fact, the hardness of martensitic carburized layers starts to decrease after exposure at temperatures of about 200 °C, whereas the nitrided layers retain their original hardness even after exposures up to about 400 °C. In both the compound and diffusion layers, residual compressive stresses are present. They originate from the precipitation of the nitrides that have a lower density, and therefore a higher specific volume, than the matrix. This also means that hardness and residual stress are correlated since they both depend on the

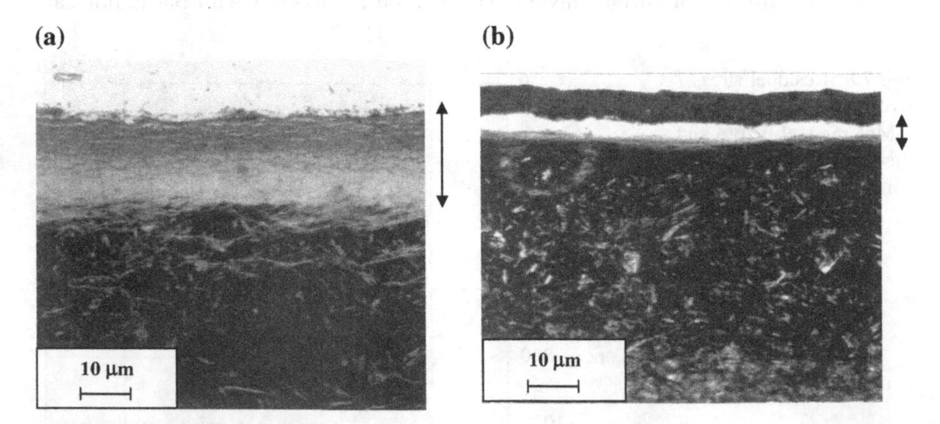


**Fig. 7.9** Cross sections of a gas (**a**) and plasma (**b**) nitrided 42CrAlMo7 steel [13]. The arrows indicate the thickness of the compound layer

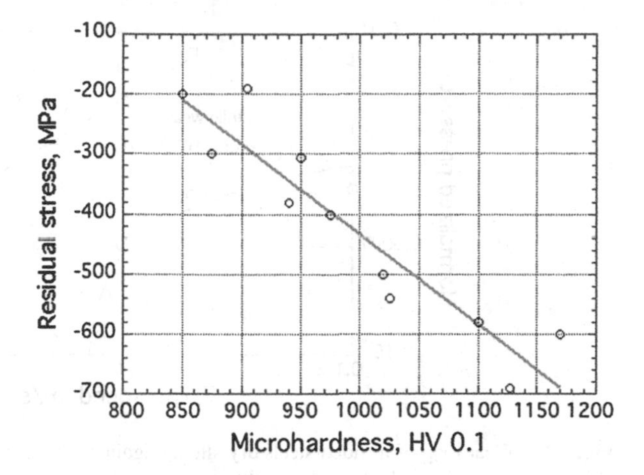


**Fig. 7.10** Experimental relationship between residual stress and microhardness (in the diffusion layer) for a nitrided AISI M2 tool steel (modified from [14])

precipitation of the nitrides. As an example, Fig. 7.10 shows the experimental close relationship between the microhardness and the residual stress in the diffusion layer, in the case of an AISI M2 tool steel gas nitrided at 570 °C for 8 h [14].

The nitriding treatment is particularly suitable to improve sliding wear resistance. In dry sliding, the white layer is removed in the run-in stage if its thickness is low and the contact pressure is sufficiently high. Therefore, wear resistance is actually provided by the diffusion layer. Wear occurs mostly by tribo-oxidation and the specific wear coefficient is very low, around $5 \times 10^{-15}$ $m^2/N$ (as reported in Table 5.2), since the diffusion layer possesses a high ability to sustain the layer of compacted oxide wear debris. Moreover, the diffusion layer displays high thermal resistance, and local temperature rises due frictional heating soften it to a lesser extent than observed in carburized or through hardened steels. This latter property produces an increase in the critical temperature for the transition from mild to severe wear. As shown in the wear map of Fig. 7.11 [15], this transition would take place at higher values of normalized pressure and sliding speed compared to what observed in normalized or quenched and tempered steels. In general, it is not necessary to remove the white layer before service, since iron nitrides possess a hardness that is quite similar to that of the diffusion layer.

In case of dry sliding with relatively low applied pressure and in presence of a quite thick compound layer, the compound layer itself provides the wear resistance. In this case the best performances are attained when the compound layer is made of just one iron nitride (preferably, ε nitride) and it contains a few pores [13]. But in case of sliding under boundary lubrication, the presence of some porosity on the outer part of the compound layer is beneficial, since pores act as lubricant reservoirs. The compressive residual stresses in the compound layer also help in closing any possible surface crack. In this respect, it has to be further considered that nitriding induces an increase in surface roughness. An increase of $R_a$ from 0.5 μm before nitriding to 1.5–2 μm after nitriding is quite common. Such an increase

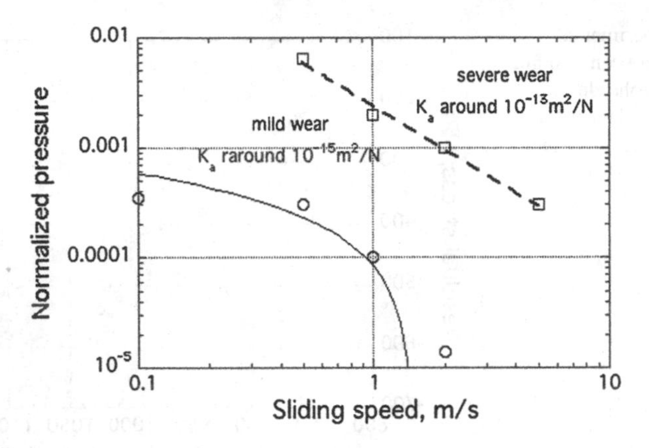


**Fig. 7.11** Wear map of nitrided steels dry sliding against a hardened steel counterface (data from [15]). For comparison, the boundary between mild and severe wear for normalized and quenched and tempered steels is also reported (from Fig. 6.1b)

clearly causes a decrease in the Λ-factor, which is only partially limited by the polishing of the higher asperities that occurs during running in [16].

Nitriding is also well suited for tribological applications characterized by abrasive wear, in particular low-stress abrasion. This is due to the quite high hardness achieved in the external regions of the nitrided layer, which can be in excess of 1250 kg/mm$^2$. As an example, Fig. 7.12 shows the specific wear coefficient for an untreated steel (H: 391 kg/mm$^2$), a nitrided steel and a chrome plated steel, obtained using a modified DSRW testing rig [17]. The test has been conducted at three temperatures under low-stress abrasion to simulate the wear process occurring inside the barrel of an injection-moulding machine of glass-filled polymers. The nitrided steel (with a surface hardness of 1284 kg/mm$^2$ compared to 1125 kg/mm$^2$ of the chromium plated steel) shows the best performances at all temperatures.

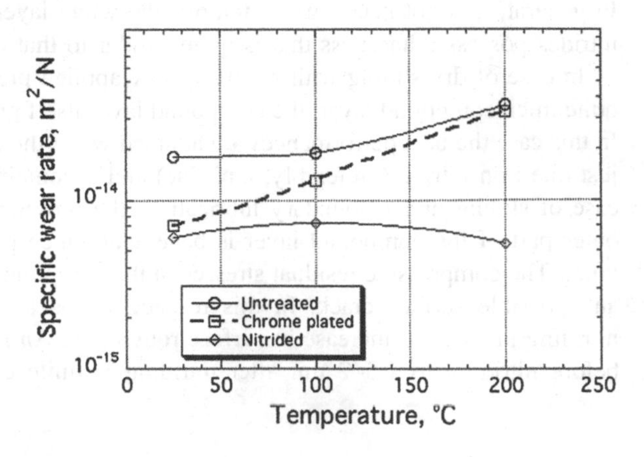


**Fig. 7.12** Mean specific wear coefficients for untreated steel, a nitrided and a chrome plated steel, under low-stress abrasive wear (modified from [17])

Finally, nitriding is also well suited to withstand wear by contact fatigue, especially in case of mixed or boundary lubrication conditions. This is because the surface hardness is very high and the case also contains compressive residual stresses. The effective thickness, however, is relatively low and this reduces the possibility of sustaining high contact pressures under well-lubricated conditions. It has to be further noticed that the increased roughness after nitriding tends to reduce the $\Lambda$-factor. As seen, such a decrease is partially reduced by some polishing during running in. In most cases, the polishing is carried out before the mechanical part is put into service. If the white layer is thick, it remains in place and contributes to the high pitting resistance of the nitrided parts [11].

### 7.3.3 Other Treatments

*Boronizing* is another diffusion treatment of steels. It involves a surface enrichment with boron. The maximum surface hardness is between 1550 and 2000 kg/mm$^2$ and is provided by the formation of quite a thick compound layer made by iron borides. The effective thickness values are similar to those obtained after nitriding.

A very high abrasion resistance can be achieved with the *Toyota Diffusion* (*TD*) process, which allows obtaining surface layers with hardness values from 2500 to 3800 kg/mm$^2$. This treatment is typically employed for tool steels. It is conducted in a salt bath and promotes the formation of very hard vanadium and titanium carbides.

The surface treatment that provides the smallest case thickness is *ion implantation*. It involves high-energy ions that strike a surface (a metal or a ceramic) and penetrate for an average depth in the 0.1 μm range, depending on the implantation conditions, like acceleration voltage and kind of ions. The treatment is typically carried out using different types of ions, including nitrogen, carbon, boron and phosphorus ions. After implantation, components are usually submitted to an annealing process to restore the crystal lattice and activate some diffusional processes. This treatment is indicated for applications characterized by adhesive and abrasive interactions under very low contact stresses.

## 7.4 Conversion Coatings

Conversion coatings are produced by a mild corrosion of a metal surface, capable to form a layer of products that are adherent to the metal. The main conversion treatments of tribological interest are *phosphating* and *anodizing*.

### 7.4.1 Phosphating

The phosphating treatment is typically made on ferrous alloys, by spraying or immersion of the parts in a bath of diluted phosphoric acid, containing specific additives. The treatment produces a uniform layer of hydrated phosphates of the metal present in the bath (typically zinc or manganese). As an example, Fig. 7.13 shows the surface morphology of a zinc phosphate layer realized on steel [18].

The thickness of the phosphate layer is difficult to be measured experimentally. In general the specific weight of the phosphating layer is considered. It is determined by weighing the part before and after the treatment. For tribological applications, heavy treatments are carried out, with specific weight in the range of 10–40 g/m$^2$. To a first approximation, 1 g/m$^2$ corresponds to a thickness of about 0.5 μm. As shown in Fig. 7.13, the phosphate layer is quite porous and the porosity may range from 3 to 13 %. Thanks to this porosity, the phosphate layers display a high capacity to take up oils or solid lubricants with clear advantages for both sliding friction and wear resistance.

*Zinc phosphate layers* are mainly used as a support for the subsequent application of oils and soaps in the manufacturing processes by plastic deformation, such as wire and tube drawing, cold stamping and deep drawing of metal sheets. *Manganese phosphate layers*, realized by immersion in a heated solution (90–98 °C), are typically used as self-lubricating and anti-wear coatings, especially in applications characterized by a low Λ-factor and hence possible scuffing problems, such heavily loaded and slow gears. In fact, the oil retained in the porosity greatly helps lubrication. In addition, the layer can slightly deform under contact and this allows for a better distribution of the involved contact stresses. In general, the layer is partially worn out after running in but the conditioning of the contacting surfaces produces long-lasting beneficial effects. The annealing of the coatings at an optimized temperature in the range of 250–450 °C, can induce the formation of

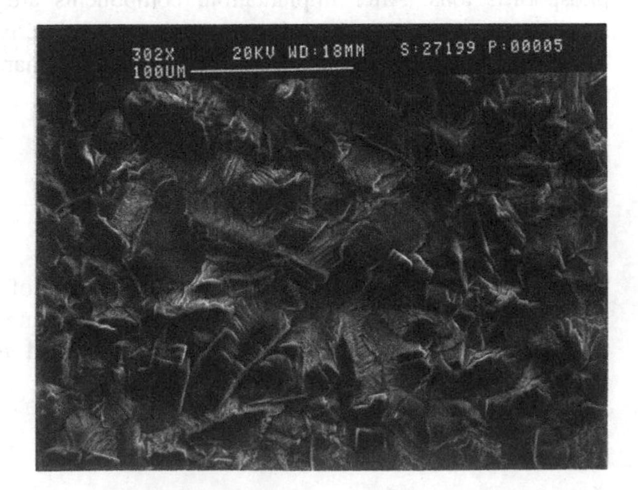


**Fig. 7.13** Morphology a zinc phosphate coating formed on a steel surface [18]

cracks perpendicularly to the surface by dehydration. Such cracks do not trigger brittle fracture but rather act as additional reservoirs for lubricating oil. This gives rise to an increased resistance to wear and scuffing [19].

### 7.4.2 Anodizing

Anodic oxidation is an electrochemical conversion process aimed at producing a hard and adherent oxide layer on a metal surface. It is mainly used for aluminium alloys (only certain Al-Cu and Al-Si alloys do not anodize well). Titanium, magnesium and zinc alloys are also anodized. The coatings can be obtained for decoration purposes (a thickness of some micrometers is sufficient), architectural purposes (a thickness of 10–20 μm is required), and for protective purposes, against corrosion and wear (a layer thickness of 50 μm is typically employed). Since the coating is formed by the dissolution of the metal surface, a layer with a thickness of 50 μm induces an increase in surface level of 25 μm only.

The anodization of aluminium alloys is typically conducted in a bath of sulphuric acid. To obtain protective coatings, the so-called *hard anodizing* process is used, which is characterized by a careful control of the process parameters that include cold electrolyte (<10 °C) and vigorous stirring of the solution. The layers are made up by a mixture of alumina and hydrated alumina, and are relatively porous. Pores are typically 200–350 nm in size and are oriented perpendicularly to the metal surface. Because of the presence of porosity and of the hydrated oxides, the hardness of the layers ranges between 350 and 600 kg/mm$^2$, i.e., well below the hardness of alumina (1600–2600 kg/mm$^2$). These layers are characterized by a quite high surface roughness that depends on the process parameters (a typical value is $R_a = 2.5$ μm). It can be reduced by grinding. Hard anodic layers are mainly used for applications that need low-stress abrasion and corrosion resistance. Examples are automotive brake callipers and pistons, pump components (operating in erosive media), pneumatic cylinder tubes and valve spools [20].

Aluminium alloys are characterized by a very low hardness (typically between 100 and 180 kg/mm$^2$) and therefore by a comparatively quite high abrasive wear. After hard anodizing, the surface hardness is considerably increased and the specific wear rate under low-stress abrasion (determined by DSRW) assumes values around $10^{-13}$ m$^2$/N (see Fig. 5.20) that may be acceptable for many applications. The resistance to solid particle erosion depends on different parameters. The residual porosity induces a brittle behaviour, and thus the erosion rate increases with the impact angle [21]. In most cases, the resistance of the untreated alloy, which behaves in a ductile manner, is better than that of the anodized one. The residual porosity influences also the dry sliding behaviour of the treated parts. The specific wear coefficients are around $10^{-14}$ m$^2$/N. However, it has to be considered that hard layers can cause severe abrasive wear of the counterface, since alumina has a high hardness and the layers may posses a quite high roughness. At the same time the friction coefficient is quite high (greater than 0.6) because of the contribution given by the abrasive interaction. In order to reduce friction and wear (of both the treated alloy and the counterface), the porous layer can be impregnated with oil, or coated with solid lubricants, such as PTFE or $MoS_2$. Different techniques can be used to apply the solid lubricants, as topcoats or as integral parts of the coating. The most commonly used solid lubricant is PTFE, which can induce a 50 % decrease in the wear of both the treated alloy and the counterface [20].

Very hard layers can be obtained using the micro arc-oxidation (MAO) technique, also know as plasma electrolytic oxidation (PAO). The deposition process is completely different from hard anodizing, and leads to the formation of surface layers with hardness in the range 1400–1800 kg/mm$^2$. Typically the hardness is lower on the external porous part of the layer, which can be eventually removed by grinding, and higher in the internal pore-free part made by melting-solidification processes of the aluminium alloy. This treatment produces a noticeable increase in the low-stress abrasion and sliding resistance of Al-alloys [21]. Since friction is still quite high under dry sliding, self-lubricating topcoats, such as PTFE or DLC layers, can be further deposited onto the MAO layers.

## 7.5 Surface Coatings

The main surface coatings are listed in Table 7.1. For the ease of discussion, they can be further classified into: metallic plating, thin coatings and thick coatings.

### *7.5.1 Metallic Plating*

The metallic deposits of greatest interest in tribology are *hard chromium* and *electroless nickel* deposits. Both can be obtained on almost all metals.

#### 7.5.1.1 Hard Chromium Plating

Hard chromium plating is produced by electrodeposition from solutions containing chromic acid and a catalytic anion. Thanks to its nanocrystalline microstructure, the hardness of the as-deposited layers is quite high, between 800 and 1200 kg/mm$^2$. The coatings can be relatively thin (about 2 μm) for decorative purposes, or have greater thickness, from 20 to 300 μm for industrial applications. In case of thick coatings, the thickness may not be uniform, and excessive built-ups on external corners have to be removed by grinding and lapping. The coatings are also characterized by multiple cracks perpendicular to the substrate surface. They form during the production process because of the co-deposition of hydrogen that result in the formation of hexagonal chromium hydride. This hydride has a tendency to release hydrogen gas that induces the formation of residual tensile stress and hence vertical cracks [22]. Hydrogen can also diffuse into the substrate inducing its embrittlement. In most cases the chromium coatings are submitted a dehydrogenation treatment (typically at 190 °C for 4 h) that also reduces the intensity of the residual stresses. Hard chromium plating is widely used to increase corrosion resistance in specific environments and to increase the wear resistance under sliding and low-stress abrasion. Typical applications include piston rings, valves, hydraulic rods, dies, moulds, rolling mills. These coatings are also employed in rebuilding operations (if the wear thickness is lower than 250 μm). A great drawback of this treatment is the presence of hexavalent chromium in the electrolyte that is environmentally harmful.

In general, hard chromium deposits are characterized by high sliding wear resistance, in particular when sliding against a steel counterface and when the coated part is under continuous contact with the counterface. In this condition, wear can be by tribo-oxidation with the formation on the coating of a protective layer of oxides coming from the steel antagonist. The recorded specific wear rates are very low, around $5 \times 10^{-16}$ $m^2/N$ [23]. The corresponding friction coefficient is around 0.6, as typical of the steel-steel pair under dry conditions. If the plating is realized on the part that is not in continuous contact with the antagonist (such as on the disc in the case of the pin-on-disc test, just to give an idea), the situation changes. The wear of the steel counterface may not be sufficient to establish a mild tribo-oxidative wear condition, and the plating wears by adhesion. Typical specific wear rates are in the $4 \times 10^{-14}$ $m^2/N$ range [24]. However, the corresponding friction coefficients are, in this case, 0.4 or lower and this can be an advantage in many applications. In case of adhesive wear, incorporating hard nanoparticles in the plating can decrease the specific wear coefficient. For example, a chromium coating containing 1.13 % of SiC nanoparticles displayed a specific wear coefficient of around $1.4 \times 10^{-14}$ $m^2/N$ [25].

To optimize the sliding wear performance of hard chromium coatings it is necessary to choose a subsurface material that is capable to sustain the plating without extensive deformation (see Fig. 7.2). In this respect, the chromium plating may be successfully realized on a martensitic steel substrate. In addition, the plating thickness should be optimized in order to avoid plating damage by brittle contact. In case of skin-pass rolling mills, an optimized thickness of about 10 μm has been obtained [26]. Hard chromium plating is particularly suitable in boundary-lubricated conditions, since the vertical cracks in the layers may act as lubricant reservoirs. However, it has to be noted that plating increases the surface roughness. For example, $R_a$ may increase from 0.1 to 1.15 μm after the deposition.

Regarding the wear behaviour under abrasion, hard chromium coatings display an excellent performance in case of low-stress abrasion, as is apparent from the analysis of the data in Fig. 5.20.

### 7.5.1.2 Electroless Nickel Plating

Electroless nickel (EN) plating is regarded, together with thermal spray coating, as a suitable candidate for the replacement of hard chromium plating that has adverse health and environmental effects. These coatings are realized by a chemical reduction of nickel ions on a catalytic surface. Since no electric current is applied, the plating is quite uniform in thickness, irrespective of the shape. EN plating can be classified into three classes: pure metallic plating, alloy plating and composite plating. The most popular coatings are binary Ni-P and Ni-B alloys. In the as-deposited condition they are almost amorphous and display high corrosion resistance. However, the wear resistance is quite low, although the hardness may be around 600 kg/mm$^2$ in Ni-P coatings and 650–750 kg/mm$^2$ in Ni-B coatings. Higher hardness values can be obtained through heat treatment at 300–400 °C that converts the amorphous layer into a microcristalline nickel matrix containing $Ni_3P$ or $Ni_3B$ precipitates. A peak hardness of 1000–1200 kg/mm$^2$, i.e., comparable to that of hard chromium coatings, can be attained [27]. This treatment induces an increase in wear resistance (both under sliding and low-stress abrasion conditions) but a decrease in the corrosion resistance. EN plating is used in valves, aluminium piston heads, components in the aircraft industry, and in automotive parts such as fuel injectors, pinion ball shafts and disk brake pistons [28].

The sliding wear resistance of properly heat-treated EN coatings is comparable to that of hard chromium coatings. As an example, Fig. 7.14 shows the results of dry sliding tests of a Ni-P plating with a thickness of about 20 μm in the as deposited and in different thermal treatment conditions that allowed obtaining hardness values from 650 to 1000 kg/mm$^2$. The tests were carried out in a disc-on-disc configuration with rotating plated disc, and counterface disc made with an AISI M2 tool steel kept fixed. The specific wear rate is seen to decrease as hardness is increased and it reaches the typical value of hard chromium coatings for a hardness of 1000 kg/mm$^2$ (the peak value). Wear was by adhesion/abrasion, and the recorded friction coefficient was around 0.3 in all cases.

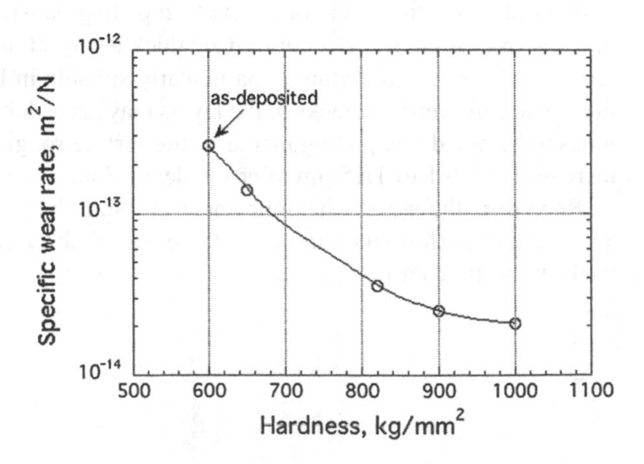


**Fig. 7.14** Influence of hardness on the dry sliding wear resistance of Ni-P coatings

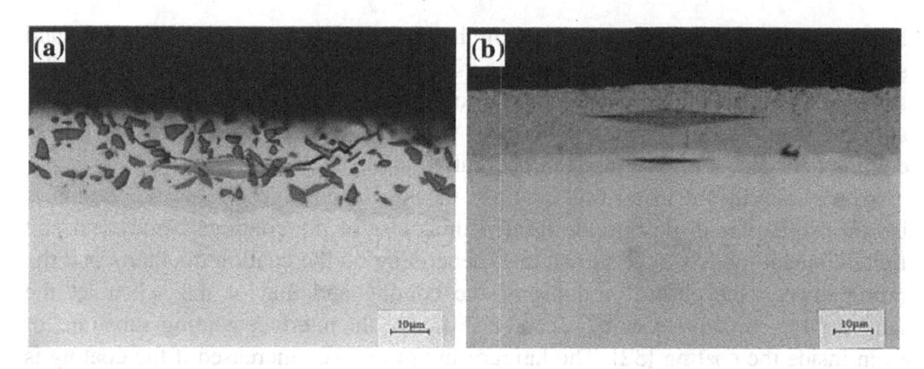


**Fig. 7.15** **a** Knoop indentation (load = 50 g) on the Ni-P/SiC layer showing the formation of brittle cracks; **b** Knoop indentation (load = 50 g) on the internal Ni-P layer and the Ni-P/PTFE layer [30]

The low-stress abrasive wear of EN coatings is generally much lower than hard chromium coatings, as shown in Fig. 5.20. However, recent proprietary Ni-B coatings have shown a wear resistance that is comparable or better than that of chromium coatings [29]. In case of abrasive interaction, the phosphorous (or boron) content and the treatment cycle should be optimized to avoid an excessive plating brittleness that would increase wear by favouring brittle contact.

The properties of EN plating can be improved with the co-deposition of hard particles, such as SiC, WC, alumina or diamond, to increase the wear resistance, or solid lubricant particles, such as Teflon, graphite or $MoS_2$, to reduce friction coefficient. The incorporation of hard particles increases hardness and sometimes also brittleness, whereas the presence of soft particles reduces hardness. Figure 7.15 shows the microstructure of two Ni-P/SiC and Ni-P/PTFE composite coatings [30]. Hard composite coatings display better wear resistance than EN coatings under sliding as well as low-stress abrasion. The hard particles, however, can abrade the counterface. This effect can be reduced, or eliminated, by employing micro- or nanoparticles, or by co-depositing SiC and graphite particles together. When using Ni-P/PTFE composite coatings, the friction coefficient can be reduced below 0.1 [31]. However, because of the lower hardness, wear resistance of these coatings is lower and this has to be taken into account in design.

## 7.5.2 Thin Coatings

It is usual to define *thin coatings* (or *thin films*) hard ceramic coatings (such as nitrides, carbides, oxides or polycrystalline diamond) obtained by two families of vapour-phase deposition techniques, called *physical vapour deposition* (PVD) and *chemical vapour deposition* (CVD). These coatings usually have a thickness between 2 and 10 μm, and are particularly suited to improve sliding wear and low-stress abrasive wear. For this reason they are widely used for cutting tools, and since the late 1980s the golden coloured TiN coatings have become very popular for this application. Thin coatings are also employed in forming dies and moulds, and are continuously proposed for new tribological systems in industry. These coatings are also currently used in decorative applications.

The possibility of using thin coatings in case of damage by contact fatigue is being also explored, despite the limited thickness of the coatings. Analytical and finite element models have shown that, depending on the coating thickness and the ratio between the elastic modulus of the coating and that of the substrate, the tangential maximum stress, $\tau_{max}$, can be shifted to the interface coating-substrate, or even inside the coating [32]. The fatigue strength is then increased if the coating is free from defects, and if the adhesion coating-substrate is high enough to withstand the high interfacial stresses. In case of mixed or boundary lubrication, fatigue crack would mainly nucleate at the surface and the use of thin films is therefore quite promising for increasing the fatigue life of the component. In general, the use of thin coatings has to be evaluated in a case-by-case basis considering also the relevant cost.

#### 7.5.2.1 Coating Technologies

In the CVD processes, gaseous chemical reactants are introduced into a closed reactor where they are thermally activated and react to form a thin coating on the part to be treated [4]. The process temperature is around 1000 °C and this technique is thus mainly used to coat hard metals that can tolerate such high temperatures. With this technique it is possible to deposit different carbides, nitrides and oxides, including multiple coatings. The most popular CVD coatings are TiN and TiC. The adhesion with the substrate is excellent because the high process temperature induce formation of chemical bonds across the interface. In the *plasma assisted CVD* technique (PA CVD), the gaseous reactants are activated by plasma ignited around the components. In this way, the process temperature can be lowered to 300–500 °C, or less, and the technology becomes suitable also for high strength steels. This technique is also used to deposit polycrystalline diamond.

The PVD processes require a lower coating temperature (500 °C or less) and are conducted under sub-atmospheric pressure. They can be used to obtain coatings on any metal and also on polymers. The main PVD processes are:

- *Evaporation*. A physical mean, such as an electron beam, promotes the evaporation of a metal (the target) and the vapour condenses on the work piece. If a reactive gas (nitrogen, oxygen, or hydrocarbon) is introduced into the reactor, the gas reacts with the ionized metal vapour to form a compound that is deposited on the work piece. For example, TiN is obtained by using a titanium target and nitrogen.
- *Sputtering*. The target is bombarded by inert gas ions (typically argon ions) and the liberated ions are attracted to the work piece. The use of reactive gases

allows deposition of compounds. Sputtered films generally display a better adhesion on the substrate than the evaporated coatings.
- *Ion-plating*. The target metal is evaporated and the atoms enter an argon plasma. Some atoms are ionized and attracted to the work piece. Others are accelerated towards the work piece by the thermal collisions. Also in this case, the presence of reactive gases allows the deposition of compounds. This process is widely used in numerous applications.

Because of the relatively low process temperatures, the adhesion between the coating and substrate may be weak. In order to favour it, a careful surface preparation of the substrate to remove the contaminants and increase the surface reactivity is required. It may include ultrasonic cleaning and sputter etching. In addition, a metallic interlayer is firstly deposited to aid adhesion. For example, in the case of TiN and TiAlN coatings a 0.1–0.2 μm titanium interlayer is usually deposited. A limitation of the PVD process is the difficulty to coat cavities deeper than their width.

#### 7.5.2.2 Main Types of Thin Coatings

Table 7.2 shows the main available commercial coatings. Some additional observations on the main coating systems are reported herewith:

- TiN. It has excellent general tribological properties. It is widely used to coat cutting and stamping tools. Its relatively low oxidation resistance at high temperatures gives a limitation to their operating conditions.
- TiCN. It possesses higher hardness than TiN and therefore better wear resistance, in particular by low-stress abrasion. TiC is very much suited for the deposition of hard sandwich coatings on hardmetals, such as $TiC/Al_2O_3/TiN$ coatings, because of its improved adhesion to the carbides of the substrate.
- TiAlN. It has greater hardness than TiN and, most of all, greater thermal resistance. During sliding at high temperature, a layer of protective aluminium oxide is formed on the surface of the coating.

**Table 7.2** Typical properties of some commercial thin coatings

| Coating type | Typical thickness (μm) | Process temperature (°C[a]) | Micro hardness (HV 0.05) | Colour | Maximum application temperature (°C) |
|---|---|---|---|---|---|
| TiN | 1–4 | 350–450 | 2400 | Gold | 600 |
| TiCN | 1–4 | 450 | 3000 | Silver | 400 |
| TiAlN | 1–3 | 450 | 3500 | Brown | 800 |
| AlCrN | 1–4 | 450 | 3000 | Grey | 1100 |
| CrN | 1–10 | 450 | 1800 | Silver | 700 |
| WC/C | 1–5 | 250 | 1000–2000 | Black | 300 |

[a]They refer to a typical PVD process

- AlCrN. It shows the best oxidation resistance.
- CrN. Widely used where it is necessary a good sliding wear resistance and protection against corrosion.
- WC/C. It has a lamellar structure, consisting of alternating layers of hard WC nanoparticles and layers of amorphous carbon. The hard particles increase the hardness of the coating, whereas the amorphous carbon induces a solid lubricant behaviour.

Thin coatings *contain residual stresses* that are formed during the manufacturing process [3, 4]. They typically have two contributions, namely the intrinsic and thermal stresses. The *intrinsic stresses* are due to the growth mechanisms of the coating. They are typically compressive in nature and their intensity decreases as the process temperature is raised. *Thermal stresses* would build up on cooling from the process temperature and are induced by the different thermal expansion coefficients between coating and substrate. The thermal residual stress, $\sigma_{rs}$, can be supposed to be biaxial, and it can be evaluated by the following relation:

$$\sigma_{rs} = \frac{E_d}{1 - \nu_d}(\alpha_d - \alpha_s)(T_d - T) \tag{7.2}$$

where $E_d$ and $\nu_d$ are the Young's modulus and the Poisson's ratio of the coating, $\alpha_d$ and $\alpha_s$ are the thermal expansion coefficient of the coating and the substrate, $T_d$ is the process temperature and T is the final temperature after cooling.

In Table 7.3 typical values of the coefficients of thermal expansion and elastic properties of some thin coatings are listed. As an example, consider the case of PCD deposited onto a hard metal substrate. Using Eq. 7.2 and the data listed in Table 7.3, a thermal residual stress of −2 GPa is obtained if the deposition temperature is 850 °C and the coating is cooled down to 25 °C. As a matter of fact, PCD coatings obtained by CVD typically contain compressive residual stresses that are only thermal in nature owing to the high processing temperature [33]. On the other hand, in PVD TiN coatings obtained at 200–300 °C, the thermal part is only 25 % of the residual stress [34].

**Table 7.3** Typical thermal expansion coefficients and elastic properties of some coatings and substrates materials (taken from different literature sources)

| Coating | Coefficient of thermal expansion (°C) | Young's modulus (GPa) | Poisson's ratio |
|---|---|---|---|
| TiN | $9.3 \times 10^{-6}$ | 500 | 0.25 |
| TiC | $8.3 \times 10^{-6}$ | 450 | 0.17 |
| CrN | $0.7 \times 10^{-6}$ | 350 | |
| PCD | $2.85 \times 10^{-6}$ | 1000 | 0.2 |
| HSS | $11.9 \times 10^{-6}$ | 215 | 0.3 |
| WC/Co | $5 \times 10^{-6}$ | 620 | 0.22 |

**Table 7.4** Experimental measurements of residual stresses in different coatings (data from [34])

| Coating/substrate | Deposition process | Typical residual stress (GPa) | Typical thickness (μm) |
|---|---|---|---|
| CrN/tool steel | PVD | −1 | 4 |
| TiN/HSS | PVD | −4 | 4–5 |
| TiN/hard metal | PVD | −4 | 4 |
| TiC/hard metal | CVD | +0.2 | 4 |
| PCD/hard metal | CVD | −2 | 10 |

The analysis of the thermal expansion coefficients listed in Table 7.3 shows that the thermal residual stresses in coatings on tool steels and hard metals are, respectively, compressive and tensile. In general, the residual stresses depend greatly on the process parameters and on the features of the coatings. Table 7.4 summarizes the results of experimental measurements of residual stresses in different coatings [34]. If the residual stress is high and the coating/substrate adhesion is low, the easy detachment of the coating, even after cooling, can be observed. This problem is particularly important in PCD coatings, characterized by poor adhesion when realized on WC-Co that has not been properly pre-treated [35].

#### 7.5.2.3 Tribological Properties of Thin Coatings

As said, thin coatings are particularly suitable to increase the wear resistance by sliding and low-stress abrasion. The lifetime, t, of a coated component under sliding can be simply obtained by Eq. 5.2:

$$t = \frac{h}{K_a \cdot p_0 \cdot v} \qquad (7.3)$$

where h is here the coating thickness, $K_a$ is the specific wear coefficient, $p_0$ and v are the nominal contact pressure and sliding velocity respectively. For a given set of $p_0$ and v values, t increases if $K_a$ decreases and h increases. However, as thickness is increased, the presence of defects in the coating also increases. Therefore, the thickness has to be optimized in order to obtain the longest possible lifetime. An interesting opportunity is given by the realization of sandwich or *multilayer coatings*. Multilayer coatings have a laminated structure, consisting of periodically repeated lamellae. The lamellae can act as crack stoppers thus reducing the damaging role of defects.

In hard coatings, $K_a$-values are typically in between $10^{-15}$ (or lower) and $10^{-14}$, i.e., very similar to the values of bulk ceramics. As an example, Table 7.5 reports the friction and wear rates of three ball counterbodies dry sliding against a TiAlN coating, deposited by PVD onto a WC-12 % Co substrate [36]. The specific wear rates of the balls and the coating are listed. In the TiAlN/steel pair, a quite severe wear of the steel ball is observed, with the formation of a transfer layer rich in iron

Table 7.5 Friction and wear rate of TiAlN sliding against steel, SiC and $Al_2O_3$ (data from [36])

| Sliding pairs | Friction coefficient | $K_a$ ball ($m^2/N$) | $K_a$ coating ($m^2/N$) |
|---|---|---|---|
| TiAlN/steel | 0.3 | $2.8 \times 10^{-13}$ | Negligible |
| TiAlN/SiC | 0.17 | $4.13 \times 10^{-16}$ | $1.7 \times 10^{-16}$ |
| TiAlN/$Al_2O_3$ | 1–0.6 (decreasing) | $5.7 \times 10^{-15}$ | $2.3 \times 10^{-15}$ |

oxides on the coating. As a consequence, the wear of the coating is almost negligible. The occurrence of material transfer is quite typical in the case of sliding of relatively soft surfaces against hard coatings. Quite often this transfer also induces an increase in friction that could favour the premature damage of the coating. In the TiAlN/$Al_2O_3$ pair, a typical mild adhesive wear of ceramic materials is observed. In the TiAlN/SiC pair, a tribolayer is formed during sliding. It firstly involves the oxidation of SiC and the reaction of $SiO_2$ with moist air to form different hydrides. As a consequence, a boundary lubrication is established and both friction coefficient and wear rates are particularly low.

If *premature failure* of the coatings takes place during sliding, the lifetime predicted by Eq. 7.3 is not attained. Three main factors may induce premature failure:

(1) a too low hardness of the substrate beneath the hard coating;
(2) the fragmentation of the coating by brittle contact;
(3) the flaking of the coating.

As shown in Fig. 7.2, if the hardness if the substrate is too low and a point or angular load is applied to the coated part, plastic deformation may be initiated at the substrate/coating interface. If the coating is unable to accommodate this deformation, it will fracture by bending. The hardness difference between coating and substrate can be minimized by depositing the coating on a hard metal or a martensitic steel (such as a tool steel). A further improvement can be achieved by the so-called *duplex treatments*, which typically comprise a nitriding process followed by the coating deposition. In general, a plasma or low-pressure nitriding is carried out, since with these techniques the formation of the compound layer is limited or avoided. In fact, the presence of the compound layer decreases the adhesion between the coating and the substrate, because it tends to decompose during the deposition. The nitrided diffusion layer provides support for the coating, and also increases the adhesion of the substrate with nitride coatings, such as TiN, TiAlN, and CrN coatings.

The fragmentation of the coating by brittle fracture, schematized in Fig. 7.16b, is also called *cohesive fracture*. It is favoured by the presence of defects in the coating that can induce brittle fracture if the applied tensile stress overwhelms a critical value (see Eq. 4.3). The flaking of the coating (Fig. 7.16c) is also called *adhesive failure*. It is favoured by a low coating/substrate adhesion and by the presence of compressive stresses that induce debonding and buckling. In both cases, the hard fragments can also act as abrasives, thus increasing the wear rate.

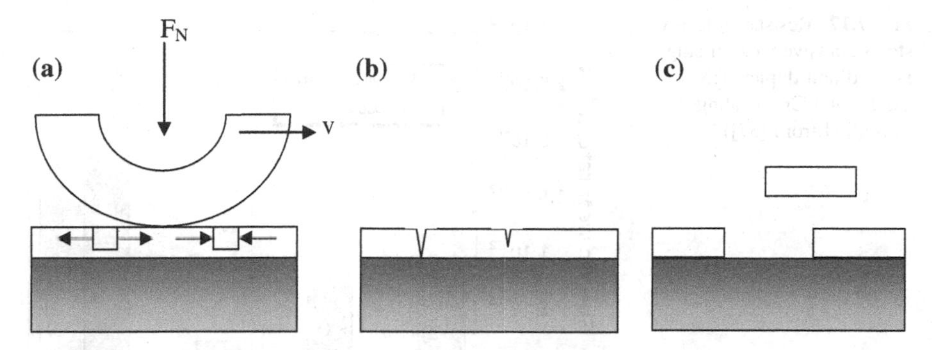


**Fig. 7.16** During sliding, friction induces a compressive stress in front of the slider and a tensile stress behind it (**a**). The tensile stress favours cohesive fracture within the coating (**b**), whereas the compressive stress favours adhesive fracture between coating and substrate (**c**)

The stress state within a coating plays then a crucial role. It is given by the sum of three contributions: the contact Hertzian stress (which may be tensile at the surface in case of point contact, see Eq. 1.7), the friction stress (Eqs. 2.1 and 2.2), and the residual stresses. Generally speaking, compressive residual stresses are beneficial with regard to cohesive fracture since they decrease any applied tensile stress. On the other hand, if the adhesion coating/substrate is weak, tensile residual stresses may be beneficial since they reduce the magnitude of any applied compressive stress that could induce flaking. However, the picture is complicated by the local temperature rises induced by frictional heating. An abrupt temperature change can induce the formation of tensile stresses that favour brittle cohesive fracture within the coating. Alternatively, the frictional heating can induce a relaxation of the residual stresses. In addition, many real components contain geometrical discontinuities, which amplify internal stresses and thus facilitate cohesive or adhesive fracture.

As regards the abrasive wear, it is clear that thin coatings are not particularly suitable in applications with high-stress abrasion, since their high brittleness would favour macroscopic brittle contact. Instead they appear more suitable in applications involving low-stress abrasion. In this case the high hardness of the coatings can provide high resistance to abrasive wear.

Figure 7.17 shows the abrasion resistance of three coatings: TiAlN, TiN and CrN, deposited on an AISI H13 tool steel (hardened to a hardness of about 600 kg/mm$^2$), with and without a prior plasma nitriding treatment [37]. The tests were conducted with a micro-abrasion tester that allows obtaining the specific wear coefficients of the coatings, using a suspension of SiC particles. For comparison, the figure also shows the wear rate of the uncoated steel. It can be noticed that TiN and, especially, TiAlN coatings display a better wear resistance than the uncoated steel, while the CrN coating resulted in a worse performance. In any case, the duplex treatment provides an increase in wear resistance. These results can be explained by considering the different hardness of the coatings in comparison with that of the abrasive particles. The TiAlN has a hardness of about 3000 kg/mm$^2$, i.e., greater than that of the SiC

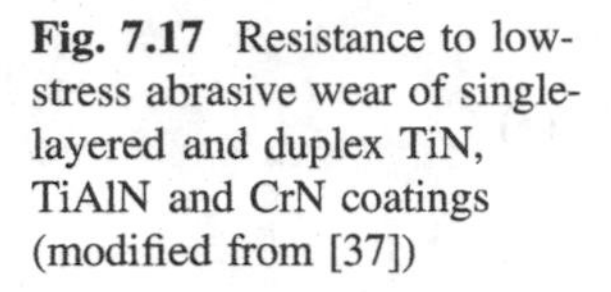

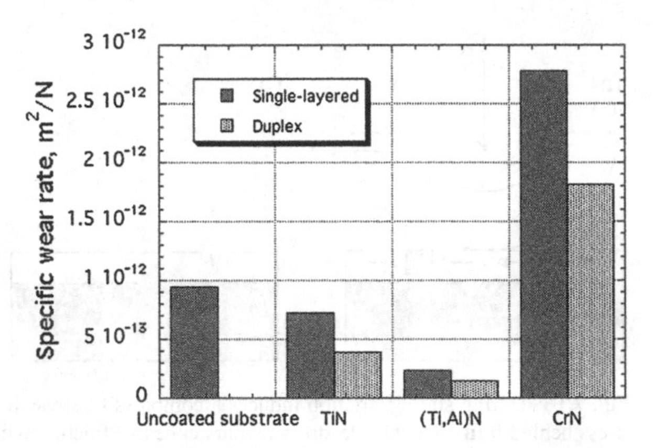


**Fig. 7.17** Resistance to low-stress abrasive wear of single-layered and duplex TiN, TiAlN and CrN coatings (modified from [37])

particles, that is about 2500 kg/mm$^2$. The TiN has a comparable hardness to that of the SiC particles. The CrN coating has a lower hardness, 2000 kg/mm$^2$, and is then more easily abraded by the SiC particles. The substrate in AISI H13 displays a lower wear than the CrN coating, despite of a much lower hardness. This can be attributed to its greater ductility and toughness that reduce abrasive wear by microcutting.

## 7.5.3 *Thick Coatings*

As shown in Table 7.1, thick coatings can be obtained by *thermal spraying* and *welding* processes. Both techniques are used for the surface engineering of new components and in the repair of worn parts.

### 7.5.3.1 Thermal Spraying Coatings

In thermal spraying the material to be deposited is melted in a torch and projected, in the form of molten or nearly molten droplets, onto the substrate. After the impact, each droplet acquires a lamellar shape and solidifies. The thickness of the coating can range from more than 20 mm to some micrometres. Typical thicknesses of coatings for tribological application are around 100 μm. These coatings are mainly used in case of sliding wear and low-stress abrasion wear.

There are various thermal spraying processes available, including flame spraying, detonation gun, arc and plasma (in air or in vacuum) spraying, and High-Velocity Oxy-Fuel (HVOF) flame spraying. The *HVOF* technique is the state-of-the-art of thermal spraying of metallic coatings. Figure 7.18 schematically shows a HVOF torch employing a liquid (kerosene) fuel. Kerosene and oxygen are burnt in a combustion chamber and the hot gases emerge as a free jet from a converging-diverging throat. The material to be deposited is injected in the hot gases in the form

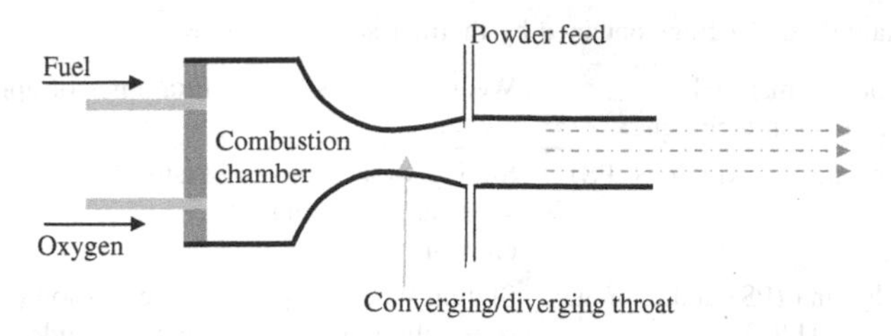


**Fig. 7.18** Schematic of the HVOF torch. It is evident that a limit of the thermal spraying technology is its capacity to treat only surfaces that the torch can 'see'

of powder. The droplets are sprayed against the surface at an extremely high speed, which may reach values of 2000 m/s as opposed to about 100 m/s in flame spraying and 1000 m/s in plasma spraying. This results in quite highly adherent coatings, with low porosity (typically lower than 1 %) and a content of oxides that is generally lower than 1 % [38].

The adhesion between the coating and the substrate is actually limited in thermal spraying coatings. For coatings realized by plasma spraying, the typical bond stresses obtained by the pull-off test (Fig. 7.3a), are between 45 and 145 MPa. They are between 145 and 360 MPa in the case of coatings produced by HVOF and detonation gun. For coatings obtained with simpler and cheaper techniques, such as flame spraying, the bond stresses are about 45 MPa. The limited coating/substrate adhesion is mainly due to the low temperature that the substrate reaches during the deposition (usually less than 200 °C), which strongly limits the interdiffusion phenomena necessary to promote the formation of strong bonds. On the other hand, the low deposition temperature has the advantage of producing very low thermal distortions. In order to improve adhesion, it is necessary to pre-treat the surface of the substrate by cleaning and sand blasting. To promote the adhesion of ceramic coatings on metal substrates, and in particular on steels, intermediate bond layers made of molybdenum or other alloys, such as NiAl, are firstly deposited.

By thermal spraying it is possible to deposit almost any type of material in the form of powder, wire or rod. Table 7.6 provides a list of some materials that are commonly employed to prevent different wear processes. Zirconia coatings, not included in the table, are widely used as thermal barriers. The microstructure of the coatings is typically inhomogeneous. The coatings are characterized by a lamellar structure containing also residual porosity, oxides, which are formed following the interaction with the environment, incompletely molten particles, and cracks. These features reduce the mechanical and tribological properties of the coatings with respect to the cast, worked or sintered materials having the same chemical composition. In general, the coatings are also characterized by a high surface roughness ($R_a$ about 5 μm) that should be reduced by grinding [39].

WC-Co and $Cr_3C_2$-NiCr hard metals coatings produced by HVOF processes are widely used in tribological applications, especially in case of sliding wear. They are widely viewed as being capable of replacing hard chromium plating. The properties of the coatings strongly depend on the spray conditions and on the properties of the

**Table 7.6** Example of coatings obtained by thermal spraying processes

| | Coating material (deposition technique) | Wear processes | Examples of applications |
|---|---|---|---|
| Ceramics | Chromium oxide (FS, PS) | Sliding wear Low-stress abrasion and erosion | Pistons |
| | Alumina (PS) and alumina/titania (FS, PS) | Sliding wear low-stress abrasion and erosion Cavitation wear | Hot extrusion dies, press punches, slides, pumps and impellers, turbine blades |
| Metals | Bronze (FS) | Sliding wear | Sliding bearings (even if high corrosion resistance is required) |
| | Molybdenum (FS, AS, PS) | Sliding wear (with additional resistance to scuffing) Fretting wear | Piston rings, cams, rocker arms |
| | High carbon iron alloys, such as tool steels, martensitic steels, and cast irons (FS, AS) | Sliding wear | |
| | Co-Mo-Cr-Si alloys (FS, HVOF) | Sliding wear Fretting wear | Moulds |
| | Ni-Cr-Mo alloys containing SiC and/or WC particles (PS, HVOF) | Abrasion Cavitation wear | Seal rings, oil drilling tools |
| Hard metals | WC-Co (PS, HVOF) | Sliding wear Low-stress abrasion and erosion | Sliding bearings, valves, turbine blades, components for the textile industry, cold rolling cylinders |
| | $Cr_3C_2$-NiCr (PS, HVOF) | Sliding and fretting wear at high temperature | Forging dies, drying cylinders in the paper industry, seal rings, valves |

Deposition techniques: *FS* flame spraying; *AS* arc spraying; *PS* plasma spraying; *HVOF* high-velocity oxy-fuel flame spraying

feedstock powder materials [40]. Coatings of the WC-Co system possess a higher hardness and wear resistance than the $Cr_3C_2$-NiCr coatings. However, these coatings cannot be used in case of temperatures in excess of 450–540 °C, because of the decarburizing of WC followed by the formation of embrittling carbides, such as $W_2C$ and complex Co-W-C carbides (the so-called η-phase). Therefore, the $Cr_3C_2$-NiCr coatings are widely used in high-temperature wear resistance applications.

As an example, Fig. 7.19 shows the results of dry sliding wear tests, performed using the block-on-ring test, using steel block and ring samples coated with three different HVOF coatings [40]:

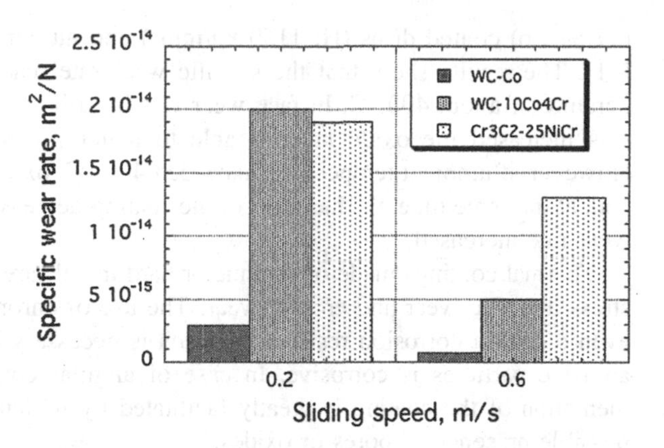


**Fig. 7.19** Specific wear rate versus sliding speed for three HVOF coatings dry sliding against a counterface of the same type in a block-on-ring configuration (modified from [40])

(1) WC-Co (12 % Co), with a hardness of 1260 kg/mm$^2$.
(2) WC-CoCr (10 % Co, 4 % Cr), with a hardness of 1255 kg/mm$^2$.
(3) $Cr_3C_2$-NiCr (25 % NiCr), with a hardness of 1060 kg/mm$^2$.

All samples were ground to a surface roughness of $R_a = 0.1$ μm. The first coating shows mild wear that is quite similar to that typically displayed by the as-sintered WC-Co hard metals (see Table 5.2). The second coating, however, displays a much larger specific wear coefficient even if its hardness is similar to that of the first one. The different behaviour was attributed to the presence of η-phase (pre-existing in the starting powder), which strongly embrittled the coating. The third coating, proposed for its better corrosion resistance, shows a specific wear coefficient similar to that of the second one. This last coating does not contain the η-phase, but its hardness is lower than that of the first coating. In fact, $Cr_3C_2$ has a hardness of about 1400 kg/mm$^2$, lower than that of WC (about 2000–2400 kg/mm$^2$).

Thermal coatings made of ceramic or hard metals are also used in high-temperature applications, since they are able to retain high hardness values at temperatures up to 600 °C and above (Fig. 6.5). Figure 7.20 shows the results of pin-on-disc sliding tests conducted at different temperatures, using HVOF WC-Co

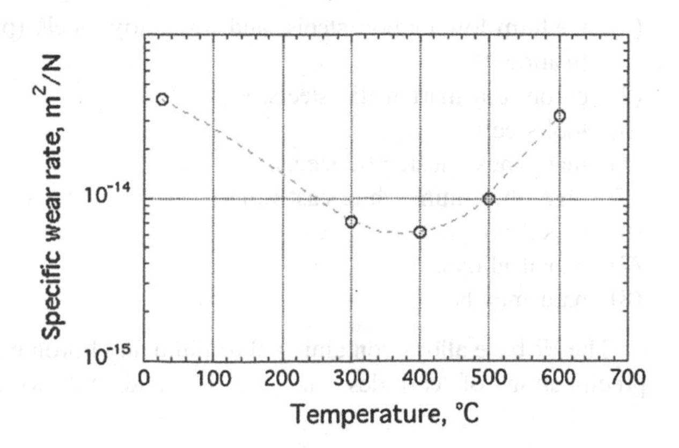


**Fig. 7.20** Specific wear rate as a function of ambient temperature for HVOF WC-Co coatings sliding against an alumina ball counterface (modified from [41])

(17 % Co) coated discs (H: 1150 $kg/mm^2$) and alumina balls (H: 1600 $kg/mm^2$) [41]. The results show that the specific wear rate reaches a minimum at a temperature of about 400 °C. In fact, wear was by tribo-oxidation and as temperature was increased the oxide layer is able to better protect the coating from wear. However, if ambient temperature exceeded 400 °C, oxidation rate became too high and, at the same time, the hardness of the coating decreased. As a result, the specific wear rate increased.

Thermal coatings made of ceramic or hard metals are also suitable to resist low-stress abrasive wear and erosive wear. The use of chromium rich coatings affords even a certain corrosion resistance, which is necessary if the fluid that carries the abrasive particles is corrosive. In case of angular contacts (Fig. 1.7), the fragmentation of the coating is greatly facilitated by its lamellar structure and by the possible presence of pores or oxides.

#### 7.5.3.2 Weld Hardfacings

Quite thick weld overlays for tribological applications are obtained using almost all the available welding technologies, such as oxyacetylene, electric arc, plasma and laser welding. In general, the obtained coatings have a thickness between 1 and 5 mm, and the bonding with the substrate is particularly strong. In fact, the significant heat input induces a dilution of the substrate and the formation of a typical weld microstructure. As usual in welding processes, appropriate preheating of the work piece to be coated or post-weld heat treatments should be carried out to optimize the microstructure and reduce the residual stresses that are typically tensile and could thus favour the cold cracking phenomenon.

Weld hardfacing is widely used in repairing worn surfaces and in the manufacture of new components that have to tolerate a large amount of sliding, abrasive and also impact wear. Typical applications include the protection of earth and rock moving equipment's in agriculture, mining, construction and steel industry, and the protection of components for the chemical, automotive and oil industries. The main hardfacing coating materials are:

(1) medium-low carbon steels, and low-alloy steels (placed in service after heat treatment).
(2) chromium martensitic steels.
(3) tool steels.
(4) manganese austenitic steels.
(5) high chromium white cast irons.
(6) nickel alloys.
(7) cobalt alloys.
(8) hard metals.

The Ni-base alloys containing chromium and boron reach high hardness after the precipitation of carbides and borides and are widely used in applications

**Table 7.7** Illustrative summary of the materials and surface treatments developed to counteract the wear mechanisms

| Wear mechanism | Suitable materials | Surface engineering |
|---|---|---|
| Adhesion/tribo-oxidation | • Hard metals versus hard metals<br>• Tool steels versus tool steels<br>• Against high-strength steels: ceramics, polymer composites, bronzes, cast irons | • Thick coatings<br>• Thin coatings<br>• Metallic plating<br>• Diffusion treatments<br>• Surface heat-treatments |
| Contact fatigue with fluid lubrication | • Bearing steels<br>• High-strength martensitic steels | • Diffusion treatments<br>• Surface heat-treatments<br>• Weld hardfacing |
| Contact fatigue with mixed or boundary lubrication | • Hard metals<br>• Bearing steels<br>• Against high-strength steels: bronzes and gray cast irons | • Diffusion treatments<br>• Weld hardfacing<br>• Phosphating<br>• Hard chromium |
| Low-stress abrasion | • Hard metals<br>• Ceramic materials<br>• White cast iron<br>• Tool steels<br>• Cr martensitic steels<br>• Mn–C steels<br>• Elastomers | • Thermal spraying and weld coatings<br>• Thin films<br>• Metallic plating<br>• Diffusion treatments<br>• Surface heat-treatments<br>• Hard anodizing (for Al alloys) |
| High-stress abrasion | • Hard metals<br>• White cast irons<br>• Tool steels<br>• Cr martensitic steels<br>• Mn–C and HSLA steels | • Weld hardfacing<br>• Surface heat-treatments |

characterized by sliding at elevated temperatures. The Co-base alloys containing chromium and tungsten (the so-called *stellites*) are also particularly suited when sliding or abrasive wear resistance at elevated temperatures is required.

## 7.6 Summary

The choice of a material for a given tribological application relies on many parameters, the most important one being the dominant wear mechanism that is expected to act in service. To counteract each damaging mechanism, many materials have been developed and they were broadly described in the previous chapter. In many cases, it is more suitable, or more convenient, to adopt a specific surface treatment, as seen in the present chapter. Table 7.7 summarizes the materials and surface treatments that have been specifically developed to counteract the different wear mechanisms. In the table, adhesion and tribo-oxidation are considered together since they are quite similar as concerns material's selection, whereas both abrasion and contact fatigue were split in two, following the considerations introduced when treating the wear processes. The list shown in the table is only illustrative. For more detailed information, reference should be made to the content of the above paragraphs, to the specialized literature (partly reported in the bibliography) and also to the catalogues and experience of the manufacturers.

## References


1. K.G. Budinski, M.K. Budinski, *Engineering Materials, Properties and Selection* (Prentice Hall, Englewood Cliffs, 2002)
2. J. Takadoum, *Materials and Surface Engineering in Tribology* (Wiley, Hoboken, 2008)
3. K. Holmberg, A. Matthews, *Coatings Tribology* (Elsevier, Amsterdam, 1994)
4. R.F. Bunshah (ed.), *Handbook of Hard Coatings* (Noyes Publications, New Jersey, 2001)
5. S. Zhang, X. Zhang, Toughness evaluation of hard coatings and thin films—critical review. Thin Solid Films **520**, 2375–2389 (2012)
6. T. Tobie, Einfluss der Einsatzhartungstiefe auf die Grubchen- und Zahnfusstragfahigkeit grosser Zahnrader, FVA Absclussbericht (2000)
7. G. Totten, M. Howes, T. Inoue (eds.), *Handbook of Residual Stress and Deformation of Steel* (ASM International, Materials Park, 2002)
8. V. Savaria, F. Bridier, P. Bocher, Measuring in-depth stress gradients: the challenge of induction hardened parts. Mater. Sci. Forum **768–769**, 158–165 (2014)
9. O.R. Lang, Berichtsband zur AWT-Tagung am 23./24. Marz 1988, Darmstadt, pp. 332–348
10. O. Asi, A.C. Can, J. Pineault, M. Belassel, The relationship between case depth and bending fatigue strength of gas carburized SAE 8620 steel. Surf. Coat. Technol. **201**, 5979–5987 (2007)
11. M. Boniardi, F. D'Errico, C. Tagliabue, Influence of carburizing and nitriding on failure of gears—a case study. Eng. Fail. Anal. **13**, 312–339 (2006)
12. M. Benedetti, V. Fontanari, B.R. Hohn, P. Oster, T. Tobie, Influence of shot peening on bending tooth fatigue limit of case hardened gears. Int. J. Fatigue **24**, 1127–1136 (2002)
13. A. Molinari, G. Straffelini, M. Pellizzari, M. Pirovano, Wear behaviour of diffusion and compound layers in nitrided steels. Surf. Eng. **14**, 489–495 (1998)
14. Z. Gawronski, Residual stresses in the surface layer of M2 steel after conventional and low pressure ('NITROVAC 79') nitriding process. Surf. Coat. Technol. **124**, 19–24 (2000)
15. H. Kato, T.S. Eyre, B. Ralph, Sliding wear characteristics of nitrided steels. Surf. Eng. **10**, 65–74 (1994)
16. G. Straffelini, G. Avi, M. Pellizzari, Effect of three nitriding treatments on tribological performance of 42CrAlMo7 steel in boundary lubrication. Wear **252**, 870–879 (2002)
17. S.J. Bull, Q. Zhou, A simulation test for wear in injection moulding machines. Wear **249**, 372–378 (2001)
18. S. Rossi, F. Chini, G. Straffelini, P.L. Bonora, F. Moschini, A. Stampali, Corrosion protection properties of Nickel/PTFE, Phosphate/$MoS_2$ and bronze/PTFE coatings applied to improve the wear resistance of carbon steel. Surf. Coat. Technol. **173**, 235–242 (2003)
19. Ph Hivart, B. Hauw, J. Crampon, J.P. Bricout, Annealing improvement of tribological properties of manganese phosphate coatings. Wear **219**, 195–204 (1998)
20. D.R. Gabe, Hard anodizing—what do we mean by hard? Met. Finish. **100**, 52–57 (2002)
21. L. Rama Krishna, A. Sudha Purnima G. Sundararajan, A comparative study of tribological behaviour of microarc oxidation and hard-anodized coatings. Wear **261**, 1095–1101 (2006)

22. L.H. Chiu, C.F. Yang, W.C. Hsieh, A.S. Cheng, Effect of contact pressure on wear resistance of AISI H13 tool steels with chromium nitride and hard chromium coatings. Surf. Coat. Technol. **154**, 282–288 (2002)
23. K.G. Budinski, The wear resistance of diffusion treated surfaces. Wear **162–164**, 757–762 (1993)
24. J.D.B. De Mello, J.L. Goncalves, H.L. Costa, Influence of surface texturing and hard chromium coating on the wear of steels used in cold rolling mill rolls. Wear **302**, 1295–1309 (2013)
25. M.A. Juneghani, M. Farzam, H. Zohdirad, Wear and corrosion resistance and electroplating characteristics of electrodeposited Cr-SiC nano-composite coatings. Trans. Nonferrous Met. Soc. China **23**, 1993–2001 (2013)
26. J. Simao, D.K. Aspinwall, Hard chromium plating of EDT mill work rolls. J. Mater. Process. Technol. **92–93**, 281–287 (1999)
27. D.T. Gawne, U. Ma, Structure and wear of electroless nickel coatings. Mater. Sci. Technol. **3**, 228–238 (1987)
28. P. Sahoo, S.K. Das, Tribology of electroless nickel coatings—a review. Mater. Des. **32**, 1760–1775 (2011)
29. Y.W. Riddle, T.O. Bailer, Friction and wear reduction via an Ni-B electroless bath coating for metal alloy. J. Met. **57**, 40–45 (2005)
30. G. Straffelini, D. Colombo, A. Molinari, Surface durability of electroless Ni-P composite deposits. Wear **236**, 179–188 (1999)
31. Y.S. Huang, X.T. Zeng, I. Annergren, F.M. Liu, Development of electroless NiP-PTFE-SiC composite coating. Surf. Coat. Technol. **167**, 207–211 (2003)
32. B. Bushan, *Introduction to Tribology* (Wiley, New York, 2002)
33. P. Hollman, A. Alahelisten, M. Olsson, S. Hogmark, Residual stress, Young's modulus and fracture stress of hot flame deposited diamond. Thin Solid Films **270**, 137–142 (1995)
34. K. Holmberg, H. Ronkainen, A. Laukkanen, K. Wallin, S. Hogmark, S. JAcobson, U. Wiklund, R.M. Souza, P. Stahle, Residual stresses in TiN, DLC and $MoS_2$ coated surfaces with regard to their tribological fracture behavior. Wear **267**, 2142–2156 (2009)
35. G. Straffelini, P. Scardi, A. Molinari, R. Polini, Characterization and sliding behaviour of HF CVD diamond coatings on WC-Co. Wear **249**, 461–472 (2001)
36. R. Ramadoss, N. Kumar, R. Pandian, S. Dash, T.R. Ravindran, Tribological properties and deformation mechanism of TiAlN coating sliding with various counterbodies. Tribol. Int. **66**, 143–149 (2013)
37. J.C.A. Batista, C. Godoy, A. Matthews, Micro-scale abrasive wear testing of duplex and non-duplex (single layerd) PVD (Ti, Al)N, TiN and Cr-N coatings. Tribol. Int. **35**, 363–372 (2002)
38. T. Sudaprasert, P.H. Shipway, D.G. McCartnay, Sliding wear behaviour of HVOF sprayed WC-Co coatings deposited with both gas-fuelled and liquid-fuelled systems. Wear **255**, 943–949 (2003)
39. L. Fedrizzi, S. Rossi, R. Cristel, P.L. Bonora, Corrosion and wear behaviour of HVOF cermet coatings used to replace hard chromium. Electrochim. Acta **49**, 2803–2814 (2004)
40. G.M. La Vecchia, F. Mor, G. Straffelini, D. Doni, Microstructure and sliding wear behaviour of thermal spray carbide coatings. Int. J. Powder Metall. **35**, 37–46 (1999)
41. L. Valentinelli, A. Loreto, T. Valente, L. Fedrizzi, Tribological behaviour at high temperature of cermet coatings, Atti del 7° Convegno AIMAT, Ancona, 29 giugno-2 luglio (2004)

# Chapter 8
# Tribological Systems

In this chapter, the main features of a number of important tribological systems will be presented. Systems from mechanical design are considered together with tribological systems involved in manufacturing processes. A brief description of the relevant design will be provided along with the analysis of the tribological damage. Materials and surface treatments currently used for each application will be also outlined. The overall goal is to provide basic tools for a preliminary design and control of the tribological system under consideration using the information provided in the previous chapters.

Of course, the tribological systems of engineering interest are much more numerous than those considered in this chapter, and on many occasions designers would be required to intervene on a case by case basis with novel solutions since no extensive bibliography is available. However, proposed cases might be also regarded as useful starting point for approaching special tribological issues for which no established procedures still exist.

## 8.1 Sliding Bearings

The most widely used sliding bearings are *thrust bearings*, capable of stand axial loads, and *journal bearings*, suitable for radial loads [1]. In this paragraph the focus is on journal bearings, consisting of a fixed cylindrical body capable to stand a rotating shaft, usually made with a high-strength or carburized steel. They are used in different applications, such as small motors, car engines and large electrical generators. As an example, Fig. 8.1 shows the main components in an *internal combustion engine* including the journal bearings supporting the conrod and the crankshaft [2].

The relative motion of the rotating shaft against the bearing is by sliding, and the two mating surfaces may undergo sliding wear. In general, sliding bearings are oil lubricated and conditions of fluid-film lubrication are attained. If the applied load is very high and the sliding speed is low, lubrication with grease is recommended. On



G. Straffelini, *Friction and Wear*, Springer Tracts in Mechanical Engineering,
DOI 10.1007/978-3-319-05894-8_8

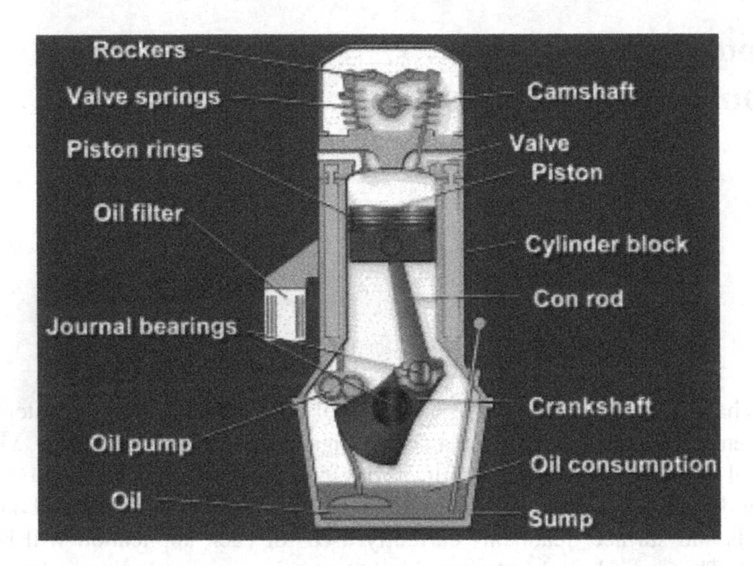


**Fig. 8.1** Main components in an internal combustion engine [2]

the contrary, if the sliding speed is particularly high, lubrication with gas can be successfully adopted.

### *8.1.1 Oil Lubricated Journal Bearings*

In oil lubricated journal bearings, the lubricant is conveyed to the contact region between the bearing and the shaft. During steady-state operation, oil is pumped in a converging gap and it exerts a pressure that balances the applied load, as shown in Fig. 8.2. The bearing design is conducted by verifying that the Λ-factor is greater than 3. A simplified approach for estimating the minimum film thickness, $h_{min}$, is provided by the design-graphs of *Raimondi and Boyd*, based on the solutions of equations for hydrodynamic oil pressure [1]. As an example, Fig. 8.3 shows the graph for a 360° journal bearing. In this approach, first it is necessary to calculate the *Sommerfeld number*, S, defined by the following relation [1, 3]:

$$S = \frac{\eta \cdot n \cdot B \cdot D}{\psi^2 \cdot F_N} \tag{8.1}$$

where η is the viscosity of the lubricant at the operating temperature, n is the rotational speed of the shaft (unit: revolutions per second), ψ is defined by s/D, where s is given by: s = D−d, and $F_N$ is the applied load (see Fig. 8.2 for the other definitions). Then, using the graph in Fig. 8.3 it is possible to evaluate the ratio 2 $h_{min}$/s (usually about 0.2). In engine bearings (Fig. 8.1), $h_{min}$ is typically lower than 1 μm, and the composite surface roughness is about 0.35 μm [4].

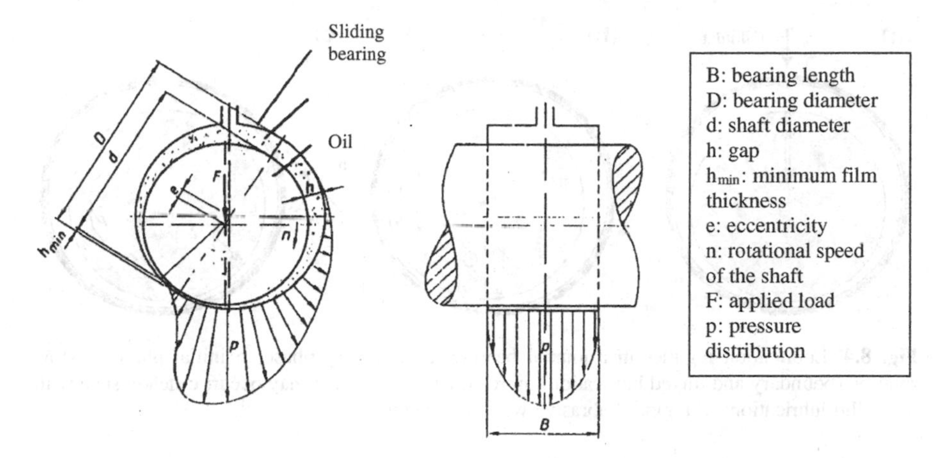


**Fig. 8.2** Part names of a journal bearing and film pressure distribution (modified from [3])

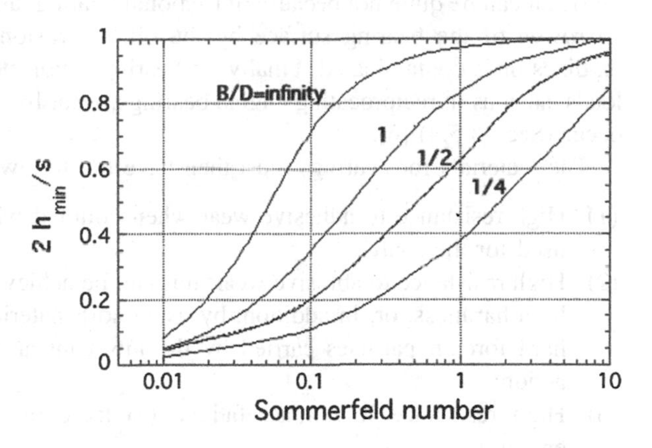


**Fig. 8.3** Raimondi-Boyd design chart to obtain the minimum film thickness (for different B/D ratios) for a 360° hydrodynamic journal bearing (modified from [1])

Even if the steady-state film thickness is correctly chosen, the journal bearing may still suffer from wear damage. Indeed, before rotation starts, bearing and shaft are in contact, as shown schematically in Fig. 8.4a, and no oil is present between the mating surfaces. At the beginning of the motion the coefficient of friction, μ, can be quite high and this causes the rolling of the shaft on the bearing surface, as shown in Fig. 8.4b. At static equilibrium: $\mu = tg\Theta$, with Θ about 30° [5]. Therefore, during this phase sliding wear by *adhesion* may occur, and the process is repeated during each start-up and shut-down. This type of wear is particularly important in case of high applied loads. After some rotations, the amount of oil dragged in the meatus increases, and the lubrication regime passes from boundary to mixed, and then to fluid-film lubrication at steady-state (Fig. 8.4c). However, surface damage by *abrasion* may still occur if the lubricant carries contaminating abrasive particles. Wear depends on the hardness of the particles, and on their mean size as compared to $h_{min}$. If the particles are larger than $h_{min}$, wear by abrasion can be quite severe.

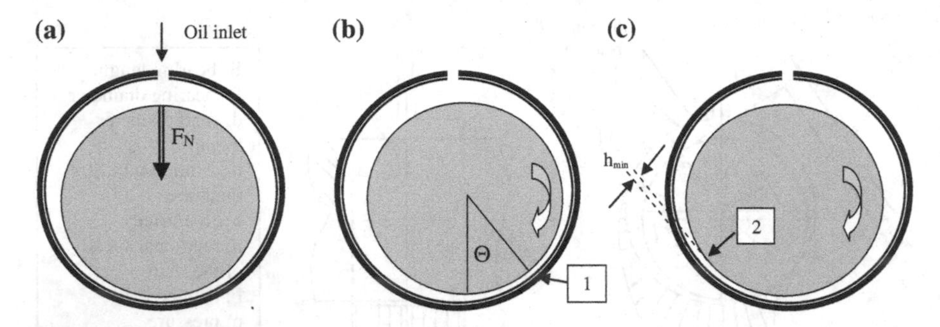


**Fig. 8.4** Lubrication regimes in a journal bearing. **a** stationary phase; **b** initial phase of slow rotation (boundary and mixed lubrication); in region 1 sliding wear may occur; **c** steady state with fluid-film lubrication; in region 2 abrasive wear may occur

Abrasion damage is mainly located in region 2 of Fig. 8.4b. In this region the lubricant can be quite hot because of frictional heating, and this may also favour the *corrosion* of the bearing surface by the oil. Corrosion can be severe if the oil oxidizes or is contaminated. Finally, in bearings subjected to high and fluctuating loads, such as in reciprocating engine bearings, wear by *cavitation erosion* can also occur (Sect. 5.5.3) [6].

The materials for bearings must then have the following requirements:

(1) High resistance to adhesive wear when coupled with steel (which is usually used for the shaft);
(2) High resistance to abrasive wear; this can be achieved by using materials with high hardness, or, in addition, by using soft materials able to incorporate the hard foreign particles carried by the lubricant and thus limit their abrasive action;
(3) High resistance to contact fatigue (to have high resistance to cavitation erosion).

To match all these requirements different materials must be used. Quite often, bearings are made of different layers over a steel shell, such as an outer layer of a soft material with low compatibility for ferrous alloys, and an intermediate layer with high hardness. In case of low nominal contact pressures (with values in the order of 1–2 MPa being the nominal pressure defined by: load/projected area, i.e., $F_N/BD$), as in the case of bearings for electric motors or centrifugal pumps, *white metals* over a steel shell are mainly used [7]. White metals are tin-based and lead-based alloys (typical compositions are: 89 % Sn, 7.5 % Sb, 3.5 % Cu, and: 75 % Pb, 15 % Sb, 10 % Sn). These materials have a low tribological compatibility towards steels, as is clear from Fig. 1.12, and then friction coefficient and adhesive wear during the initial stages under boundary lubrication are particularly low. Furthermore, these materials are soft and they easily incorporate any abrasive foreign particle. However, because of their low hardness they also display a relatively low resistance to wear by adhesion and contact fatigue.

Journal bearings for automobile engines operate under high nominal pressure (typically up to 60 MPa [4]). Therefore they must be made of more resistant materials [7]. In general, such bearings consist of a steel support coated with two layers: an intermediate layer of a copper alloy with a thickness of about 300 μm (made by casting), and an outer thin layer (with a thickness of about 25 μm) in white metal (obtained by electrochemical deposition). Another type of bearings is made of aluminium alloys (for example: Al-20 % Sn); such bearings are obtained by rolling an aluminium sheet on a steel shell. If the nominal pressure is very high, bearings made by heat-treated steel or carburized steels are also employed [8]. From the point of view of corrosion resistance, white metals, copper alloys and aluminium alloys afford the best performances.

### *8.1.2 Unlubricated Sliding Bearings*

In unlubricated or *dry bearings*, sliding wear is clearly present for all the servicing duration. The materials for dry bearing must then give rise to a sufficiently low friction coefficient when sliding against the shaft (usually made of steel). The design is then based on the verification that the bearing wear is sufficiently low. In Table 5.3 some materials commonly used for dry bearings are listed. The design procedure is generally based on the $PV_{limit}$-concept, outlined in Sect. 5.1.4, and on the maximum $PV_{limit}$-concept, outlined in Sect. 6.8.1 for polymers.

## 8.2 The Piston Ring/Cylinder Liner System

The piston ring/cylinder liner is also quite an important component for internal combustion engines (Fig. 8.1), since the friction and wear losses in this system influence to a great extent the energy and fuel consumption, and the corresponding environmental emissions [2]. In general, three or more rings are placed in specific circumferential grooves at the top of the piston, and they are forced against the cylinder liner by their own elasticity. They form a moving labyrinth seal, and thus maintain a gas seal between the combustion chamber and the crankcase. This primary role of the piston assembly is mainly played by the top rings, as schematized in Fig. 8.5. The bottom ring is an oil-control ring that limits the upward lubricant flow from the crankcase.

The elastic pressure exerted by the rings against the cylinder wall is comparatively low, whereas the additional cylinder gas pressure is much higher, about 70 MPa in a gasoline engine [4]. The system is lubricated, and the top ring has a barrel-shape profile that form a converging gap thus allowing oil to exert an EHD pressure able to counteract the applied load. As a matter of fact, during each stroke the velocity changes, and the ring/cylinder contact experiences pure sliding under boundary, mixed and fully film lubrication. Friction and wear may be particularly

Fig. 8.5 Scheme of a compression piston ring/cylinder liner system

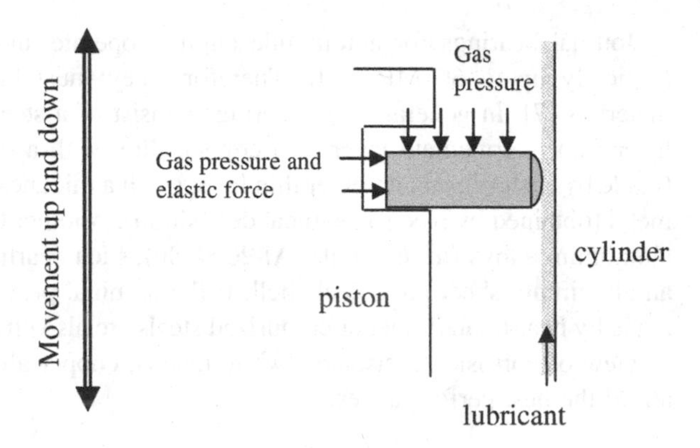


high at the *top dead centre*, where the relative velocity between the ring and the cylinder wall approaches zero and the applied gas pressure is maximum. In this position a boundary lubrication regime is present. As an example, Fig. 8.6a shows the profile of the cylinder wall at top dead centre after long-term running [4]. A deep wear scar is observed at the reversal position of the top compression ring.

Piston rings and skirts, and cylinder liners are commonly made of cast iron (grey, malleable or nodular cast iron) and steel [2, 9]. Different materials and surface treatments have been recently introduced: some of them are listed in Table 8.1. In general, the specific wear rate of materials for piston rings should be lower than $5 \times 10^{-17}$ $m^2/N$, and lower than $10^{-16}$ $m^2/N$ for the liner materials. It should be noted that thermal sprayed molybdenum coatings and also gas nitrided parts contain some porosity on the outer layers that act as oil reservoirs and avoid the possible failure by scuffing during sliding contact under boundary lubrication (Fig. 8.6b). Chromium plated rings are widely used against cast iron cylinder liners. Surface treatments are generally limited to the periphery of the rings. The composite roughness of the piston ring/cylinder liner system is about $R_a = 0.2$ μm. Surface finishing is often optimized by a specific texturing aimed at improving oil retention at the surface.

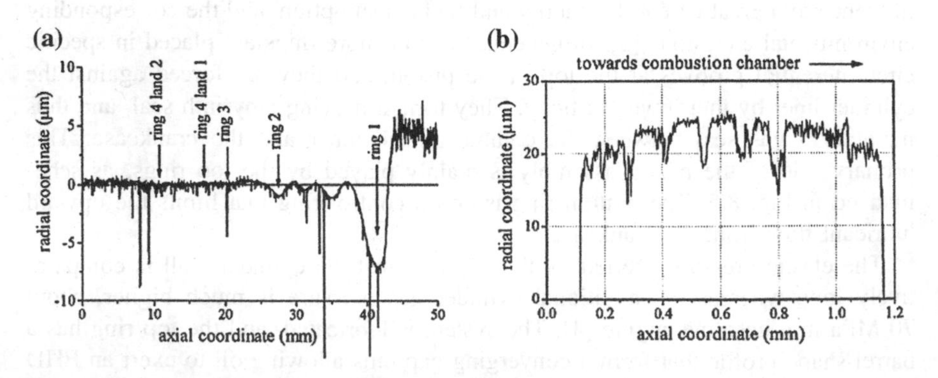


**Fig. 8.6** **a** Wear profile in an engine cylinder wall at *top* dead centre. Ring 1 is the *top* compression ring; **b** profile of a *top* ring coated with molybdenum. Note the barrel-faced shape and the deep valleys that are pores in the coating [4]

**Table 8.1** Materials and surface treatments used for piston rings, piston skirts and cylinder liners (modified from [9])

| | Piston ring | Piston skirt | Cylinder bore/liner |
|---|---|---|---|
| Conventional materials | • Grey cast iron<br>• Malleable/nodular cast iron<br>• Steel | Grey cast iron | Grey cast iron |
| Recently used materials | Tool steel | Copper and aluminium alloys | • Si-containing aluminium alloy<br>• Cast Al-base metal matrix composite<br>• Steel<br>• Compacted graphite iron |
| Surface treatments | • Chromium plating<br>• Thermal sprayed molybdenum on cast iron<br>• DLC on steel<br>• Gas nitriding of steel | • Chromium plating on cast iron<br>• Ni-P reinforced with SiC | Chromium plating on cast iron |

## 8.3 Cam/Follower System

Cam/follower systems are widely used in different machineries to transform the rotary motion into linear reciprocating motion, and vice versa. They are also used in the valve train in internal combustion engines (Fig. 8.1), where a cam, driven from by the crankshaft, controls the opening and closure of the valves. Figure 8.7 schematizes the non-conformal contact between a cam and a *flat follower*. In this case, friction and wear are mainly due to sliding. The cam/follower system can be also designed using a *roller follower*. In this case motion is mainly by rolling (although some sliding is present due to possible rapid changes in the roller speed), and friction coefficient is reduced by an order of magnitude as compared to a pure sliding coupling.

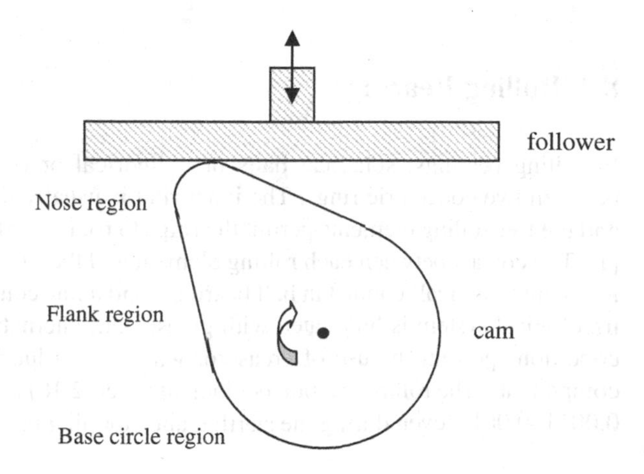


**Fig. 8.7** Schematization of a cam/follower system with a flat follower

**Table 8.2** Roughness change (in $R_a$) in cam surfaces due to 1000-h laboratory tests (data from [11])

| Nose region | | Base circle region | |
|---|---|---|---|
| Before | After | Before | After |
| 0.28 | 0.17 | 0.28 | 0.27 |
| 0.31 | 0.22 | 0.48 | 0.48 |
| 0.96 | 0.34 | 2.41 | 2.4 |

Cam/follower systems typically operate in lubricated conditions under large variations of load, speed and temperature [10]. Generally speaking, in the contact over the flank and base circle lubrication is mixed, whereas over the nose region a boundary lubrication regime is attained, and this limits the system lifetime. For this reason, lubricants with EP additives are typically employed. In the nose region of cam/follower systems of internal combustion engines, the Hertzian pressure is about 600 MPa and the lubricant film thickness is about 0.1–0.15 μm [4]. Since the composite roughness is typically 0.3 μm, it turns out that $\Lambda$ is about 0.3. As an example, Table 8.2 shows the experimental roughness values of different cams before and after 100 h tests in a laboratory simulator [11]. The roughness values in the base circle of the cam were almost unaffected by testing, showing that an adequate lubricant film was present to avoid any contact between the asperities. On the contrary, over the nose region the surface roughness has changed. The contact between the asperities produced a reduction of roughness and then a surface flattening.

The most popular materials used for cams and followers are grey cast irons and steels. The cast irons are heat-treated or chilled to achieve high hardness values necessary for the requested sliding resistance. The followers are also made by Fe–Cr–Mo–C alloys obtained by powder metallurgy. When wear is quite severe because of high sliding contact and/or poor lubrication, proper surface treatments can be adopted. Steel camshafts may be submitted to induction hardening or carburizing, whereas steel followers may be chromium plated, carbonitrided or coated with thin films, such as TiN, CrN or DLC [9]. Hard coatings are particularly beneficial since they reduce friction and wear.

## 8.4 Rolling Bearings

In rolling bearings, spherical balls or cylindrical or conical rollers are enclosed between two concentric rings. The inner ring is usually fastened to a rotating shaft, and the free rolling elements permit the rings to rotate whilst supporting a radial load [1]. The contact between each rolling element and the ring races is non-conformal. It is a point (elliptical) contact in ball bearings, and a line contact in roller bearings. The tribological system is lubricated with grease and, alternatively, with oil, if working conditions prevent the use of grease or when an oil lubrication is used for nearby components. The rolling friction coefficient (Sect. 2.10) is typically low, in the range 0.0011–0.0015, even during the starting and stopping phases, when sliding bearings

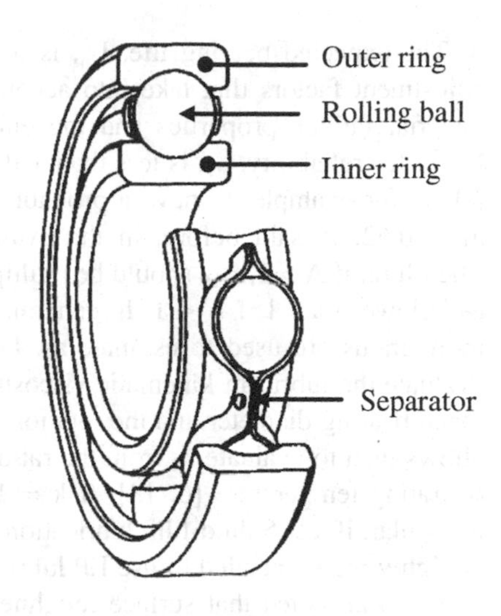


**Fig. 8.8** Schematization of a deep-groove ball bearing (modified from [12])

display high friction, as mentioned in Sect. 8.1. As an example, Fig. 8.8 shows a *deep-groove ball bearing*, where the balls fit into deep grooves making possible to support radial loads as well as axial loads [12].

The surfaces of the rolling elements and the raceways may undergo wear by contact fatigue. Typically, all loaded parts are made with high-strength heat-treated steels, described in Sect. 6.1.2. For the estimation of bearings life, a simplified design procedure is usually followed, since rolling bearings are available in different standardized configurations. The basic bearing life, $L_{10}$, is obtained by the following relation, similar to Eq. 5.8 [6]:

$$L_{10} = \left(\frac{C}{F_N}\right)^p \tag{8.2}$$

where:

- $L_{10}$ (in bearing life calculations, L is usually employed in place of N to indicate the number of revolutions) is the *basic bearing life*, i.e., the number of revolutions required to have contact fatigue damage with a probability of 10 %;
- $F_N$ is the equivalent radial load;
- C is the *dynamic load capacity*, which represents the load to have contact damage with a 10 % probability (90 % reliability) after one million of revolutions; it depends on the material's resistance and the geometry of the bearing assembly (compare to Eq. 5.8). Design values for C are provided by the bearing's manufacturers and are obtained assuming that the lambda factor is 1, i.e., lubrication is mixed;
- p is a constant set equal to 3 in case of spherical bearings, and 10/3 in case of roller bearings ($\Lambda = 1$).

The expected bearing life, $L_n$, is then obtained by multiplying $L_{10}$ by proper adjustment factors that take into account the required reliability level, the actual material fatigue properties and the effective lubrication regime. The adjustment factor for reliability, $a_1$, is less than 1 if the required failure probability is less than 10 %; for example, to have a probability of failure of 5 %, it is necessary to put $a_1 = 0.62$. A said before, in the evaluation of $L_{10}$, it is assumed that $\Lambda = 1$. Therefore, if $\Lambda > 1$, $L_{10}$ should be multiplied by a factor, $a_2$, greater than 1. Clearly, $a_2$ is lower than 1 if $\Lambda < 1$. In general, $\Lambda$ is not directly evaluated, and simplified monograms are used to estimate $a_2$. For example, Fig. 8.9a shows a diagram to evaluate the lubricant kinematic viscosity, $\nu_1$, required to have $\Lambda = 1$, knowing the mean bearing diameter and the rotation velocity (units: $rev^{-1}$) [12, 13]. Figure 8.9b allows then to evaluate $a_2$ from the ratio $k = \nu/\nu_1$, where $\nu$ is the oil viscosity at the operating temperature [6, 13]. If $k = 1$, $a_2 = 1$ and $\Lambda = 1$. If $k > 1$, $a_2 > 1$; in particular, if $k > 5$ fluid-film lubrication is attained. If $k < 1$, $a_2 < 1$; when $k < 0.4$ it is highly recommended to use EP lubricants to help lubrication and avoid scuffing. It has to be noted that surface roughness is not included in the calculation, since modern bearings have all similar roughness values.

In some applications it is difficult, or impossible, to replenish the lubricant during the operation. In such cases it is suggested to increase the contact fatigue life by using self-lubricating coatings (such as soft metals or $MoS_2$ coatings), or hard PVD or CVD coatings. Soft coatings reduce the adhesion during the starting and the stopping phases when lubrication is less effective, and allow a more homogeneous distribution of contact pressures. As already mentioned, the improvement in contact resistance achieved by thin hard coatings is due to the difference in modulus of

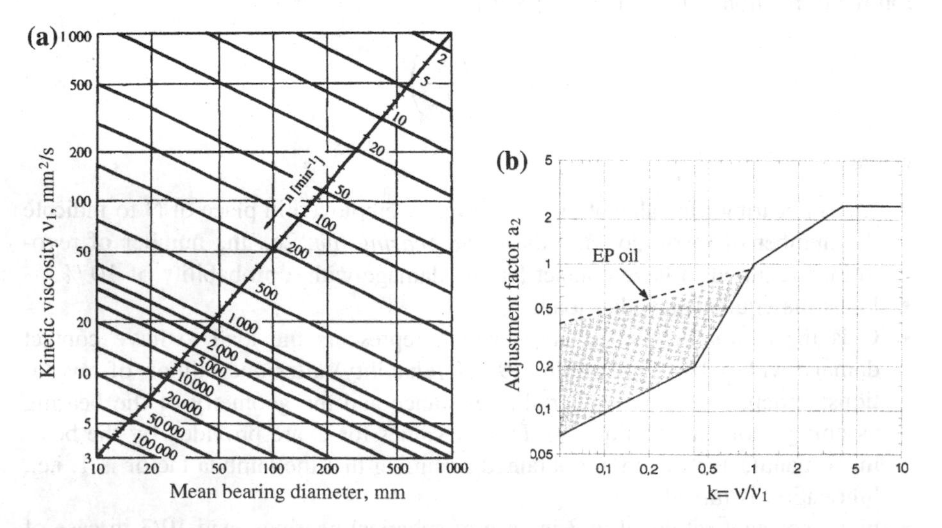


**Fig. 8.9** **a** Monogram for the evaluation of the kinematic viscosity, $\nu_1$, required to ensure lubrication with a lambda factor of 1; **b** monogram to evaluate the adjustment factor $a_2$ knowing $k = \nu/\nu_1$. The *solid line* refers to a normal lubricating oil while the *dashed line* refers to an EP oil (modified from [12, 13])

elasticity between the coating and the substrate. Such a difference shifts the maximum tangential stress inside the coating or at the coating/substrate interface. As already mentioned, the performance of the coated components is very sensitive to the coating quality and to the coating thickness. It has been shown that an optimum thickness is generally quite low, between 0.1 and 1 μm [14].

## 8.5 Gears

Gears transmit rotational motion and power from one shaft to another. They are essential parts of many mechanical devices. There are various types of gears [1]. *Spur gears*, for example, have teeth parallel to the axis of rotation of the shaft. Gears are typically lubricated (in a oil bath or by spraying) even if in some cases they also run in dry conditions.

Figure 8.10 schematically shows the interaction of spur gear teeth during contact. The relative motion is pure rolling only when the contact occurs at the pitch point. Otherwise the contact is by rolling-sliding. In particular it is observed that a *negative sliding* (with sliding and rolling having opposite directions, compare with Fig. 5.7) occurs in the contact area between the pitch point and tooth root. On the contrary, positive sliding occurs between the pitch circle and tooth tip. In spur gears a line contact is established, and the contact stresses may be evaluated using the Hertz theory. The contact stresses may be evaluated in all points from the beginning to the end of contact by considering the local values of the radii of curvature. The maximum Hertzian pressure is attained at the pitch point. In case of lubricated gears, the minimum film thickness can be evaluated using Eq. 3.11 (specific simplified relations are also available [10]). This allows evaluation of the $\Lambda$-factor and thus of the lubrication regime. The velocity of the pitch line provides a simple indication of the lubrication regime, given the role of velocity in determining the hydrodynamic thrust force [10]. In particular, gears running at low velocity, lower

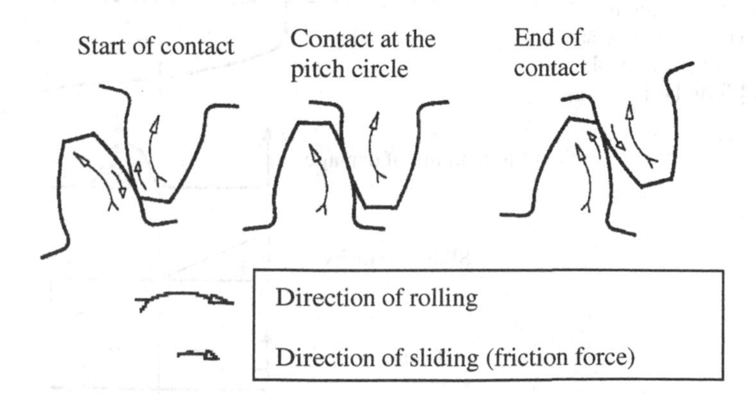


**Fig. 8.10** Schematization of the gear tooth contact. The *lower* wheel is the driver

than about 0.1 m/s, are characterized by boundary lubrication; gears running at medium velocity (in the range 2–20 m/s) are characterized by mixed lubrication, and high velocity gears (with velocity over than 20 m/s) feature fluid film lubrication with $h_{min}$-values of the order of 1 μm. The analysis of the relative motion between the faces of the mating teeth shows that tribological damage is by rolling-sliding. Damage by scuffing may also occur in service and this should be checked for in designing using available relationships [10]. If foreign hard particles contaminate the lubricant, abrasive wear may also intervene.

In most cases, gears are made of high-strength steel, cast iron (spheroidal) or bronze. Quite often different materials are used for producing the two mating wheels. In gears that have to transmit a high power, suitable surface treatments are adopted such as surface hardening, carburizing or nitriding. To improve lubrication, the surface roughness is reduced to a minimum. In case of relatively low contact loads, non-ferrous gears made by aluminium or copper alloys, or polymeric materials can be used.

## *8.5.1 Damage by Contact Fatigue*

In well lubricated gears (with Λ-factors greater than, say, 1.2), wear damage is mainly by contact fatigue. Damage by pitting or spalling is concentrated in the region between the pitch point and the teeth root, as schematized in Fig. 8.11a [15]. As an example, Fig. 8.12 shows the contact fatigue damage in a carbonitrided and carbonitrided/shot peened steel gear [16]. As seen, the Hertzian pressure is maximum at the pitch point where pure rolling is present. Moving away from the pitch point, the Hertzian pressure decreases but the sliding contribution increases (Fig. 8.11b). Damage by rolling-sliding is thus maximum where the combination of applied pressure and sliding is most damaging, which is between the pitch point and the tooth root where a negative sliding velocity is present and the tensile stress at

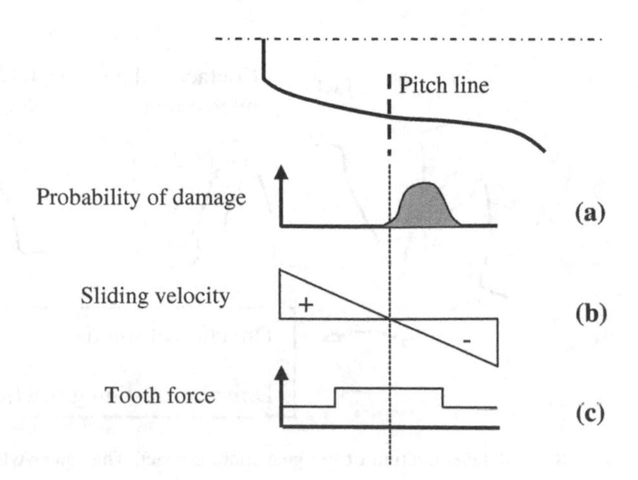


**Fig. 8.11** Probability of damage, sliding velocity and tooth force on a gear tool (modified from [15])

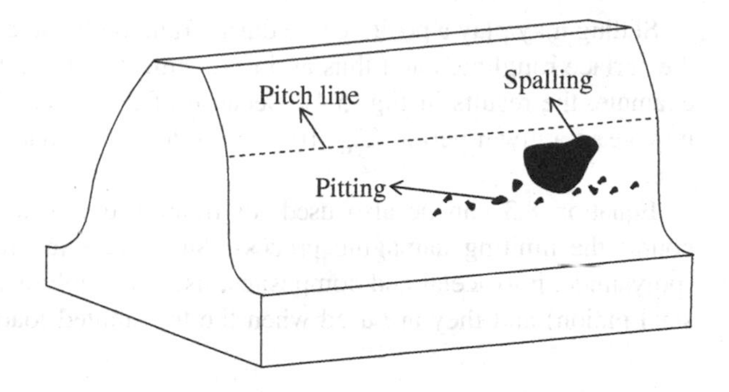


**Fig. 8.12** Schematic of damage observed in a carbonitrided gear (modified from [16])

the beginning of contact favours the formation of surface cracks and then open them, allowing the lubricant to pour into them thus activating the oil pumping effect (Fig. 4.15). It can be also observed that in the regions at the beginning and the ending of contact, the tooth force is lower than around the pitch point, since the load is shared with the tooth leaving the contact and then with the next tooth entering the contact (Fig. 8.11c).

To check if the expected gear life is sufficient for the intended use, it is therefore necessary to use relation 5.9, with the appropriate values for $p_{end}$ and n, which depend on the gear materials and the lubrication regime (Tables 5.5 and 5.6). A common practice is based on standardized procedures, such as that proposed by the AGMA or ISO [1]. The Hertzian pressure at the pitch point is calculated and, if needed, modified considering manufacturing, assembly and loading variability. The obtained value is then compared with a *tabulated allowable strength* that may be also adjusted to account for the required lifetime or survival reliability. In any case, the stress situation on the contact surface of gears is very complex, and reference to test results obtained from actual gears is advisable to obtain the best design.

### 8.5.2 Damage by Sliding and Its Control

If the lambda factor is less than about 1, sliding wear may play an overwhelming role. In this case the use of lubricants with EP additives is strongly recommended. The evolution of wear depth, h, can be evaluated by using Eq. 5.14. To a first approximation the ratio $v_s/v_t$ can be set to 1.45 for the pinion and the gear [14]:

$$h \cong 1.45 \cdot \frac{K}{H} \cdot \frac{F_N}{L} \cdot N \tag{8.3}$$

where $F_N$ is the tooth force, L is the tooth width, N is the number of rotations, H is the hardness of the material which constitutes the tooth, and K is the wear coefficient. As described in Sect. 5.1.5, K depends on the lubrication regime, and typically ranges between $10^{-10}$ and $10^{-6}$ in the case of mixed lubrication and between $10^{-6}$ and $10^{-5}$ for boundary lubrication.

Sliding may play a positive role during running in, since it induces a reduction of the surface roughness and thus an improvement of the lubrication regime (see, for example, the results in Fig. 5.4). Because of this, it is often recommended to run new gears a given period (e.g. 10 h or so) to a load that is one half of the nominal one.

Equation 8.3 can be also used for running dry gears, for which sliding is of course the limiting damaging process. Such gears are mostly made of polymers (polyamide, polyacetal and composites; usually a polymeric gear is coupled with a steel pinion) and they are used when the transmitted loads are low.

## 8.6 Contact Seals

Radial shaft seals are used to prevent fluid leakage from a mechanical system, and also to avoid dust or dirt to enter into the system. These seals can operate with or without contact. Contactless seals, such as labyrinth seals, are often used to protect bearings. From a tribological point of view, contact seals are particularly important. They run in sliding conditions and may therefore suffer from sliding wear. Among the contact seal, of particular interest are *lip seals* and *mechanical face seals*, schematized in Fig. 8.13 [6].

The tips of lip seals are generally made with an elastomer [17]. They are sufficiently compliant to be elastically pushed, possibly under the action of a spring, against the shaft surface. The shaft is typically made of high-strength steel,

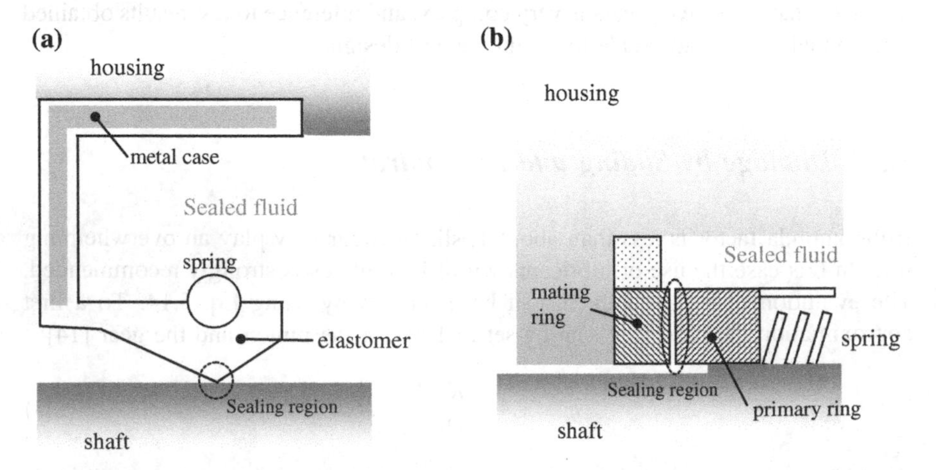


**Fig. 8.13** Schematics of a lip seal (**a**) and a mechanical face seal (**b**). The sealing regions also represent the sliding regions (modified from [6])

carburized steel, or steel coated with ceramic material (e.g. alumina) by thermal spraying, to resist the abrasive action of any possible contamination particle. The shaft surface must have a controlled roughness (not above $R_a = 0.8$ μm) to prevent abrasive wear of the lip by the mating hard asperities. During operation, the pressurized fluid (oil or water, typically) penetrates between the lip and the shaft inducing a mixed or fluid-film lubrication that reduces both friction and wear.

Rotating mechanical face seals are used when the pressure difference, Δp, between the inner and outer region is particularly high. They are therefore widely used in centrifugal pumps, compressors, and turbines, just to give a few examples. As shown in Fig. 8.13b, there are two rings in conformal contact: the *primary ring* is fixed to the shaft and rotates with it; the *mating ring* is fixed to the housing. The sliding region between the two rings is also the sealing region and thus the leakage path for the sealed fluid. The fluid enters the contact and acts as a lubricant, exerting an opening force on the seal faces. The mating faces reach an equilibrium position, determined by this opening force and the closing forces exerted by the spring and, above all, by the pressure of sealed fluid acting on the backside of the primary ring. This pressure ($p_{fluid}$) is given by [6, 14]:

$$p_{fluid} = \Delta p \cdot (b - k) \qquad (8.4)$$

where b is the so-called *balance ratio*, defined by the ratio between the effective area of the backside of the primary ring and the annular area of contact between the two rings; k is a constant that depends on the pressure distribution in the sealing region (b is typically maintained between 0.65 and 0.75, a k is set to 0.5 assuming a linear pressure distribution).

The fact that, during operation, the fluid enters the sealing region has two consequences:

(1) to induce a fluid leakage;
(2) to promote a (hydrostatic) lubrication.

Most mechanical seals work with mixed lubrication in order to minimize leakage rate and also minimize wear and friction (and the associated temperature rise) [18]. The wear design is carried out using Eq. 5.4, where $p_o$ is the contact pressure given by the sum of the pressure exerted by the spring (quite low, in the range of 69–276 kPa [6]), and the pressure exerted by the fluid given by Eq. 8.4, v is the sliding velocity (connected with the rotational speed), $\dot{h}_{al}$ is the allowable linear wear rate (for example, 0.02 mm after 1000 h), and $K_a$ is the specific wear rate determined by the adopted material and lubrication regime. In most applications, the rotating ring is made of carbon-graphite while the fixed ring is made of harder materials, such as ceramics, hard metals, stellites, or, in special cases, martensitic stainless steels. Wear is generally limited to the carbon-graphite ring and is characterized by some transfer of graphite onto the harder counterface.

## 8.7 Automotive Disk Brakes

Friction brakes are employed in a variety of applications, including winches, elevators, movers, washing machines and, of course, in the automotive industry. The automotive brakes have the main purpose of slowing down or stopping the vehicle. In modern cars *caliper disc brakes* are widely used and the main parts are shown in Fig. 8.14. The disc is fixed to the rotating wheel and two brake pads are placed in a caliper and fixed on the chassis. During braking a piston pushes the pads against the disc and the kinetic energy of the vehicle is converted into thermal energy by friction. Since the pads extend over only a small sector on the disc, most part of the disc is cooled by convection with the ambient atmosphere. This effect is further increased by special vanes in the disc (in ventilated discs).

Indicating with m the mass of the vehicle, the braking force, $F_b$, is given by: $F_b = ma$, where a is the deceleration, usually between 2.5 and 5 m/s$^2$. The friction force, $F_f$, acting on each braking pad is given by: $F_f = F_b/n$, where n is the number of pads (usually n = 8). Therefore, the average pressure, p, acting onto the pad-disc contact for having the required braking torque is given by:

$$p = \frac{F_f}{A_{pad} \cdot \mu} \tag{8.5}$$

where μ is the friction coefficient (typically between 0.4 and 0.5) and $A_{pad}$ is the nominal area of contact between each pad and the disc. Table 8.3 lists the typical values of the contact pressure in case of normal and heavy braking conditions (note that Eq. 8.5 is actually simplified since it neglects the aero-dynamic drag and the rolling resistance).

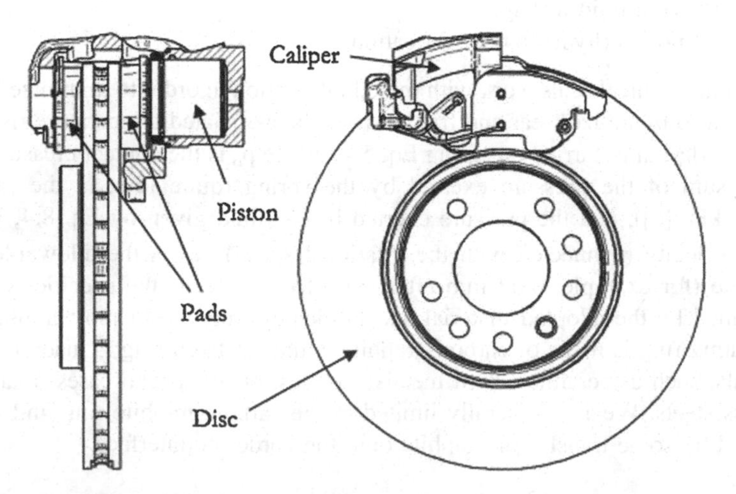


**Fig. 8.14** Schematic of a disk brake assembly with a single-piston floating caliper and a ventilated disc [19]

**Table 8.3** Summary of typical automotive braking conditions

| | Normal braking (normal urban and high-way cycles) | Heavy braking (heavy urban cycles, downhill and sport drives) |
|---|---|---|
| Typical decelerations during braking, m/s$^2$ | <2 | >5 |
| Nominal contact pressure, MPa | 0.2–1.5 | 1.1–4 |
| Average surface temperature | <300 °C | >300 °C |
| Pad wear, $K_a$ in m$^2$/N | $5 \times 10^{-15}$–$4 \times 10^{-14}$ | Increases with temperature |
| Friction coefficient | 0.4–0.6 | Decreases with temperature |

The brake disc is typically made of pearlitic grey cast iron, featuring quite good sliding wear resistance and high thermal conductivity, required to reduce the contact temperature reached during braking. The friction pads are usually made of organic friction materials (Sect. 6.8.1) with quite a low thermal conductivity (about 1–2 W/mK). The average surface temperature rise can be evaluated using the approach described in Sect. 2.11.1, considering that most of the frictional heat (about 95 %) is dissipated by the rotating disc (the Pe number is mostly greater than 5) that is in turn cooled by convection [1]. The surface temperature reached during braking depends on the geometry of the system and the braking conditions, as outlined in Table 8.3. Sliding wear of the pad is mild as long as the surface temperature is lower than a critical value, which is about 250–300 °C. This critical temperature is mainly determined by the characteristics and content of the phenolic resin, and, in particular, by its thermal degradation tendency. Above this value wear is severe and it increases almost exponentially with surface temperature [10]. The counterface cast iron disc undergoes tribo-oxidative wear (with some abrasive contribution) and its wear rate is about 20 % that shown by the friction pad.

## 8.8 The Wheel/Rail System

The tribological performance of the wheel/rail system is paramount in the rail transportation system [20]. Figure 8.15 shows a pair of wheels, fixed on their axis, in contact with two rails. Since the two wheels rotate at the same angular velocity, large wheel-rail sliding occurs on a curved track, depending on the radius of the curve. Sliding velocities up to 1 m/s have been reported. The whole wear process is then by rolling-sliding, and the sliding contribution is particularly important in the curves, especially if their radius of curvature is lower than about 600 m. Wheels and rails are usually made of pearlitic steel containing between 0.6 and 0.8 % of carbon. Wheels have a hardness in the range 260–300 kg/mm$^2$, and rails have a somewhat greater hardness that may reach 390 kg/mm$^2$ in case of curved tracks that may undergo severe sliding wear. Bainitic or martensitic steels are also used. A martensitic

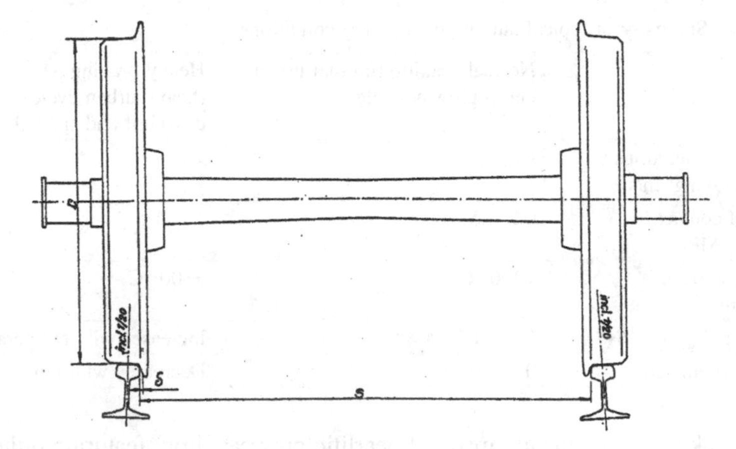


**Fig. 8.15** Pair of railway wheels in contact with the rails. The distance between the rails, denoted by s, is the track gauge. In a *curve* the outside rail is raised and the internal rail is lowered to compensate for the action of the centrifugal force

microstructure can be obtained in the contact region of the rails by a special *head-treating* cycle.

In the wheel/rail system, conditions for both free rolling and tractive rolling exist (Sect. 2.10). The coefficient $\mu_v$ under tractive rolling depends on the operational conditions. It amounts typically to 0.2 and is decreased in wet conditions [21]. The static friction between the rotating wheel and the rail is necessary to control the acceleration phases (especially in locomotives) and for braking purposes.

On each wheel acts a vertical load of several tonnes due to the mass of the train. On straight tracks and on curves with large radius, the contact involves the wheel tread and the rail head (Fig. 8.16a), and the contact stresses can be evaluated using the Hertz theory considering a contact between two cylinders (in a simplified approach, a line contact can be assumed). The Hertzian pressure is typically about 600–1000 MPa [20]. The ratio between the Hertzian pressure and the shear yield strength ($p_{max}/\tau_Y$) is between 2 and 5 [22]. The comparison with Fig. 2.4 shows that the materials response to repeated contact loading can be easily characterized by cyclic plasticity. The wear damage is thus mainly by contact fatigue, and it may be

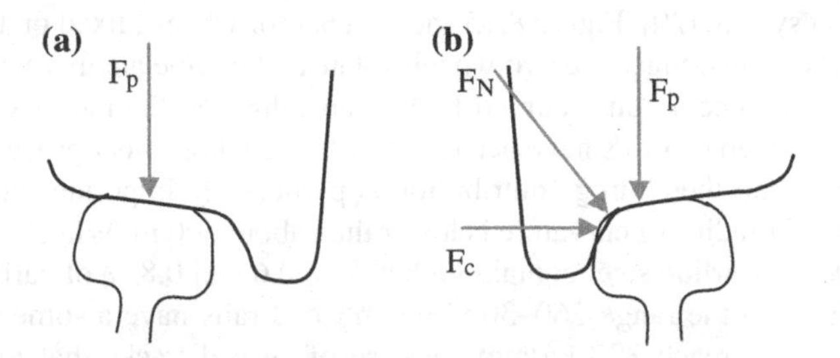


**Fig. 8.16** Schematization of rail head/wheel tread contact rail gauge/wheel flange contact (**b**)

by *low-cycle fatigue* [23]. The possible presence of ratcheting increases the severity of damage evolution. As an example, Fig. 8.17a shows the surface of a wheel damaged by contact fatigue. The cross section in Fig. 8.17b shows the presence of numerous surface nucleated cracks [24].

On sharp curved tracks, the contact involves the wheel flange and the rail gauge (Fig. 8.16b). In addition to the vertical load, $F_p$, a sideways centrifugal force is

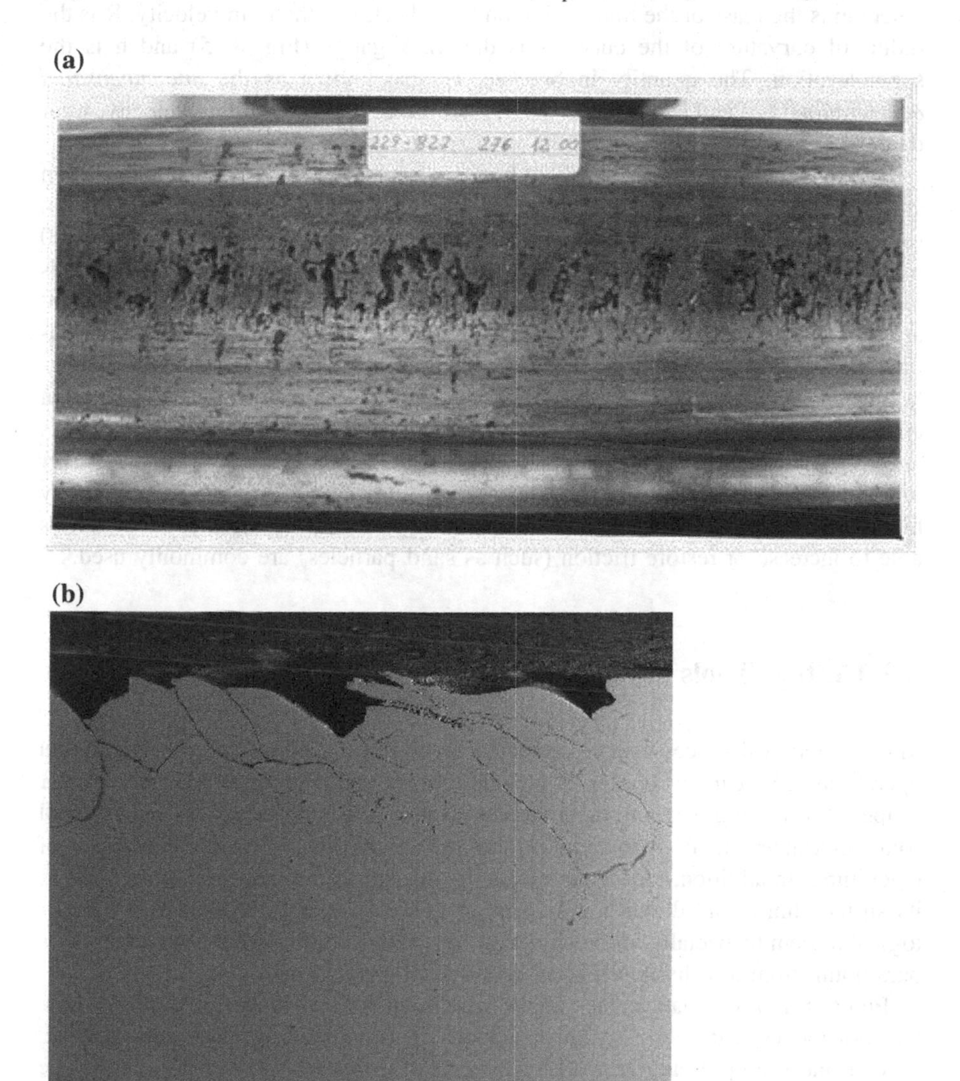


**Fig. 8.17** Damage by contact fatigue of a railway wheel. **a** Surface appearance; **b** Cross section showing the surface damage and subsurface cracks [24]

present, $F_c$, and the resulting applied force, $F_N$, is therefore inclined. $F_c$ is simply given by:

$$F_c = m\left(\frac{v^2}{R} - \frac{hg}{s}\right) = ma_{nc} \tag{8.6}$$

where m is the mass of the train acting on the wheel, v is the train velocity, R is the radius of curvature of the curve, s is the track gauge (Fig. 8.15) and h is the superelevation. The quantity in brackets is also known as the *uncompensated acceleration* ($a_{nc}$) and is 0 if $hg/s = v^2/R$. Because of the large sliding velocity, wear damage is mainly by sliding and the wear mechanism is typically adhesion.

The rails and wheels are reprofiled by grinding and eventually substituted when surface damage is excessive. The wheel life is typically 300,000/1,200,000 km (including two to five reprofilings) and the rail life is in between 100 and 2500 million gross tonnes [21]. The best way to reduce wear damage by rolling-sliding is to increase lubrication. However, lubrication decreases the static friction coefficient and this may pose serious problems with regard to traction and braking. The situation is also complicated by the possible presence of water and external contaminants, such as snow, ice and leaves. Therefore, oils or greases are often employed to lubricate the wheel flange/rail gauge interface and the lubricant is applied when the train enters a curve thus reducing sliding wear [20]. However lubrication should not involve the wheel tread/rail head contact, where a relatively high friction is required for traction and braking. In this region, friction modifiers able to increase or restore friction (such as sand particles) are commonly used.

## 8.9 Cutting Tools

Material-removal processes are conducted by cutting, grinding and non-traditional operations [25]. *Cutting processes* include turning and drilling to produce circular shapes, and milling, shaping or broaching to produce more complex shapes. Tool wear determines the life span and it is therefore an important item in the total cost of machining. In addition, tool wear affects the quality of the machined part, such as its surface finish and dimensional accuracy. Therefore, optimization of the tribological system (especially in the choice of tool materials and cutting parameters) is paramount from a technological and economical point of view.

Figure 8.18a schematizes the simple two-dimensional *orthogonal cutting*, where the tool moves, with constant linear velocity, perpendicularly to its cutting edge. Two surfaces are present: the work surface, i.e., the surface of the work piece to be cut, and the machined surface, i.e., the newly produced surface. In a *turning* operation (Fig. 8.18b) a third transient surface is present, e.g. the surface being cut positioned between the work and machined surface.

Figure 8.18a also shows the forces acting on the tool. $F_c$ is the *cutting force*, acting in the direction of the cutting velocity, and $F_t$ is the *thrust force* acting

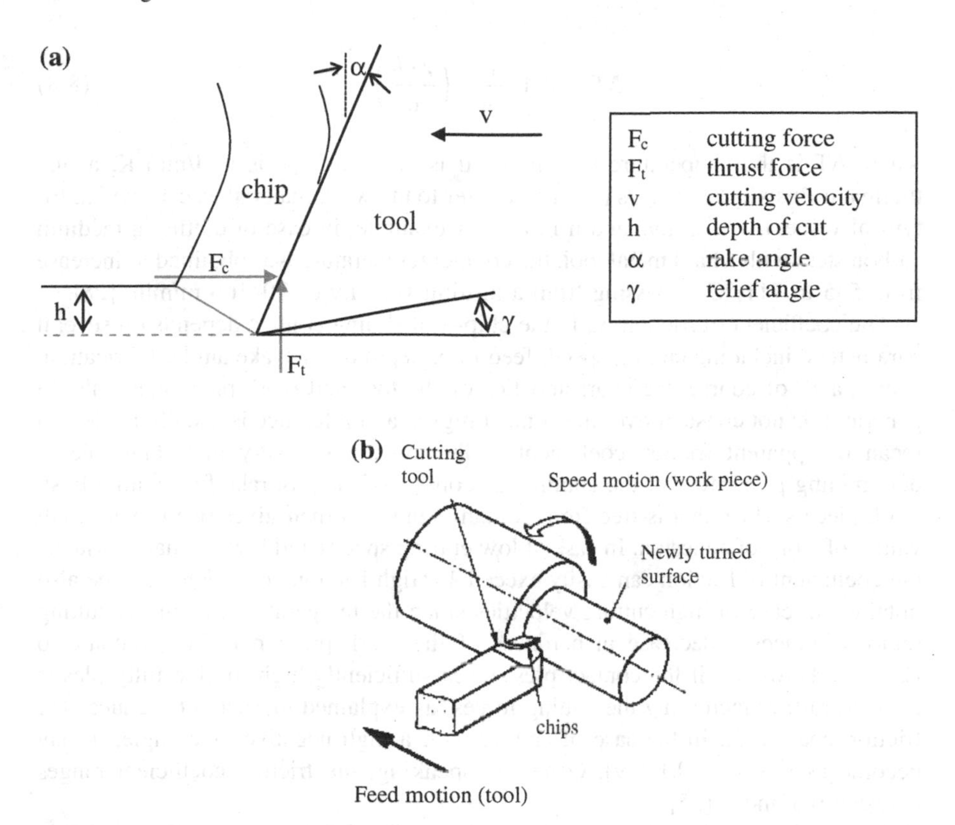


**Fig. 8.18** **a** Schematic diagram and terminology of orthogonal cutting. The cutting tool moves from *right* to *left*; **b** Scheme of turning

perpendicularly to the cutting velocity and to the work piece. $F_c$ supplies the power for cutting that is given by: $F_c$ v. $F_c$ is roughly given, at least at the beginning of cutting, by the following relation [26]:

$$F_c = u \cdot b \cdot h \tag{8.7}$$

where b is the width of cut, h is the depth of cut and u is the specific energy for cutting. Typical values for u are listed in Table 8.4. Such values approximately coincide with the hardness of the work piece material, in substantial agreement with the analysis carried out in Sect. 2.8 and, in particular, with Eq. 2.13 (setting $u \approx p_Y$). The thrust force is influenced by the rake angle, α, and the coefficient of friction at the tool-chip interface; when $\alpha = 0$, $F_t = \mu F_c$.

The power dissipated during cutting (to form the chip by plastic shearing, and also by friction due to the sliding at the chip-tool contact) produces a considerable temperature rise in the cutting zone. The highest temperature is achieved at the interface between the tool and the chip, and it can be estimated to a first approximation using the following experimental relationship [27]:

$$\Delta T = 0.4 \cdot \frac{u}{\rho c} \cdot \left(\frac{v \cdot h}{a}\right)^{0.33} \quad (8.8)$$

where ΔT is the temperature rise in °C, u is in J/mm$^3$, ρc is in J/mm$^3$K, a (the thermal diffusivity) is in m$^2$/s (ρ, c and a refer to the work material; see Table 2.2 for typical values), v is in m/s and h in m. For example, in case of cutting a medium carbon steel with a hard metal tool, the contact temperature was obtained to increase from 535 to 711 °C in passing from a turning velocity of 70–160 m/min. [28].

The coefficient of friction, μ, in the chip-tool sliding contact depends on several parameters, including cutting speed, feed rate, depth of cut, rake angle, lubrication, if any, and, of course, the characteristics of the tool and work piece materials. In general, μ is not constant over the contact region and reference is usually made to a mean or apparent friction coefficient. Adhesion plays a very important role in determining μ, because of the continuous contact of the tool rake face with a fresh work piece surface that is free from contaminants and then gives rise to very high values of work of adhesion. In case of low cutting speeds and low normal pressures, the coefficient of friction can easily exceed 1. High friction coefficients can be also obtained in case of high cutting velocities since the temperature rise in the cutting regions induces a decrease in hardness of the work piece material (compare to Eq. 2.8). However, if the contact pressure is sufficiently high to give fully plastic contact, further increasing the cutting forces, as explained in Sect. 3.6, reduces the friction coefficient. In the case of cutting with a high negative rake angle, μ may become as lower as 0.1 [29]. Generally speaking, the friction coefficient ranges between 0.5 and 2 [25].

Lubrication is reported to reduce friction. Cutting lubricants are often *emulsions* of oils and water. They mainly exert a cooling action thus reducing distortions, wear and friction as mentioned before. It is believed that they are able to penetrate, driven by the capillary forces, into the regions between the contacting asperities in the rear of the contact, where the real area of contact is lower than the nominal one. The lubrication effect of cutting fluids increases with the penetration depth, which is inversely proportional to the chip velocity (and hence to the cutting velocity) [30]. A sort of solid lubrication is attained by machining free-cutting steels, containing manganese sulphides (around 0.3 % sulphur) and/or lead (about 0.25 %) in their microstructure, and grey cast irons containing graphite. During cutting, these inclusions form a transfer layer on the tool surface that decreases the friction coefficient.

**Table 8.4** Typical values of the specific energy for cutting, u (*units* N/mm$^2$. Data from different literature sources)

| | | | |
|---|---|---|---|
| Aluminium alloys | 400–1100 | Nickel alloys | 4900–6800 |
| Copper alloys | 1400–3300 | Steels | 2700–9000 |
| Magnesium alloys | 500 | Stainless steels | 3500 |
| Titanium alloys | 3000–4000 | Grey cast iron | 1380 |

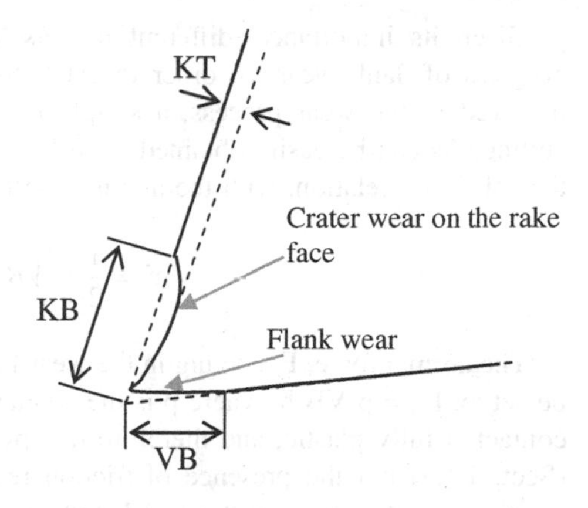


**Fig. 8.19** Schematic drawing showing the crater and flank wear

The high adhesion at the chip-tool interface may also promote formation of a transfer layer on the tool rake face. Such a layer is also known as *built-up edge* (BUE). When the BUE is too large, it changes the depth of cut, and when it breaks, it favours the wear of the tool and reduces the quality of the work piece surface. The BUE mainly forms at low cutting velocities and when cutting relatively soft alloys. Its formation should be avoided to improve the quality of cutting.

The sliding between the chip and the tool rake face produces a wear damage known as *crater wear*. As schematized in Fig. 8.19, it is mainly quantified by the crater depth (KT) and the crater width (KB). It is promoted by the cutting force ($F_c$) and the chip sliding velocity; the high contact temperatures at the tool-chip interface also accentuate it. Wear is mainly by adhesion, but also by abrasion if the work piece material contains hard abrasive particles in its microstructure (such as carbides or hard alumina oxides). In case of tools made of hard metals, the high contact temperature may weaken the tool and favour its wear through the interdiffusion of cobalt and tungsten into the machined work piece (and possibly iron from the work piece to the tool).

The most important form of wear in metal cutting is *flank wear* since it directly determines the finishing of the machined surface (Fig. 8.19). At the beginning of cutting, the contact area between the tool and the work piece is very small. Then, a running in stage characterized by a quite high sliding wear rate follows up, which gives rise to the formation of a *wear land* in the tool flank. In Fig. 8.19 the length of the flank wear land is indicated, as usual, with VB. After running in, VB increases in a uniform way almost linearly with cutting time. Flank wear has detrimental effects on the surface finish of the work piece. In addition, it increases the cutting force and also the contact temperature. As a consequence, when it becomes too severe, i.e., when VB reaches a critical value of about 0.3 mm, the tool has to be resharpened or replaced. The wear mechanism responsible for flank wear is mainly adhesion (and abrasion in presence of hard particles). Interdiffusion plays a minor role because of the lower contact temperature (estimated to be about 273 °C lower than the maximum contact temperature at the chip-tool interface [31]).

Given its importance, different models have been proposed to evaluate the progress of flank wear. In order to catch the main physical parameters that are involved in the wear process, a simple model will be presented for orthogonal cutting. As can be easily obtained from Fig. 8.19, the wear volume V is given by the following relation, with the assumption that α (the rake angle) is zero:

$$V = \frac{1}{2} b \cdot VB^2 \cdot tg\gamma \tag{8.9}$$

The normal force, $F_N$, acting in the wear land is very difficult to evaluate. It may be set to: $F_N = p\ VB\ b$, where p is the contact pressure. It can be assumed that the contact is fully plastic, and therefore $p = p_Y$ [30]. In absence of sliding, $p_Y = H$ (Sect. 1.1.3) but the presence of friction reduces the pressure required for fully plastic contact, as seen in Sect. 2.1 and, in particular, in Fig. 2.4. When $\mu = 0.5$, fully plastic contact is attained at $p/\tau_Y \approx 2$, which gives $p_Y \approx \tau_Y$. Considering that $H = 6\ \tau_Y$ (using the Tresca yield criterion), from the Archard equation (Eq. 4.1), it is easily obtained that:

$$VB = \frac{K_{ad}}{6 \cdot tg\gamma} \cdot s = \frac{K_{ad}}{6 \cdot tg\gamma} \cdot v \cdot t \tag{8.10}$$

where $K_{ad}$ is the coefficient of adhesive wear of the tool material, s is the sliding distance, and v and t are the cutting speed and time.

As an example, Fig. 8.20a shows the evolution of VB as a function of testing time in case of dry turning of a medium carbon steel (H: 196 kg/mm$^2$), using a commercial hard metal tool (with b = 2.5 mm, $\gamma = 8°$ and feed rate = 0.25 mm/rev) [31]. Tests at different cutting speeds were conducted. It is observed that after an initial running in (particularly evident at low cutting speeds), wear increases almost linearly with time, as predicted by Eq. 8.10. The wear rate in the steady state, given by VB/t, increases also with cutting velocity, passing from about 0.12 μm/min. at 80 m/min. to about 5.6 μm/min. at 250 m/min. However, the dependence of VB/t to

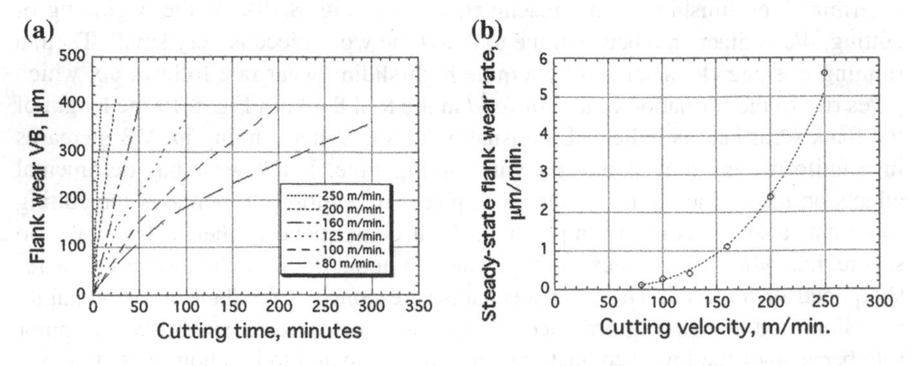


**Fig. 8.20** Flank wear VB as a function of cutting time in case of turning a medium carbon steel using an uncoated carbide tool (modified from [31])

v is not linear, as it would be predicted by Eq. 8.10, but rather follows a power law relationship, as shown in Fig. 8.20b. This is a general result and may mostly attributed to the fact that $K_{ad}$ also depends on v; in fact, $K_{ad}$ depends on the hardness of the tool material, which decreases as contact temperature increases, i.e., as cutting speed increases.

To overcome such difficulties, the wear life, $t_c$, for a chosen VB-value is thus determined using simplified empirical relations, such as the *Taylor equation*:

$$v \cdot t_c^n = C \tag{8.11}$$

where n and C are two experimental constants. Using the data of Fig. 8.20a for VB = 0.3 mm, it is obtained that n = 0.36 and C = 580 (for t in min. and v in m/min.). In Table 8.5 representative values of n for different cutting materials are shown. By comparing Eqs. 8.10 and 8.11 follows that the *life-index* n is larger in materials characterized by a hot-hardness that is less affected by temperature.

In addition to sliding wear, cutting tools can experience other damage processes in service, such as plastic deformation, brittle fracture, oxidation and thermal fatigue, depending on the materials and cutting process. Therefore, the most important requirements for tool materials are: high hot hardness and oxidation resistance; low tribological compatibility with the work piece material; adequate fracture toughness (when impact loads or vibrations are involved) and thermal shock resistance. The materials most widely used are:

- Cold-work tool steels, mainly used at low cutting speeds when contact temperatures are below about 300 °C;
- High-speed steels, HSS, used in a variety of applications, also coated with hard ceramics; they possess high fracture toughness but soften when the tool temperature becomes too high;
- Hard metals. They feature a balanced combination of hardness and fracture toughness and therefore excellent performances especially in turning operations. Most carbide materials used in industry are coated with TiN, TiCN, TiAl(C)N and CrN [32], and are mainly used as inserts clamped on the tool shank;
- Ceramics, such as alumina and silicon nitride, and super-hard ceramics, such as cubic boron nitride (cBN) and PCD. These latter are mainly employed at high cutting speeds; however, since the extremely high hardness is accompanied by a low fracture toughness, these materials are then not commonly employed in processes such as drilling and milling.

Ceramics and HSS or carbides coated with $MoS_2$ or WC/C are often used in dry machining operations, thus avoiding the use of cutting fluids that may have adverse

**Table 8.5** Representative values of n for the Taylor equation

| | |
|---|---|
| High-speed steels | 0.08–0.2 |
| Hard metals | 0.2–0.5 |
| Ceramics | 0.5–0.7 |

**Table 8.6** Recommended tool and work material combinations

| | Soft non ferrous alloys (Al, Cu) | Carbon and low alloy steels | Hardened too and die steels | Cast iron | Nickel-based alloys | Titanium alloys |
|---|---|---|---|---|---|---|
| HSS | 2, 3 | 2, 3 | 4 | 3, 4 | 3, 4 | 3, 4 |
| WC-Co (inc. coated) | 2 | 1, 2 | 3 | 1, 2 | 1 | 2 |
| Ceramic | 4 | 1, 2 | 2 | 1 | 1, 2 | 4 |
| cBN | 3, 4 | 4 | 1 | 1, 2 | 2 | 2 |
| PCD | 1 | 4 | 4 | 4 | 4 | 1 |

*1* good; *2* all right in some conditions; *3* possible but not advisable; *4* to be avoided. Modified from [30])

environmental effects. The use of tools coated with solid lubricants is particularly recommended when machining materials with a high work of adhesion.

Table 8.6 lists some recommended tool and work material combinations.

## 8.10 The Grinding Process

Grinding is a material-removal process for obtaining mechanical parts with high dimensional accuracy and low surface roughness (Table 1.3). In this process, abrasive wear is profitably exploited for part shaping. In fact, material removal is obtained by small and hard abrasive particles (*grits*) fixed to a grinding wheel. The wheels are made of composite materials, consisting of hard particles (typically $Al_2O_3$, SiC, cBN, PCD) held together by either a metallic, polymeric or vitreous binder. As an example, Fig. 8.21a shows sharp and blocky cBN grits that are approximately 125 μm in size, and Fig. 8.21b shows the surface of a vitrified cBN-grinding wheel [33]. The grinding surface has a randomly structured topography and each grit removes a small chip by microcutting. This explains the better surface finish achieved by grinding with respect to cutting operations.

The most important parameters that govern the grinding process are:

- The abrasive type. To exert an abrasive interaction, the grits have to be harder than the work piece. To grind hard materials like hard metals or ceramics, super-abrasives, like diamond or cBN are used.
- The grit size. Large particles are more effective in material removal since they are characterized by higher values of the attack angle (see Sect. 4.3.1). However, fine particles remove less material at each interaction and give rise to a better surface finish. They are also more suited to machine very hard materials, for which the removal of a large amount of material at each interaction might be difficult.
- The bond properties. Metallic bonds are very tough and able to retain the grits in place during grinding. Vitreous bonds are more flexible; the possibility of

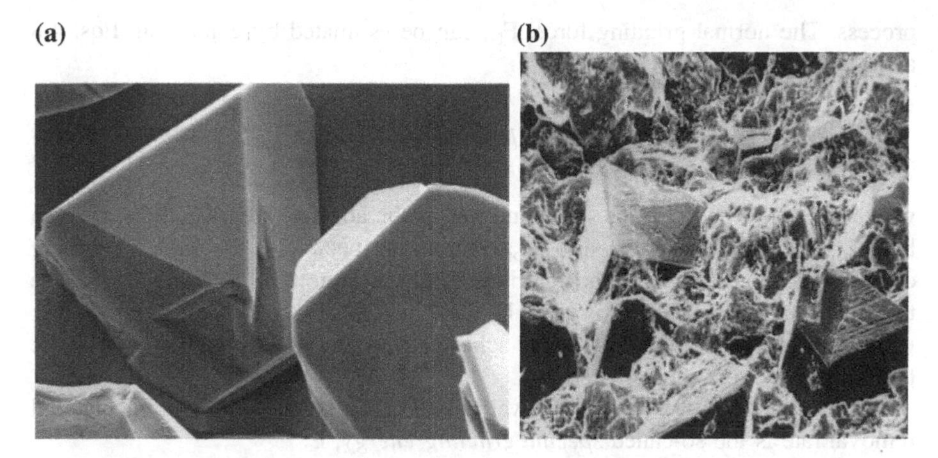


**Fig. 8.21** **a** Sharp and blocky cBN grains approximately 125 μm in diameter; **b** Surface of a vitrified cBN grinding wheel [33]

modifying their strength by controlling the residual porosity allows obtaining different removal rates and surface finish characteristics [34]. Phenolic and polyamide resins easily expel the damaged grains during grinding but have a lower life. In any case, the conditioning of worn wheels by adequate *dressing* processes is necessary to maintain the grinding efficiency of the wheel.

- Lubrication. The use of grinding fluids (water-base emulsions) is important to reduce friction and also to reduce the temperature rise, which could damage the work piece surface by inducing tempering effects or, in case of steels, by favouring the local formation of hard and brittle martensite.

Figure 8.22 schematizes the surface grinding process. The process performance is given by the *material removal rate*, $\dot{V}$:

$$\dot{V} = h \cdot b \cdot v_f \tag{8.12}$$

where h is the depth of cut, b is the grinding width and $v_f$ is the work piece speed. The estimation of the grinding forces is very important in the control of the entire

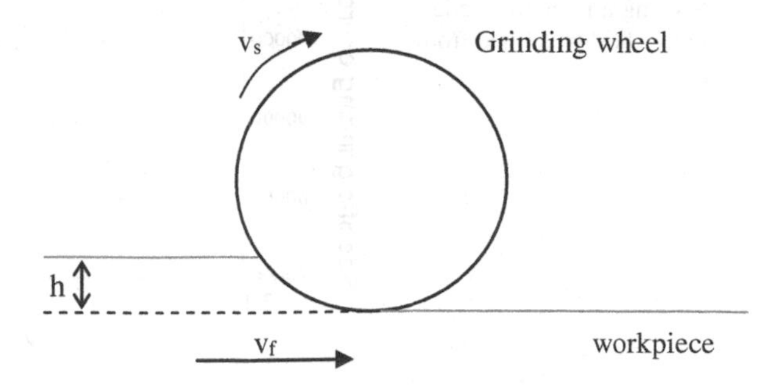


**Fig. 8.22** Schematization of a surface grinding process

process. The normal grinding force, $F_N$, can be estimated by combining Eqs. 4.8 and 8.12:

$$F_N = h \cdot b \cdot \frac{H}{K_{abr}} \cdot \frac{v_f}{v_s} \tag{8.13}$$

where $v_s$ is the wheel surface speed and $K_{abr}$ is the abrasive wear coefficient, given by: $\Phi\ \mu_{abr}$, where $\Phi$ (< 1) and $\mu_{abr}$ express, respectively, the microploughing contribution to abrasion and the contribution of abrasion to friction (Eq. 2.14). The tangential grinding force, $F_T$, is given by $\mu\ F_N$. $\mu$ is the friction coefficient given by: $\mu = \mu_{ad} + \mu_{abr}$, where $\mu_{ad}$ refers to the adhesive interaction between the grits, or the bond, and the work piece material.

The ratio between the grinding power (given by $F_T\ v_s = \mu\ F_N\ v_s$) and the material removal rate is the so-called *specific grinding energy*, E:

$$E = \frac{\mu}{K_{abr}} \cdot H \tag{8.14}$$

The knowledge of E allows for a quick evaluation of $F_T$:

$$F_T = h \cdot b \cdot E \cdot \frac{v_f}{v_s} \tag{8.15}$$

and then of $F_N$ (= $F_T/\mu$). Typical values of E are listed in Table 8.7. However, as shown by Eq. 8.14, E is not constant but depends on the phenomena occurring during the grits-work piece interactions, such as the adhesive and abrasive friction, the ploughing contribution, the hardness of the work material that in turn depends on contact temperature and the local plastic strain rate. In general, E increases noticeably as the ratio $v_f/v_s$ decreases. As an example, Fig. 8.23 shows the experimental values of E as a function of $v_f/v_s$ for a surface grinding operation of a mild steel work piece (SAE 1018, with H about 130 kg/mm$^2$) using a medium grade

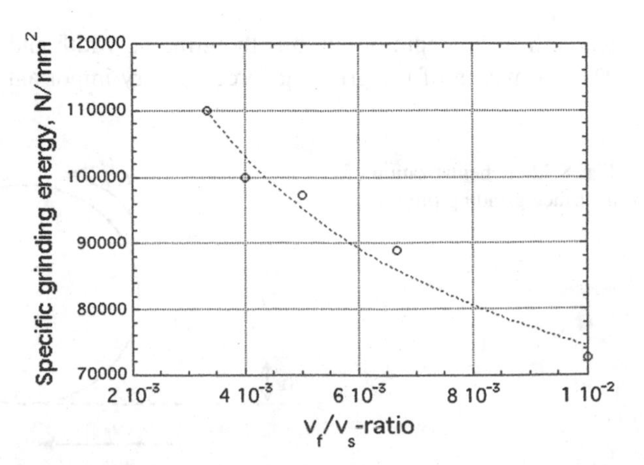


**Fig. 8.23** Experimental values of the specific grinding energy for a surface grinding operation of a mild steel work piece using a medium grade alumina wheel (modified from [35])

**Table 8.7** Typical values of the specific grinding energy (from different literature sources)

| Material | Hardness, kg/mm$^2$ | Specific grinding energy, N/mm$^2$ |
|---|---|---|
| Aluminium | 150 | 6800–27000 |
| Grey cast iron | 215 | 12,000–60,000 |
| Low-carbon steel | 110 | 13,700–60,000 |
| Tool steel | 67 HRC | 17,700–82,000 |
| Titanium alloys | 300 | 16,400–55,000 |

alumina wheel [35]. The interpolating curve is: E = 14,481 $(v_f/v_s)^{-035}$ (units: N/mm$^2$). It follows that $F_T$ depends on $(v_f/v_s)^n$ with n = 0.64. In general n ranges between 0 and 1, and tends to 1 in materials that are easy to grind [36]. The experimental friction coefficient, μ, was found to be about 0.54 irrespective of the $v_f/v_s$-ratio. Setting $\mu_{ad} = 0.2$ for the $Al_2O_3$/steel couple, $\mu_{abr}$ turns out to be 0.34 and the average attack angle of the $Al_2O_3$ grits is thus 28° (from Eq. 2.14). Finally, assuming $\Phi \approx 0.25$ (from Fig. 4.12), it is obtained that $K_{abr} = \Phi\ \mu_{abr} = 8.5 \times 10^{-2}$, that is a quite high value, as expected.

Grinding is the most widely used operation for shaping ceramic materials and hard metals. Diamond particles are used as abrasives. Since ceramics are quite brittle, abrasive wear follows the Lawn and Swain mechanism (Sect. 4.3.2). The formation of surface microcracks by brittle fracture may deteriorate the surface quality. In order to obtain a mirror-like surface, grinding is usually followed by a *lapping* stage with a decreasing size of the abrasives that are free to rotate during the contact, and, in the end, by a *polishing* stage using very fine abrasive particles.

The experimental values for the specific grinding energy of ceramics are quite similar to those displayed by metals. For example, in case of alumina E-values ranging from 10,000 to 60,000 N/mm$^2$ are reported [33, 37]. This means that sliding friction and plastic deformation during the local indentation stages (see Fig. 1.7) play an important role in the grinding process. This latter observation opens the possibility of the so-called *ductile machining* of ceramics. A ductile, damage-free machining requires the use of low applied load to have local indentations stresses below the critical value for brittle contact. The grinding process may be quite slow but it allows obtaining optical mirror-like surfaces without using the lapping/polishing stages.

## 8.11 Hot Forging Dies

In a forging process, a work piece is shaped by compressive plastic deformation. The process is carried out either in open die, as in the case of *upsetting* (schematically shown in Fig. 8.24), or in closed (shaped) dies. In this latter case, metal is not free to deform but it is constrained to flow in the die cavities. In hot forging the billets are preheated and the plastic deformation is greatly favoured by the activated creep and recrystallization phenomena.

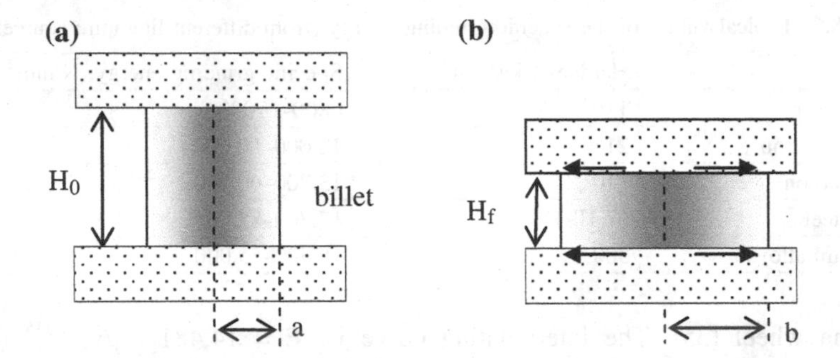


**Fig. 8.24** Scheme of the upsetting process of a cylindrical billet. $H_0$ and a indicate the initial height and radius of the billet (**a**) whereas $H_f$ and b indicate the height and radius after forging. The *horizontal arrows* in (**b**) indicate the tool areas subjected to sliding due to the plastic deformation of the billet

Steel billets are heated to temperatures between 1000 and 1300 °C and the dies are maintained at about 300 °C. During the forging cycle, the intense heat exchange between the billet and the dies, favoured by the fact that the real area of contact tends to the nominal one, causes a temperature rise at the tool surface in the 600–900 °C range. Both work piece and dies undergo oxidation, and the thermo-mechanical cycle induces three main interacting damaging processes:

(1) Oxidation-assisted thermal fatigue (or *heat checking*), with the appearance of a network of surface cracks due to the building up of alternating tensile and compressive stresses, particularly in the outer layers of the contacting surfaces;
(2) Plastic deformation, when the contact pressure exceeds the yield strength of the tool material. Most tools are made by hot-work tool steels, which are characterized by high hot-hardness and adequate fracture toughness. However, if the tool temperature becomes too high, the material can soften and may undergo plastic deformation during the forging impacts;
(3) Sliding wear, due to the sliding between the work piece on the tool surface caused by plastic deformation (see Fig. 8.24). It is the dominant factor affecting the die life [38].

The progression of sliding wear can be modelled using Eq. 5.2. The worn die is replaced or repaired (it is typically re-machined) when the depth of wear, h, reaches a critical value (that is in the range 0.3–0.5 mm), in correspondence of which the quality of the forged part is impaired. In order to use Eq. 5.2, the contact pressure must be evaluated. Different relations are available. In the simple upsetting process of a cylindrical billet (Fig. 8.24), the average contact pressure, p, is given by [25]:

$$p = \bar{\sigma}\left(1 + \frac{2\mu a}{3H_0}\right) \tag{8.16}$$

where:

- $\bar{\sigma}$ is the plastic flow strength of the work piece material, which depends on its hardness and decreases as the working temperature is increased and the strain rate (due to the forging speed) is decreased;
- $\mu$ is the coefficient of friction due to the billet-die sliding;
- $H_0$ and a are the height and radius of the undeformed billet.

Proper solid lubricants, such as graphite, $MoS_2$ or glass particles, may of course reduce the coefficient of friction. At high process temperatures, glass acquires a low viscosity and actually behaves like a liquid lubricant.

The sliding distance, s, is given by: s = N $\Delta x$, where N is the number of forging operations and $\Delta x$ is the sliding for each operation. $\Delta x$ is maximum in correspondence of the initial radius of the billet (i.e., at a distance a from the billet axis, see Fig. 8.24a), and is given by [10]:

$$\Delta x \cong \frac{1}{2} a \cdot \varepsilon \tag{8.17}$$

where $\varepsilon$ is the billet plastic deformation: $\varepsilon = \ln(H_0/H_f)$. To summarize, the maximum depth of wear in case of upsetting is given by:

$$h = K_a \cdot \bar{\sigma} \cdot \frac{1}{2} a \cdot \varepsilon \cdot N \tag{8.18}$$

The wear mechanism is typically tribo-oxidation and, in particular, oxidation-scrape-reoxidation (Fig. 4.6) [39]. Abrasion can also play a role if hard particles are trapped in the contact region. Low values of the specific wear coefficient, $K_a$, are attained using properly heat-treated *hot work tool steels*, such as the AISI H13 steel, able to retain high hardness and thus high wear resistance also at relatively high temperatures. Typical values of $K_a$ of hot work tool steels during hot forging are about $10^{-14}$ $m^2/N$ [40]. In order to increase the tool wear resistance, it is possible to adopt specific surface treatments, such as nitriding, coating with thin films (such as TiAlN) and hardfacing with stellite. For example, when using a TiN coating of 10 μm in thickness, $K_a$ was found to be about $1.8 \times 10^{-15}$ $m^2/N$ [40]. Of course, in order to evaluate the performance of the treatments, the die service life and the cost per part have to be calculated. The work piece temperature plays a particular role. Upon increasing it up to about 1100 °C, wear is increased. However, above 1100 °C wear is found to decrease with temperature because of the overwhelming effect of the decrease in $\bar{\sigma}$ [41].

In case of closed-die forging, the evaluation of the wear process is much more complex. To a first approximation, the identification of the die regions where wear is more intense can be made by considering Eq. 8.17, and therefore by determining the locations where the product of the billet plastic deformation, $\varepsilon$, and the distance from the billet axis is maximum. If this region lies in correspondence with a geometric discontinuity, the intensity of wear is typically greater than expected, because of the interplay between wear and thermal fatigue that is particularly intense in presence of high local stress intensifications.

## 8.12 Rolling Rolls

Rolling is a manufacturing process quite similar to forging. The basic *flat rolling* operation is schematized in Fig. 8.25. The initial work piece thickness, $H_o$, is reduced by a compressive plastic deformation exerted by two counter-rotating rolls to a final thickness $H_f$. The rolls are typically made of high chromium iron, high speed steel (HSS), indefinite-chilled cast iron or forged steels. In hot rolling, the rolls are sequentially heated by contact with the work piece and drastically cooled by water spraying. The rolls surfaces are then oxidized and they are exposed to thermal fatigue and rolling-sliding wear. Sliding occurs because of the difference between the roll surface and the work piece velocity in the roll bite. As an example, in finishing stands for steels, the surface roll velocities and the sliding velocities are in the 2–20 m/s and 0.06–2 m/s range respectively [42].

In dry steel hot rolling, the coefficient of friction typically ranges between 0.3 and 0.6, depending on the rolling parameters. Several empirical formulas have been published. Robert's formula, for example, relates the friction coefficient to the work piece temperature for well-descaled steel strips [43]:

$$\mu = 0.00049 \cdot T - 0.071 \tag{8.19}$$

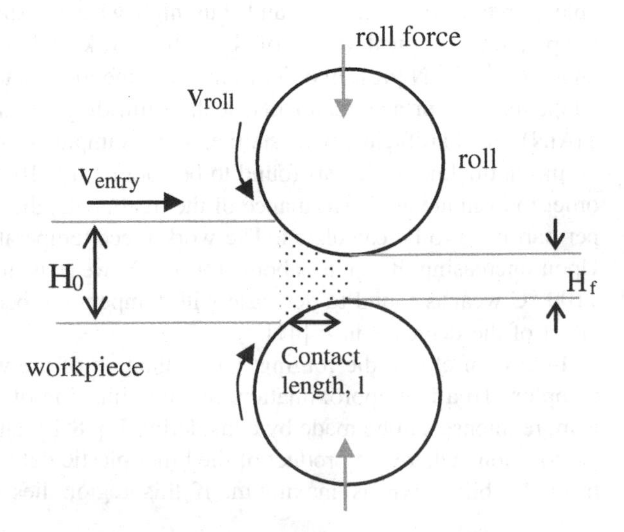


**Fig. 8.25** Scheme of the flat rolling process

(T in °C). As seen, the friction coefficient is predicted to increase with temperature. On the contrary, Galeji's formula indicates that friction coefficient decreases with temperature and surface roll speed [44, 45]:

$$\mu = 0.82 - 0.0005 \cdot T - 0.056 \cdot v \tag{8.20}$$

(T in °C and v in m/s; the relation was obtained for ground steel rolls). The diverging findings might depend on alterations in the contact conditions (in particular, in the real contact temperature). In the roll bite, the real area of contact is almost equal to the nominal one ($A_r \approx A_n$) and hence (see Sect. 3.6):

$$\mu = \frac{\tau_m}{p_o} \approx \frac{\tau_m}{\sigma'_Y} \tag{8.21}$$

where $\sigma'_Y$ is the plane stress yield stress of the work piece, and $p_o \approx \sigma'_Y$. Therefore, as the work piece temperature increases, $\sigma'_Y$ decreases and $\mu$ increases, as predicted by Eq. 8.19. But this is not the whole story. In fact, both rolls and work piece are covered with an oxide layer that influences $\tau_m$. As seen in Sects. 2.6 and 6.1.1, if contact temperature increases, the coverage degree of the oxide is increased and the oxide may behave as a solid lubricant. The corresponding decrease in $\tau_m$ may overwhelm the decrease in $\sigma'_Y$ thus leading to a decrease in the friction coefficient, as predicted by Eq. 8.20.

In order to decrease the friction coefficient, a lubricant may be used. In hot rolling of steels oil-in-water emulsions are adopted, and a boundary lubrication regime is attained. The coefficient of friction is typically lowered to 0.15–0.4. In the hot rolling of aluminium alloys, lubrication is required, to avoid a large transfer of aluminium particles onto the roll surface, which would strongly deteriorate the surface quality of the product [46].

In case of cold working, like cold rolling of thin strips, a lubricant is always used and the mixed lubrication regime is typically attained. The lubricant thickness can be evaluated by adopting Eq. 3.13 [44]:

$$h = \frac{3 \cdot \eta_0 \cdot (v_{entry} + v_{roll}) \cdot R}{l \cdot \left[1 - e^{-\alpha \cdot \sigma'_Y}\right]} \tag{8.22}$$

where $v_{roll}$ is the roll surface velocity, $v_{entry}$ is the entry velocity of the work piece, R is the roll radius and l is the projected contact length (given by $l = \sqrt{\Delta H \cdot R}$; the other parameters were already defined). Equations 8.22 and 3.1 are helpful for evaluating the influence of the different rolling parameters on the friction coefficient. Generally speaking, the coefficient of friction decreases as the roll surface roughness is decreased, the roll surface velocity is increased, and the lubricant viscosity is increased. Of course, the friction coefficient should not be too low since the friction force pulls the work piece into the roll byte.

In cold rolling of aluminium, titanium, zirconium alloys and stainless steel, insufficient lubrication may give rise to excessive transfer phenomena on to the roll surface that may lead to a sharp increase in the friction coefficient and a deterioration of the surface finish of the rolled product. As an example, Fig. 8.26 shows the recorded dependence of the *unit rolling force* on reduction, in case of cold rolling of a ferritic stainless steel with three mineral oils [47]. The additive content (ester) in the oils increases in passing from A to C. At low reductions, all three lubricants are effective but as the reduction increases above 20 %, oil A shows a sharp increase in the unit force due to the increase in the friction coefficient. This increase was due to the deterioration of the lubrication regime (decrease in Λ-factor) with the attainment of boundary lubrication that prompted the onset of transfer phenomena. As expected, the increase in additive content delays this transition to higher reduction values.

The evolution of the roll wear depth, h, can be estimated by using Eq. 5.14 where, in the present case, $v_s$ is the average sliding velocity and $v_t$ is the surface roll velocity. In flat rolling the unit rolling force, $F_N/L$, is given by [48]:

$$\frac{F_N}{L} = \sigma'_Y \cdot l \cdot \exp\left(\frac{\mu \cdot l}{H_0 + H_f}\right) \tag{8.23}$$

As an example, Fig. 8.27 shows the experimental dependency of the depth of wear as a function of the product of the unit rolling force and the number of rolling cycles for three types of rolls [49]. As expected, an almost linear experimental dependency is observed. The slope of the curves depends on the specific wear coefficient, $K_a$. In general, the $K_a$-values for high chromium cast iron rolls are in the $10^{-14}$ $m^2/N$ range. The indefinite-chilled cast iron rolls show a lower wear resistance and the CPC-type cast iron rolls a much greater wear resistance than the high chromium cast iron rolls.

Surface engineering is also used to reduce the friction coefficient or extend the roll life. For example, chromium plated rolls are often used in cold rolling mill rolls.

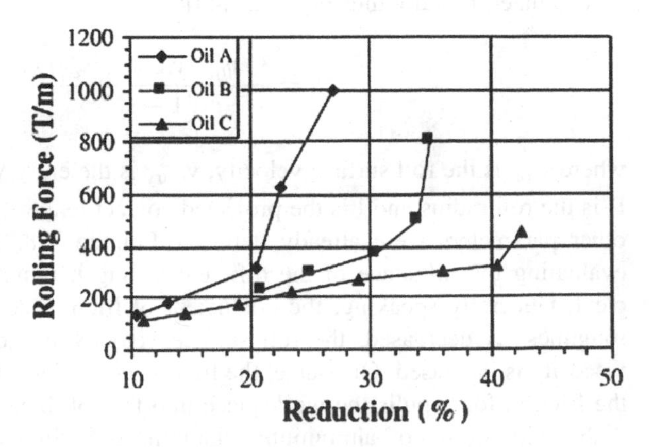


**Fig. 8.26** Experimental unit rolling force (i.e., roll separating force per unit length in the roll axial direction) versus reduction for a ferritic (FeCr17) stainless steel [47]

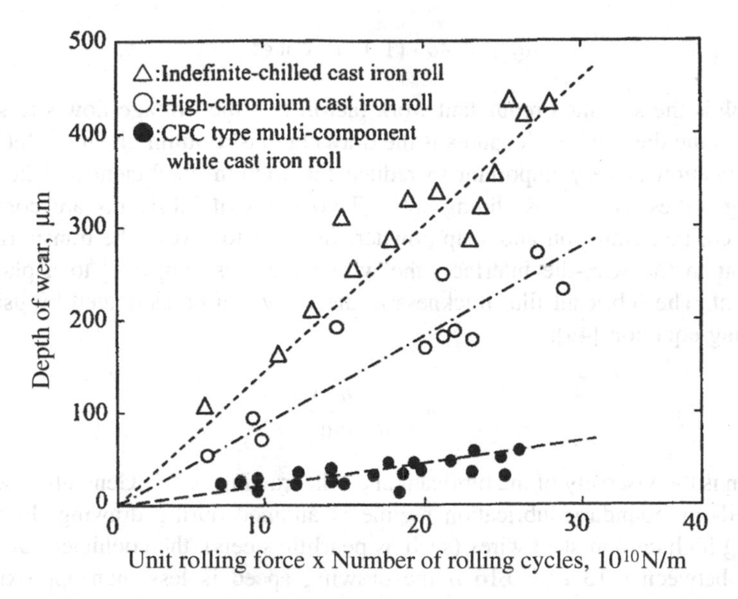


**Fig. 8.27** Wear depth versus the product of the unit rolling force and the number of rolling cycles for three types of rolls (modified from [49])

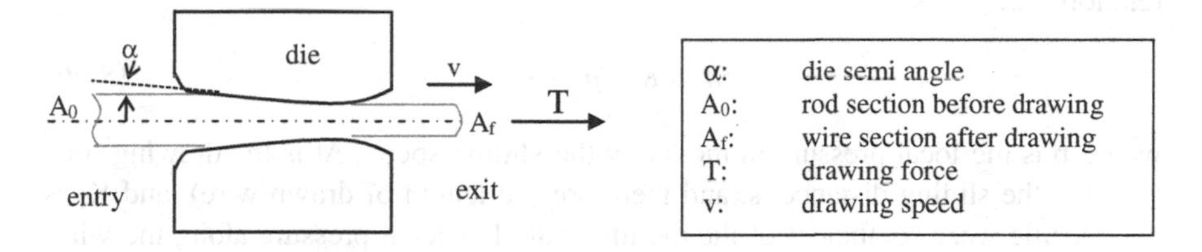


**Fig. 8.28** Scheme of the wire drawing process

## 8.13 Wire Drawing Dies

In wire drawing a rod produced by hot rolling is reduced in diameter by pulling it through a die with a conical hole by means of a tensile force applied to the exit side of the die, as shown in Fig. 8.28. The process is typically conducted in cold conditions. During drawing, sliding occurs at the wire-die interface. Therefore, friction and wear must be carefully controlled because they determine the *drawing force* and the final shape and surface quality of the product.

The drawing force, T, is given by: $\sigma_t A_f$, where $\sigma_t$ is the drawing stress that may be expressed by [48]:

$$\sigma_t = \phi \cdot \bar{\sigma} \cdot (1 + \mu \cdot \cot g\alpha) \cdot \varepsilon \tag{8.24}$$

where $\Phi$ is the so-called redundant work factor, $\bar{\sigma}$ is the average flow stress in the wire, $\alpha$ is the die semi angle, and $\varepsilon$ is the drawing plastic strain given by: $\ln(A_0/A_f)$.

Lubrication is very important to reduce the friction coefficient and hence the drawing forces as well as sliding wear. Two types of lubricants are commonly employed, i.e., emulsion and soap powder. In order to favour the transport of the lubricant to the wire-die interface, the wire surface is subjected to a phosphate treatment. The lubricant film thickness at die entry can be estimated by using the following equation [44]:

$$h = \frac{6 \cdot \eta \cdot v}{\bar{\sigma} \cdot \tan \alpha} \tag{8.25}$$

where $\eta$ is the viscosity of the lubricant and v is drawing speed. Generally speaking, a mixed- or boundary lubrication regime is attained during drawing. In case of drawing high carbon steel wires (such as pearlitic steels), the coefficient of friction ranges between 0.13 and 0.16 if the drawing speed is less than approximately 0.01 m/s. The friction coefficient decreases to 0.2–0.02 when the drawing velocity is about 1 m/s, and to about 0.01 when the drawing velocity is 20 m/s [44]. The increase in velocity increases the lubricant film thickness, and hence the $\Lambda$-factor.

The depth of wear, h, throughout the drawing channel can be evaluated using the relation 5.2:

$$h = K_a \cdot p \cdot v \cdot \Delta t \tag{8.26}$$

where p is the local pressure in the die, v the sliding speed, $\Delta t$ is the drawing time (v $\Delta h$ is the sliding distance, s, and therefore the length of drawn wire), and $K_a$ is the specific wear coefficient of the die material. The local pressure along the wire-die contact is not constant but a peak pressure is observed near the die entry. Pressure then decreases moving towards the die exit. As a result, wear depth is maximum close to the wire entry where the so-called *wear ring* is formed, and deceases moving towards the die exit. The control of the total die wear is very important in the process optimization [50]. The wear ring may render quite difficult the lubricant entry, and the wear at the die exit directly affects the diameter of the drawn wire. In general the die has to be substituted when the diameter increase of the drawn wire exceeds a given value, which is typically 1 μm. This means that a maximum wear depth of about 0.5 μm at exit die is tolerated. To assure such low wear values, materials with very high sliding resistance are used for the dies. Hard metals (typically: WC-6 % Co) and diamond (PCD, natural or synthetic diamond) are widely employed in the drawing industry. Ceramic composites have also a great potential as die materials for drawing. In a wide investigation carried out using WC-6 % Co dies and different drawing conditions, $K_a$-values around $5 \times 10^{-19}$ $m^2/N$ were obtained [51]. The comparison with the data listed in Table 5.2 shows that these values are lower by a factor of $10^{-3}$ than the typical values obtained in dry conditions. This means that the lubrication regime was in between mixed and boundary lubrication, as already evidenced.

## 8.14 Hot Extrusion Dies

Hot extrusion is used in the production of long products such as beams, tubes, rods and complicated profile shapes. In a *direct extrusion process*, schematized in Fig. 8.29a, a hot billet is pushed through the profile opening of a die. Sliding occurs between the billet and the container walls and also in the die lands. Lubrication is therefore important to reduce friction coefficient, and hence the *extrusion force*, and also to reduce the sliding wear of the die material. For example, glass powder is commonly employed in steel hot extrusion. During extrusion the glass softens and melts slowly, providing continuous lubrication.

Direct hot extrusion is commonly used to produce aluminium parts. The billet is heated to about 500 °C and the metal reaches a temperature 550–620 °C over the die land in the bearing channel (see Fig. 8.29b) [52], and the die material must thus possess high hardness and adequate fracture toughness at these temperatures. Therefore, the dies are made by hot work tool steels, typically the AISI H13 grade, generally after nitriding. In general, no lubricant is used in the extrusion of aluminium alloys in order to obtain complex shapes safely. As a consequence, transfer phenomena are quite common at the inlet of the bearing channel, as schematized in Fig. 8.30a. The friction coefficient is correspondingly quite high, typically in the range 1–1.5 [52].

The bearing surface undergoes sliding wear that, if excessive, changes the geometrical tolerances of the extrudate and worsens its surface finish. It has been observed that wear is mainly localized in the region where an unstable transfer film is formed. Wear is adhesive in nature and it is manifested by the intermittent detachment of this layer. Because of the high local temperatures at the interface between the transferred material and the die surface, interdiffusion processes and chemical reactions may occur with the possible formation of embrittling phases.

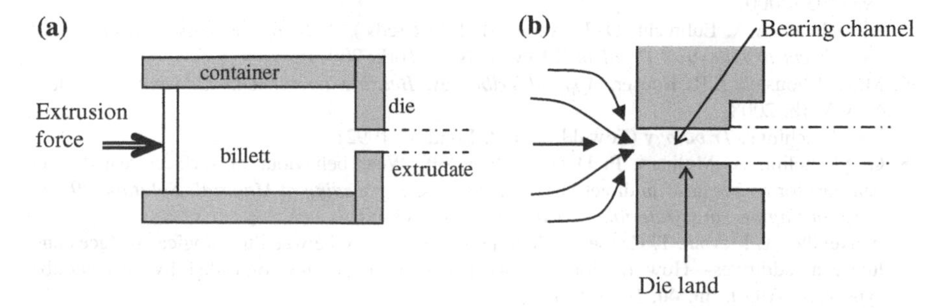


**Fig. 8.29** **a** Scheme of direct extrusion (with flat die); **b** die opening with metal flow during extrusion

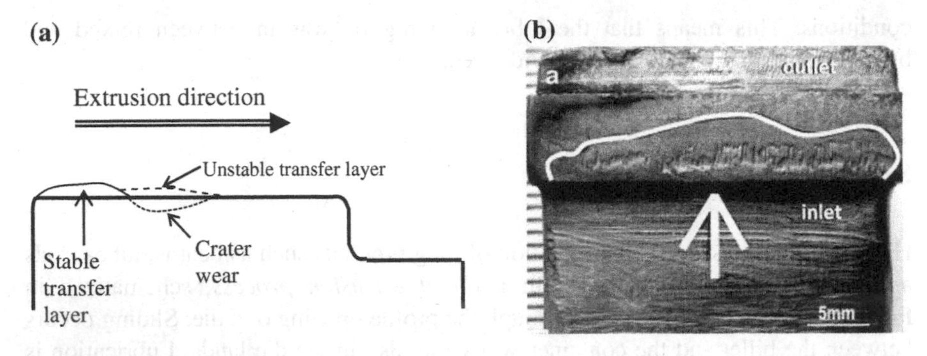


**Fig. 8.30** **a** Wear damage in the die land (modified from [51]); **b** example of worn bearing surface of a nitrided die [53]

These phenomena may help the adhesion-aided detachment of the transfer layer. In Fig. 8.30b an example of a wear crater forming on the die land is shown. In nitrided dies, such craters are typically 20–100 μm in depth. They may also form at the inlet or at the exit of the die land [53].

In general, a die is used to extrude 20–100 km of profiles. Because of wear, nitrided dies are re-nitrided five to eight times during their service life. Dies coated with hard thin films (such as AlTiN or CrN) display a greater sliding wear resistance and their actual lifetime is similar to that of nitrided dies that are re-nitrided several times [54, 55].

## References

1. J.A. Collins, H. Busby, G. Staab, *Mechanical Design of Machine Elements, A Failure Prevention Perspective*, 2nd edn. (Wiley, New York, 2010)
2. S.C. Tung, M.L. McMillan, Automotive tribology overview of current advances and challenges for the future. Tribol. Int. **37**, 517–536 (2004)
3. K.H. Czichos, K.H. Habig, *Tribologie Handbuch* (Vieweg, Reibung und Verlschleiss, 1992)
4. M. Priest, C.M. Taylor, Automobile engine tribology—approaching the surface. Wear **241**, 193–203 (2000)
5. G. Dalmaz, A.A. Lubrecht, D. Dowson, M. Priest (eds.), *Tribology Research: from Model Experiment to Industrial Problem* (Elsevier, New York, 2001)
6. M.M. Khonsari, E.R. Booser, *Applied Tribology, Bearing Design and Lubrication* (Wiley, New York, 2008)
7. I.M. Hutchings, *Tribology* (Edwald Arnold, London, 1992)
8. G. Straffelini, A. Molinari, P. Detassis, P. Groff, Wear behaviour of surface treated steel bearings for rocker arms in direct injection systems, *Proceeding of Materials Solutions '97 on Wear of Engineering Materials* (ASM, 1997), pp. 85–90
9. A. Neville, A. Morina, T. Haque, M. Voong, Compatibility between tribological surfaces and lubricant additives—How friction and wear reduction can be controlled by surface/lube synergies. Tribol. Int. **40**, 1680–1695 (2007)
10. M.B. Peterson, W.O. Winer (eds.), *Wear Control Handbook* (ASME, New York, 1980)

11. C.M. Taylor, Automobile engine tribology—design considerations for efficiency and durability. Wear **221**, 1–8 (1998)
12. J. Williams, *Engineering Tribology* (Cambridge University Press, Cambridge, 2004)
13. SKF, *I cuscinetti volventi*, Catalogo generale (1989)
14. K. Holmberg and A. Matthews, *Coatings Tribology* (Elsevier, New York, 1994)
15. M. Widmarka, A. Melander, Effect of material, Heat treatment, grinding and shot peening on contact fatigue life of carburized steels. Int. J. Fatigue **21**, 309–327 (1999)
16. A.C. Batista, A.M. Dias, J.L. Lebrun, J.C. LeFlour, G. Inglebert, Contact fatigue of automotive gears: evolution and effects of residual stresses introduced by surface treatments. Fatigue Fract. Eng. Mater. Struct. **23**, 217–228 (2000)
17. D. Frölich, B. Magyar, B. Sauer, A comprehensive model of wear, friction and contact temperature in radial shaft seals. Wear **311**, 71–80 (2004)
18. Q.F. Wen, Y. Liu, W.F. Huang, S.F. Suo, Y.M. Wang, The effect of surface roughness on thermal-elasto-hydrodynamic model of contact mechanical seals. Sci. China **56**, 1920–1929 (2013)
19. J. Wahlstrom, A study of airborne wear particles from automotive disc brakes, Doctoral thesis, Royal Institute of Technology (KTH), Stockholm, Sweden, 2011
20. U. Olofsson, Y. Zhu, S. Abbasi, R. Lewis, S. Lewis, Tribology of the wheel-rail contact—aspects of wear, particle emission and adhesion. Veh. Syst. Dyn. **51**, 1091–1120 (2013)
21. T. Lewis, U. Olofsson, *Wheel-Rail Interface Handbook* (Woodhead Publishing Limited, London, 2009)
22. K.D. Vo, H.T. Zhu, A.K. Tieu, P.B. Kosasih, FE method to predict damage formation on curved track for various worn status of wheel/rail profiles. Wear **322**, 61–75 (2015)
23. A. Ekberg, B. Akesson, E. Kabo, Wheel/rail rolling contact fatigue—probe, predict, prevent. Wear **314**, 2–12 (2014)
24. A. Ghidini, M. Diener, A. Gianni, J. Schneider, *Superlos, Innovative steel by Lucchini RS for high-speed wheel application* (Lucchini RS, Lovere, 2012)
25. S. Kalpakjian, *Manufacturing Processes for Engineering Materials* (Addison-Wesley, Essex, 1984)
26. M.C. Shaw, *Metal Cutting Principles* (Clarendon Press, Oxford, 1989)
27. M. P. Groover, *Fundamentals of Modern Manufacturing* (Prentice-Hall, Englewood Cliffs, 1996)
28. L.J. Yang, Determination of the wear coefficient of tunngsten carbide by a turning operation. Wear **250**, 366–375 (2001)
29. W. Grzesik, J. Rech, K. Zak, Determination of friction in metal cutting with tool wear and flank face effects. Wear **317**, 8–16 (2014)
30. T.H.C. Childs, K. Maekawa, T. Obikawa, Y. Yamane, *Metal machining, Theory and Applications* (Arnold, London, 2000)
31. Z. Palmai, Proposal for a new theoretical model of the cutting tool's flank wear. Wear **303**, 437–445 (2013)
32. A. Inspektor, P.A. Salvador, Architecture of PVD coatings for metalcutting applications: a review. Surf. Coat. Technol. **257**, 138–153 (2014)
33. M.J. Jackson, C.J. Davis, M.P. Hitchiner, B. Mills, High-speed grinding with CBN grinding wheels-applications and future technology. J. Mater. Process. Technol. **110**, 78–88 (2001)
34. J. Kopac, P. Krajnik, High-performance grinding-A review. J. Mater. Process. Technol. **175**, 278–284 (2006)
35. U.S.P. Durgumahanti, V. Singh, P. Venkateswara Rao, A new model for grinding force prediction and analysis. Int. J. Mach. Tools Manuf. **50**, 231–240 (2010)
36. G. Werner, Influence of work material on grinding forces. Ann. CIRP **27**, 243–248 (1978)
37. S. Jahanmir, M. Ramulu, P. Koshu (eds.), *Machining of Ceramics and Composites* (Marcel, New York, 1999)
38. C. Chen, Y. Wang, H. Ou, Y. He, X. Tang, A review on remanufacture of dies and moulds. J. Cleaner Prod. **64**, 13–23 (2014)

39. S.Q. Wang, M.X. Wei, F. Wang, H.H. Cui, C. Dong, Transition of mild wear to severe wear in oxidative wear of H21 steels. Tribol. Lett. **32**, 67–72 (2008)
40. E. Doege, G. Andreis, M. Guld, Improving tool life in hot massive forming by coating, in *Proceeding of 5th International Conferente on Tooling*, University of Leoben, 335–348, 1999
41. ASM Handbook, *Friction, Lubrication and Wear Technology 18* (ASM, 1992)
42. M. Pellizzari, A. Molinari, G. Straffelini, Tribological behaviour of hot rolling rolls. Wear **259**, 1281–1289 (2005)
43. W.L. Roberts, *Hot Rolling of Steel* (Marcel, New York, 1983)
44. J.G. Lenard, *Metal Forming Science and Practice*. Elsevier, New York (1983)
45. J.G. Lenard, L. Barbulovic-Nad, The coefficient of friction during hot rolling of low carbon steel strips. J. Tribol. **124**, 840–845 (2002)
46. J.H. Beynon, Tribology of hot metal forming. Tribol. Int. **31**, 73–77 (1998)
47. P. Montmitonnet, F. Delamare, B. Rizoulieres, Transfer layer and friction in cold metal strip rolling parameters. Wear **245**, 125–135 (2000)
48. G.E. Dieter, *Mechanical Metallurgy* (McGraw Hill, New York, 1988)
49. M. Hashimoto, Development of multi-component white cast iron rolls and rolling technology in steel rolling, *Proceedings of ABRASION 2008,* University of Trento, 2008 pp. 1–23, ed. by M. Pellizzari
50. R.N. Wright, *Wire Technology—Process Engineering and Metallurgy* (Butterworth-Heinemann, Oxford, 2011)
51. P. Gillstrom, M. Jarl, Wear of die after drawing of pickled or reverse bent wire rod. Wear **262**, 858–867 (2007)
52. T. Björk, J. Bergstrom, S. Hogmark, Tribological simulation of aluminium hot extrusion. Wear **224**, 216–225 (1999)
53. Y. Birol, Analysis of wear of gas nitrided H13 tool steel die in aluminium extrusion. Eng. Fail. Anal. **26**, 203–210 (2012)
54. T. Björk, R. Westergard, S. Hogmark, Wear of surface treated dies for aluminium extrusion—a case study. Wear **249**, 316–323 (2001)
55. M. Pellizzari, M. Zadra, A. Molinari, Tribological properties of surface engineered hot work tool steel for aluminium extrusion dies, Surf. Eng. **23**, 165–168 (2007)

# Index




G. Straffelini, *Friction and Wear*, Springer Tracts in Mechanical Engineering, DOI 10.1007/978-3-319-05894-8

**D**



**E**

Printed in the United States
By Bookmasters